Recent Advances and Issues in

ENVIRONMENTAL SCIENCE

Recent Advances and Issues in
ENVIRONMENTAL SCIENCE

Edited By
William Hunter III
Researcher, National Science Foundation, U.S.A.

Apple Academic Press
TORONTO NEW YORK

International Standard Book Number: 978-1-926692-70-8 (Hardback)

Printed in the United States of America

10 9 8 7 6 5 4 3 2 1

Library and Archives Canada Cataloguing in Publication

Recent advances and issues in environmental science/William Hunter, editor.

Includes index.
ISBN 978-1-926692-70-8
1. Environmental sciences. I. Hunter, William, 1971- II. Title.

GE105.R44 2011 363.7 C2011-905429-9

Apple Academic Press also publishes its books in a variety of electronic formats. Some content that appears in print may not be available in electronic format. For information about Apple Academic Press products, visit our website at **www.appleacademicpress.com**

Preface

The field of environmental science is relatively new, but its importance cannot be overestimated. As with many scientific fields, the work of environmental science draws attention on many other sciences to solve problems that we face today. Due to its very broad and cross-disciplinary nature, there is not a consensus as to its correct definition. However, the definition for an environmental science program of study provided by the National Center for Education Statistics (part of the United States Department of Education) encapsulates this diverse field well: "A program that focuses on the application of biological, chemical, and physical principles to the study of the physical environment and the solution of environmental problems, including subjects such as abating or controlling environmental pollution and degradation; the interaction between human society and the natural environment; and natural resources management. Includes instruction in biology, chemistry, physics, geosciences, climatology, statistics, and mathematical modeling."

As seen in the above description, environmental science is truly interdisciplinary, fostering wide-ranging relationships across sciences and math. However, the value of a solid, diverse, liberal-arts background to start, with emphasis in the social sciences, is also important. Communication of results with the general public, lawmakers, and other stakeholders and the creation of government policy will be critical. Therefore, understanding the functioning of government at the local, state, national, and global level will be important in helping to create policies that effectively deal with specific environmental issues.

Environmental science seeks answers to some of our biggest environmental challenges. Many of these tie into a greater concept of using the earth's resources sustainably. For example, reducing greenhouse emissions that lead to global warming can be accomplished in a number of ways, such as developing more efficient cars and industrial processes while also developing renewable energy. Actively saving rainforest and other wild areas can also be invaluable in reducing greenhouse gases, as they will sequester large amounts of carbon while also preserving biodiversity. Looking at interactions such as these and producing models of their effects is all in the purview of environmental scientists.

A broad background in many sciences makes an effective environmental scientist and would include course work in ecology, biology, geology, statistics, and planning. The career field is equally broad and would include work in local municipalities (such as sewage treatment plant operators) and consulting with national governments in a specialty area such as geophysics. Employment can be found with local, state, and federal governments, private companies, colleges and universities, and lobbying groups.

The relevance of environmental science is visible all around us every day. Often the world news media is focused on large-scale environmental problems such as global warming. However, if you think about everyday local life at your community level, it does not take long to develop a list of questions relevant to environmental science.

How does your community get fresh drinking water? Where does your community dispose of solid waste? What level of treatment does community sewage receive before being discharged into natural waterways? Is your community welcoming to alternative forms of transit such as buses, bicycles, and pedestrians, or is it totally reliant on automobiles? Is your food grown in the local community or shipped from distant areas? Is your home safe from toxins and contaminants? Although seemingly cliché now, the saying "Think Global, Act Local" really can have large-scale beneficial results, and directly seeing local environmental issues makes them more tangible.

Due to its relevance on the local and global level and its diverse number of careers, the future of environmental science is strong. Employment will be found at a wide range of levels and for people with highly variable skills. The top levels of this field will need people with the ability to integrate wide-ranging data sets and understand their relevance, and those who have strong communication skills to translate these results to the general public. All of the science and resources that we can bring to bear on an environmental problem will mean nothing in the end if we cannot get people to act.

— William Hunter III

List of Contributors

Karina Acevedo-Whitehouse
Institute of Zoology, London, UK.

Benjamin Arnold
Centers for Occupational and Environmental Health and Family and Community Health, School of Public Health, Berkeley, University of California, Berkeley, CA, USA.

Gianni Astolfi
IMER Registry, Department of Reproduction and Growth, St. Anna Hospital, Ferrara, Italy.

Nigel W. Beebe
School of Biological Sciences, University of Queensland, St. Lucia, Queensland, Australia.
CSIRO Entomology, Long Pocket Laboratories, Indooroopilly, Queensland, Australia.

Maria Blettne
Institute of Medical Biostatistics, Epidemiology and Informatics (IMBEI), Johannes Gutenberg-University Mainz, Germany.

Julien Bonnet
Laboratoire de Lutte contre les Insectes Nuisibles (LIN), Institut de Recherche Pour le Développement (IRD) Montpellier, France.

Wim Van Bortel
Department of Parasitology, Institute of Tropical Medicine, Antwerp, Belgium.

Cecile Brengues
Vector Research group, Liverpool School of Tropical Medicine, Liverpool, UK.

Monica Daigl Cattaneo
Department of Public Health and Epidemiology, Swiss Tropical Institute (STI), University of Basel, Switzerland.

Myriam Cevallos
Department of Public Health and Epidemiology, Swiss Tropical Institute (STI), University of Basel, Switzerland.

Andri Christen
Department of Public Health and Epidemiology, Swiss Tropical Institute (STI), University of Basel, Switzerland.

John M. Colford Jr.
Centers for Occupational and Environmental Health and Family and Community Health, School of Public Health, Berkeley, University of California, Berkeley, CA, USA.

Robert D. Cooper
Australian Army Malaria Institute, Gallipoli Barracks, Enoggera, Queensland, Australia.

Vincent Corbel
Laboratoire de Lutte contre les Insectes Nuisibles (LIN), Institut de Recherche Pour le Développement (IRD) Montpellier, France.

Frederic Darriet
Laboratoire de Lutte contre les Insectes Nuisibles (LIN), Institut de Recherche Pour le Développement (IRD) Montpellier, France.

Jean-Philippe David
Laboratoire d'Ecologie Alpine (LECA, UMR 5553 CNRS—Université de Grenoble), Grenoble, France.

C. Patrick Doncaster
School of Biological Sciences, University of Southampton, Southampton, UK.

Luc Duchateau
Department of Physiology and Biometrics, University of Ghent, Ghent, Belgium.

Agustín Estrada-Peña
Facultad de Veterinaria, Universidad de Zaragoza, Zaragoza, Spain.

Sara Fabbi
LARMA—Laboratory of Environmental Analysis, Surveying and Environmental Monitoring, Department of Mechanical and Civil Engineering, University of Modena and Reggio Emilia, Modena, Italy.

Joseph Fargione
Central Region, the Nature Conservancy, Minneapolis, MN, USA.

Matt Finer
Save America's Forests, Washington, DC, USA.

Curtis L. Fritz
California Department of Public Health, Center for Infectious Diseases, Division of Communicable Disease Control, Infectious Diseases Branch, Vector-Borne Disease Section, Sacramento, CA, USA.

José de la Fuente
Instituto de Investigación en Recursos Cinegéticos IREC (CSIC-UCLM-JCCM), Ciudad Real, Spain.
Department of Veterinary Pathobiology, Center for Veterinary Health Sciences, Oklahoma State University, Stillwater, USA.

Livia Garavelli
Department of Paediatrics, Santa Maria Nuova Hospital, Reggio Emilia, Italy.

Solomon Gebre-Selassie
Department of Microbiology, Immunology and Parasitology, Addis Ababa University, Addis Ababa, Ethiopia.

Peng Gong
Environmental Science, Policy, and Management Department, University of California, Berkeley, CA, USA.

Gaël Paul Hammer
Cancer Registry of Rhineland-Palatinate, Registration Office, Mainz, Germany.
Institute of Medical Biostatistics, Epidemiology and Informatics (IMBEI), Johannes Gutenberg-University Mainz, Germany.

Jan Hattendorf
Department of Public Health and Epidemiology, Swiss Tropical Institute (STI), University of Basel, Switzerland.

Ashley C. Holt
Environmental Science, Policy, and Management Department, University of California, Berkeley, CA, USA.

Po-Shin Huang
Meiho Institute of Technology, Pingtung and Department of Resource Engineering, National Cheng Kung University, Tainan, Taiwan.

Gabriele Husmann
Cancer Registry of Rhineland-Palatinate, Registration Office, Mainz, Germany.

Mercedes Iriarte
Centro de Aguas y Saneamiento Ambiental (CASA), Facultad de Tecnología, Universidad Mayor de San Simón (UMSS), Cochabamba, Bolivia.

Clinton N. Jenkins
Nicholas School of the Environment, Duke University, Durham, North Carolina, USA.

Brian Keane
Land is Life, Somerville, MA, USA.

Joe Kiesecker
Rocky Mountain Region, the Nature Conservancy, Fort Collins, Colorado, USA.

Kirk R. Klausmeyer
The Nature Conservancy, San Francisco, CA, USA.

Helmut Kloos
Department of Epidemiology and Biostatistics, University of California Medical Center, San Francisco, CA 94143, USA.

Katherine M. Kocan
Department of Veterinary Pathobiology, Center for Veterinary Health Sciences, Oklahoma State University, Stillwater, USA.

Jochem König
Institute of Medical Biostatistics, Epidemiology and Informatics (IMBEI), Johannes Gutenberg-University Mainz, Germany.

Anne Krtschil
Cancer Registry of Rhineland-Palatinate, Notification Office, Mainz, Germany.

Charles P. Larson
International Centre for Diarrhoeal Disease Research, Bangladesh (ICDDR, B), Dhaka, Bangladesh.
Centre for International Child Health, British Columbia Children's Hospital and Department of Pediatrics, University of British Columbia, Vancouver, Canada.

Worku Legesse
School of Environmental Health, Jimma University, Jimma, Ethiopia.

Carlotta Malagoli
CREAGEN—Environmental, Genetic and Nutritional Epidemiology Research Center, Department of Public Health Sciences, University of Modena and Reggio Emilia, Reggio Emilia, Italy.

Atilio J. Mangold
Instituto Nacional de Tecnología Agropecuaria, Estación Experimental Agropecuaria Rafaela, Santa Fe, Argentina.

Sébastien Marcombe
Laboratoire de Lutte contre les Insectes Nuisibles (LIN), Institut de Recherche Pour le Développement (IRD) Montpellier, France.

Daniel Mäusezahl
Department of Public Health and Epidemiology, Swiss Tropical Institute (STI), University of Basel, Switzerland.

Robert I. McDonald
Worldwide Office, the Nature Conservancy, Arlington, VA, USA.

William M. Miller
Department of Chemical and Biological Engineering, Northwestern University, Evanston, IL, USA.

Pipi Mottram
Communicable Diseases Branch, Queensland Health, Brisbane, Queensland, Australia.

Victoria Naranjo
Instituto de Investigación en Recursos Cinegéticos IREC (CSIC-UCLM-JCCM), Ciudad Real, Spain.
Department of Veterinary Pathobiology, Center for Veterinary Health Sciences, Oklahoma State University, Stillwater, USA.

Hazera Nazrul
International Centre for Diarrhoeal Disease Research, Bangladesh (ICDDR, B), Dhaka, Bangladesh.

Barry Nicholls
School of Biological Sciences, University of Aberdeen, Aberdeen, UK.

Gonzalo Duran Pacheco
Department of Public Health and Epidemiology, Swiss Tropical Institute (STI), University of Basel, Switzerland.

Stuart L. Pimm
Nicholas School of the Environment, Duke University, Durham, NC, USA.

Rodolphe Poupardin
Laboratoire d'Ecologie Alpine (LECA, UMR 5553 CNRS—Université de Grenoble), Grenoble, France.

Jimmie Powell
Worldwide Office, the Nature Conservancy, Arlington, VA, USA.

Paul A. Racey
School of Biological Sciences, University of Aberdeen, Aberdeen, UK.

Hilary Ranson
Vector Research group, Liverpool School of Tropical Medicine, Liverpool, UK.

Stéphane Reynaud
Laboratoire d'Ecologie Alpine (LECA, UMR 5553 CNRS—Université de Grenoble), Grenoble, France.

Francesca Rivieri
IMER Registry, Department of Reproduction and Growth, St. Anna Hospital, Ferrara, Italy.

Rossella Rodolfi
Local Health Unit of Reggio Emilia, Reggio Emilia, Italy.

Carl Ross
Save America's Forests, Washington, DC, USA.

Unnati Rani Saha
International Centre for Diarrhoeal Disease Research, Bangladesh (ICDDR, B), Dhaka, Bangladesh.

Daniel J. Salkeld
Environmental Science, Policy, and Management Department, University of California, Berkeley, CA, USA.

Irene Schmidtmann
Institute of Medical Biostatistics, Epidemiology and Informatics (IMBEI), Johannes Gutenberg-University Mainz, Germany.

Andreas Seidler
Federal Institute for Occupational Safety and Health (BAuA), Berlin, Germany.
Cancer Registry of Rhineland-Palatinate, Registration Office, Mainz, Germany.
Institute of Medical Biostatistics, Epidemiology and Informatics (IMBEI), Johannes Gutenberg-University Mainz, Germany.

M. Rebecca Shaw
The Nature Conservancy, San Francisco, CA, USA.

Li-Hsing Shih
Department of Resource Engineering, National Cheng Kung University, Tainan, Taiwan.

Thomas A. Smith
Department of Public Health and Epidemiology, Swiss Tropical Institute (STI), University of Basel, Switzerland.

Niko Speybroeck
Department of Animal Health, Institute of Tropical Medicine, Antwerp, Belgium.
Public Health School, Université Catholique de Louvain, Brussels, Belgium.

Clare Strode
Vector Research group, Liverpool School of Tropical Medicine, Liverpool, UK.

Anthony W. Sweeney
School of Biological Sciences, University of Queensland, St. Lucia, Queensland, Australia.

Sergio Teggi
LARMA—Laboratory of Environmental Analysis, Surveying and Environmental Monitoring, Department of Mechanical and Civil Engineering, University of Modena and Reggio Emilia, Modena, Italy.

Fidel Alvarez Tellez
Project Concern International (PCI), Cochabamba, Bolivia.

James R. Tucker
California Department of Public Health, Center for Infectious Diseases, Division of Communicable Disease Control, Infectious Diseases Branch, Vector-borne Disease Section, Sacramento, CA, USA.

Marco Vinceti
CREAGEN—Environmental, Genetic and Nutritional Epidemiology Research Center, Department of Public Health Sciences, University of Modena and Reggio Emilia, Reggio Emilia, Italy.

André Yébakima
Centre de démoustication, Conseil général de la Martinique, Fort de France, Martinique, France.

Delenasaw Yewhalaw
Department of Biology, Jimma University, Jimma, Ethiopia.

Maria E. Zapata
Instituto de Investigaciones Biomédicas (IIBISMED), Facultad de Medicina, Universidad Mayor de San Simon, Cochabamba, Bolivia.

List of Abbreviations

AChE	Acetylcholinesterase
AOGCMs	Atmosphere-ocean general circulation models
ARP	Acid regeneration plant
AUC	Area under the curve
BoliviaWET	Bolivia water evaluation trial
CCEs	Carboxy/cholinesterases
CCS	Carbon capture and storage
CDPH	California Department of Public Health
CIs	Confidence intervals
CMIP	Coupled Model Intercomparison Project
CNS	Central nervous system
COD	Chemical oxygen demand
CSC	China Steel Corporation
DE	Design effect
DEM	Digital elevation model
dxs	D-1-Eoxyxylulose 5-phosphate synthase gene
ECOD	7-Ethoxycoumarin-O-deethylase
EIA	Energy Information Administration
EIP	Extrinsic incubation period
EIS	Environmental impact studies
EKA	Environmental knowledge accumulation
EKC	Environmental knowledge creation
EKCP	Environmental knowledge circulation process
EKI	Environmental knowledge internalization
EKM	Environmental knowledge management
EKS	Environmental knowledge sharing
EKU	Environmental knowledge utilization
EM	Environmental management
EMF	Electromagnetic field
EMIS	Environmental management information system
EMS	Environmental management systems
ENM	Ecological niche modeling
EPD	Environmental Protection Department
EPI	Expanded program for immunization
ESRI	Environmental Systems Research Institute

FBR	Feeding buzz ratio
FPIC	Free, prior, and informed consent
GEEs	Generalized estimating equations
GIS	Geographic information system
GLMMs	Generalized linear mixed models
GSCM	Green supply chain management
GSTs	Glutathione S-transferases
HNT	Hubbell's 2001 neutral theory
ICC	Intracluster correlation coefficient
ICDDR, B	International Centre for Diarrheal Disease Research, Bangladesh
IPCC	Intergovernmental Panel on Climate Change
IQR	Interquartile range
IR	Incidence rate
ITT	Ishpingo–Tiputini–Tambococha
KC	Knowledge creation
KCP	Knowledge circulation process
kdr	Knock down resistance
KM	Knowledge management
MCE	Mediterranean climate extent
MOHFW	Ministry of Health and Family Welfare
MSP	Major surface protein
MSWI	Municipal solid waste incinerator
NAC	National Advisory Committee
NDVI	Normalized Difference Vegetation Index
NGO	Nongovernmental organization
NHL	Non-Hodgkin lymphoma
NREL	National Renewable Energy Laboratories
NSW	New South Wales
OR	Odds ratio
ORS	Oral rehydration salt
ORT	Oral rehydration therapies
PCA	Principal components analysis
PCDD/F	Polychlorinated dibenzo-p-dioxins and dibenzofurans
PCI	Project Concern International
PCI	Pulverized coal injection
PCR	Polymerase chain reaction
PDCA	Plan-do-check-action
PET	Polyethyleneteraphtalate
PNLTs	Pirbright-Miniature light-suction traps

PR	Prevalence ratio
RF	Radio frequency
RR	Relative rate
RRs	Rate ratios
SADs	Species-abundance distributions
SARs	Species-area relationships
SD	Severe diarrhea
SDS	Sodium dodecyl sulfate
SEA	Strategic Environmental Assessment
SIRs	Standardized cancer ratios
SODIS	Solar drinking water disinfection
SSs	Suspended solids
TCDD	2,3,7,8-Tetrachlorodibenzo-p-dioxin
TTC	Thermotolerant coliforms
TW	Terawatt
US EPA	United States Environmental Protection Agency
UV	Ultraviolet
WCRP	World Climate Research Programme
WHO	World Health Organization

Contents

Chapter 1

Ecological Equivalence: Niche Theory as a Testable Alternative to Neutral Theory

C. Patrick Doncaster

INTRODUCTION

Hubbell's 2001 neutral theory (HNT) unifies biodiversity and biogeography by modeling steady-state distributions of species richness and abundances across spatio-temporal scales. Accurate predictions have issued from its core premise that all species have identical vital rates. Yet no ecologist believes that species are identical in reality. Here I explain this paradox in terms of the ecological equivalence that species must achieve at their co-existence equilibrium, defined by zero net fitness for all regardless of intrinsic differences between them. I show that the distinction of realized from intrinsic vital rates is crucial to evaluating community resilience.

An analysis of competitive interactions reveals how zero-sum patterns of abundance emerge for species with contrasting life-history traits as for identical species. I develop a stochastic model to simulate community assembly from a random drift of invasions sustaining the dynamics of recruitment following deaths and extinctions. Species are allocated identical intrinsic vital rates for neutral dynamics, or random intrinsic vital rates and competitive abilities for niche dynamics either on a continuous scale or between dominant-fugitive extremes. Resulting communities have steady-state distributions of the same type for more or less extremely differentiated species as for identical species. All produce negatively skewed log-normal distributions of species abundance, zero-sum relationships of total abundance to area, and Arrhenius relationships of species to area. Intrinsically identical species nevertheless support fewer total individuals, because their densities impact as strongly on each other as on themselves. Truly neutral communities have measurably lower abundance/area and higher species/abundance ratios.

Neutral scenarios can be parameterized as null hypotheses for testing competitive release, which is a sure signal of niche dynamics. Ignoring the true strength of interactions between and within species risks a substantial misrepresentation of community resilience to habitat loss.

The HNT unifies the disciplines of biodiversity and biogeography by modeling steady-state distributions of species richness and relative species abundance across spatio-temporal scales [1]. Surprisingly accurate predictions have issued from its core premise that all species are exactly identical in their vital rates. As a null hypothesis to explain what should be observed if all species were perfectly equal with respect to all ecologically relevant properties, it has proved hard to refute [2]. Yet no ecologist, including Hubbell, believes that species are equivalent in reality [3, 4]. The challenge presented by HNT is to justify invoking anything more complex than ecological drift to define community structure [5]. Its extravagant simplicity has had an explosive impact on ecology (>1,100 citations, rising exponentially), because it appears to discount 100 years of traditional conventions on niche differentiation. If biodiversity encompasses the great richness of differently attributed species that constitutes the natural world, how can ecological equivalence yield such predictive power about the numbers of species [6]? If HNT is based on a ludicrous assumption [7], then our conceptual understanding is thrown into disarray by its fit to empirical patterns [8]. Here I explain this paradox in terms of the ecological equivalence realized by coexisting species at demographic equilibrium. Analyses and simulations of co-existence equilibria demonstrate the emergent property of ecological equivalence among species with a rich diversity of attributes, leading to novel predictions for a quantifiable gradation in species-area relationships between neutral and niche models.

A neutral model of empirical relationships eliminates "the entire set of forces competing for a place in the explanation of the pattern" [9]. Accordingly, HNT assumes that all species behave identically in a zero-sum game such that the total density of individuals in a trophically similar community remains constant regardless of species composition. The defining image of this ecological equivalence is a tropical forest canopy, with remarkably constant total densities of trees regardless of large regional variations in constituent species [1]. Interpretations of zero-sum equivalence routinely omit to distinguish between the equal vital rates realized at the system carrying capacity approximated in this image (and most datasets), and the intrinsic vital rates that define the heritable character traits of each species. Models of HNT consistently prescribe identical intrinsic rates and niche dimensions. Hubbell [1] anticipated the disjuncture between realized and intrinsic rates by comparing ecological equivalence to the fitness invariance achieved at carrying capacity, allowing for different trade-off combinations in life-history traits. The prevailing convention, however, remains that ecological equivalence explicitly requires symmetric species with identical per capita vital rates, thereby promulgating the notion that HNT is built on an unrealistic foundation [3].

Theoretical studies have sought various ways to reconcile neutral patterns with niche concepts. Intrinsically similar species can coexist under niche theory [7], and niches add stabilizing mechanisms that are absent under the fitness equivalence of intrinsic neutrality [10]. Comparisons of niche to neutral simulations in a saturated system of fixed total abundance have shown that they can predict similar species-abundance distributions (SADs) and species-area relationships [11], demonstrating that neutral patterns need not imply neutral processes [12]. Even neutral processes of intraspecific competition and dispersal limitation cannot be distinguished in principle for species-abundance predictions [13-16]. Here I use an analysis and simulation of Lotka–Volterra dynamics to model zero-sum ecological drift as an emergent property of stochastic niche structures at dynamic equilibrium. I explain its appearance in the steady-state distributions even of extremely dissimilar species in terms of the trivial expectation that species must achieve ecological equivalence at their co-existence equilibrium, which is defined by equal realized fitness for all. Although the predictions are standards of Lotka–Volterra analysis for a homogeneous environment, they drive a simulation that for the first time spans across dispersal-limited neutral to stochastic niche scenarios without fixing the total abundance of individuals.

The neutral simulation developed here is consistent with the models of Solé et al. [17] and Allouche and Kadmon [18] in having total species, S, abundance of individuals, N, and zero-sum dynamics as emergent properties (in contrast to refs. [1, 11, 12, 19]). The S species are identical in all respects including interspecific interactions equal to intraspecific (in contrast to refs [13, 16]). Non-neutral simulations developed here extend the model of Chave et al. [11] by allowing competitive differences to vary stochastically on a continuous scale, as in Purves and Pacala [12]. They extend both these models by allowing pre-emptive recruitment and emergent zero-sum dynamics, and the model of Calcagno et al. [20] by adding dispersal limitation. They are consistent with Tilman's niche theory [21, 22] in their population abundances being a function of species-specific vital rates.

These simulations confirm the previously untested prediction [12] that colonization-competition trade-offs with stochastic colonization will exhibit zero-sum ecological drift and produce rank abundance curves that resemble neutral drift. Truly neutral dynamics should nevertheless sustain a lower total density of individuals at density-dependent equilibrium. This is because intrinsically identical species must interact as strongly between as within species. They therefore experience no competitive release in each others' presence, contrasting with the net release to larger populations obtained by segregated niches. The simulations demonstrate this fundamental difference, and I discuss its use as a signal for dynamic processes when predicting species-area relationships.

Analysis of Abundance Patterns for Two-niche Communities

Species characterized by extremely different intrinsic attributes can achieve ecological equivalence in a zero-sum game played out at dynamic equilibrium. Take for example a two-species community comprising a dominant competitor displacing the niche of a fugitive (e.g., [23]). The fugitive survives even under complete subordination, provided it trades competitive impact for faster growth capacity [24]. Figure 1 illustrates the equal fitness, zero-sum outcome at density-dependent equilibrium under this most extremely asymmetric competition. The carrying capacity of each species is a function of its intrinsic lifetime reproduction (detailed in Materials and Methods Equation 1), and equilibrium population sizes are therefore a function of the species-specific vital rates. Regardless of variation in the ratio of dominant to fugitive carrying capacities, $0 \leq k_D/k_F \leq 1$, the system density of individuals is attracted to the stable equilibrium at $N = n_F + n_D = k_F$. Knocking out the fugitive reduces N to the smaller k_D, but only until invasion by another fugitive. This may be expected to follow rapidly, given the fugitive characteristic of fast turnover. The steady-state scenario is effectively neutral by virtue of the dominant and fugitive realizing identical vital rates and constant total density at their co-existence equilibrium despite contrasting intrinsic (heritable) rates. The reality that species differ in their life history traits therefore underpins the assumption of ecological equivalence, which then permits fitting of intrinsically neutral models with vital rates set equal to the realized rates. In the next section, these predictions are extended to simulate the drift of species invasions that sustains the dynamics of recruitment following deaths and extinctions among multiple species of dominants and fugitives.

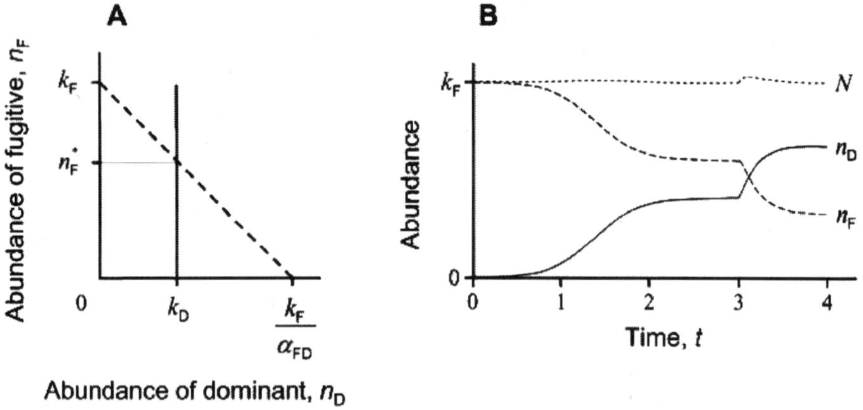

Figure 1. Equilibrium co-existence of a fugitive species invaded by a competitive dominant. With competition coefficients $\alpha_{DF} = 0$, $\alpha_{FD} = 1$, the fugitive persists provided it has the greater carrying capacity: $k_F/k_D > 1$. (A) Lotka–Volterra phase plane with steady-state abundance at the intersection of the isoclines for the fugitive (dashed line) and the dominant (solid line). (B) Equilibration of abundances over time given by Runge–Kutta solutions to Equation 1, with a 20% drop in the dominant's intrinsic death rate, d_D, imposed at $t = 3$ (equivalent to a rightward shift in its isocline) to illustrate the constancy of $N = n_F + n_D$.

The same principle of trade-offs in character traits conversely allows a sexually reproducing species to withstand invasion by highly fecund asexual mutants [25, 26]. A 2-fold advantage to the mutant in growth capacity resulting from its production of female-only offspring is canceled by even a small competitive edge for the parent species (Figure 2). Sexual and asexual types coexist as ecological equivalents to the extent that each invades the other's population to symmetric (zero) net growth for all. Although the dynamics are not zero-sum if the mutant has some competitive impact on the parent species, they approach it the higher the impact of parent on mutant and the faster its growth capacity (albeit half the mutant's). Attributes such as these accommodate greater similarity between the types in their carrying capacities and competitive abilities, which aligns the two isoclines. A consequently reduced stability of the co-existence equilibrium may result in the sexual parent ousting the asexual mutant over time, for example if the latter accumulates deleterious mutations [26, 27].

These local-scale dynamics apply equally at the regional scale of biogeography, reconfiguring individual death as local extinction, and birth as habitat colonization [24]. Equally for regional as for local scales, rate equations take as many dimensions as species in the community, with their coupling together defining niche overlap [24, 28]. Co-existence of the species that make up a community is facilitated by their different heritable traits, which is a fundamental premise of niche theory. Ecological equivalence, and hence modeling by neutral theory is nevertheless possible by virtue of the co-existence equilibrium leveling the playing field to zero net growth for all.

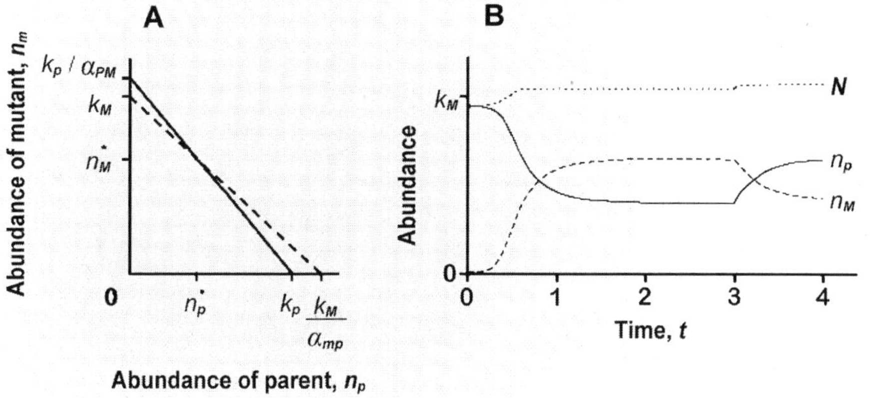

Figure 2. Equilibrium co-existence of a sexually reproducing parent population n_P invaded by an asexual mutant, n_M. With the mutant having identical vital rates except for twice the intrinsic propagation rate per capita: $b_M = 2 \cdot b_P$, the parent species persists if $\alpha_{PM} < k_P/k_M$. (A) Phase plane. (B) Equilibration of abundances over time given by Equation 1, with a 50% drop in the parent's intrinsic death rate imposed at t = 3 to illustrate approximate constancy of $N = n_M + n_P$.

The above examples of dominant versus fugitive and sexual versus asexual were illustrated with models that gave identical realized rates of both birth and death at co-existence equilibrium. Fitness invariance and zero-sum dynamics, however, require only that species have identical net rates of realized birth minus death. The simulations in the next section show how neutral-like dynamics are realized for communities of coexisting species with trade-offs in realized as well as intrinsic vital rates.

Comparison of Simulated Neutral and Multi-niche Communities with Drift

Figure 3 illustrates the SADs and species-area relationships of randomly assembled S-species systems under drift of limited immigration and new-species invasions (protocols described in Simulation Methods). From top to bottom, its graphs show congruent patterns between an intrinsically neutral community with identical character traits for all species (equivalent to identically superimposed isoclines in Figures 1 and 2 models), and communities that trade growth capacity against competitive dominance increasingly starkly. The non-neutral communities sustain more total individuals and show greater spread in their responses, reflecting their variable life-history coefficients. Their communities nevertheless follow qualitatively the same patterns as those of neutral communities. For intrinsically neutral and niche-based communities alike, Figure 3 shows SADs negatively skewed from log-normal (all $P < 0.05$, every $g_1 < 0$), and an accelerating decline in rank abundances of rare species (cf. linear for Fisher log-series) that is significantly less precipitous than predicted by broken-stick models of randomly allocated abundances among fixed S and N; Figure 4 shows constant densities of total individuals regardless of area (unambiguously linear), and Arrhenius relationships of species richness to area (unambiguously linear on loglog scales).

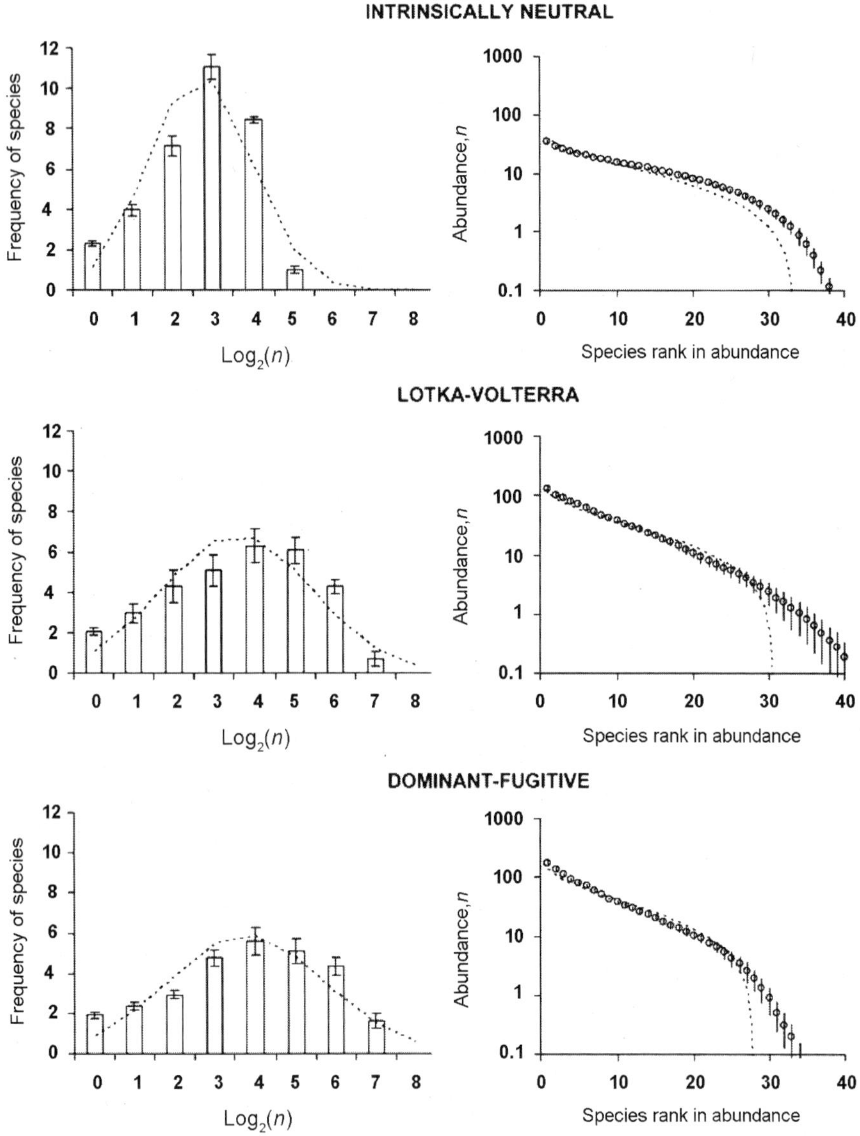

Figure 3. Simulated steady-states of species-abundance distributions (SADs). From top to bottom, graphs show average patterns for intrinsically neutral, Lotka–Volterra, and dominant-fugitive communities. The SADs each show mean ± s.e. of six replicate communities with carrying capacity K = 1000 habitable patches. Frequencies are compared to log-normal (left-hand column) and MacArthur's broken-stick (right-hand column). See Materials and Methods for input parameter values and the process of random species assembly.

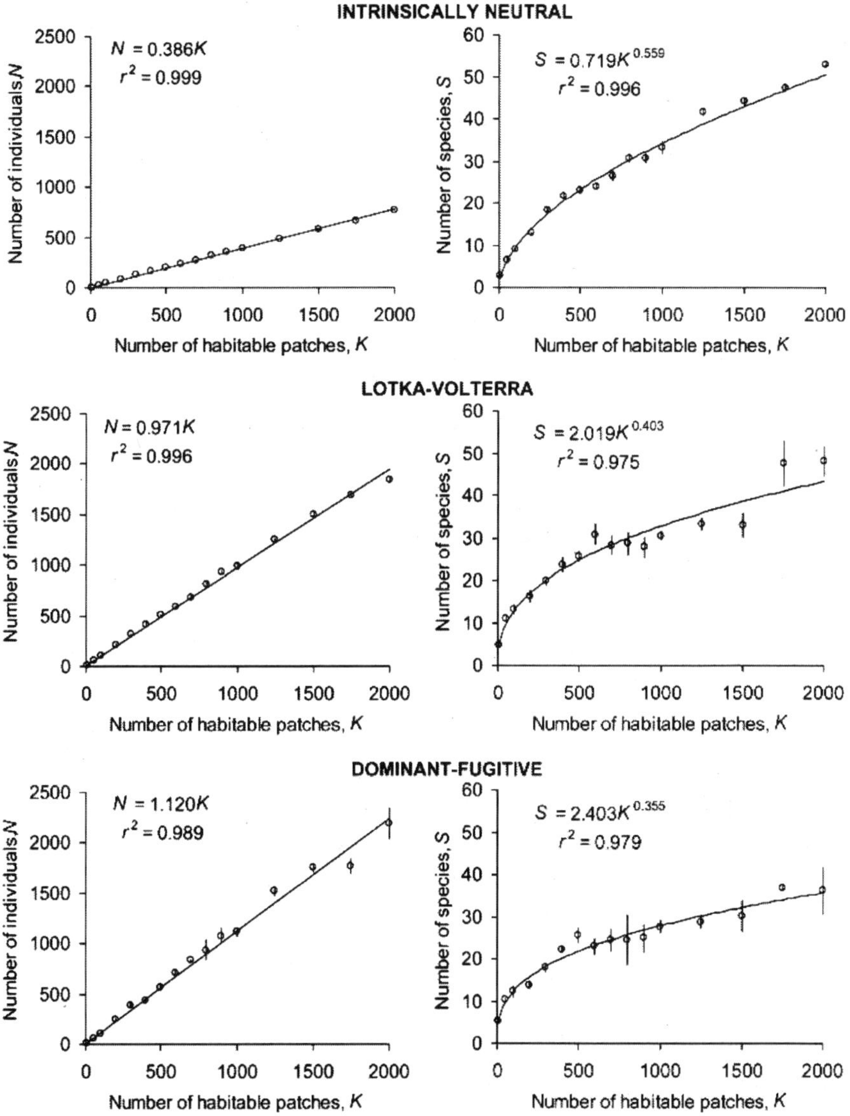

Figure 4. Simulated steady-states of species-area relationships (SARs). The SARs each show mean ± s.e. of three replicate communities. See Materials and Methods for input parameter values and the process of random species assembly.

The extended tail of rare species seen in the Figure 3 SADs is caused by single-individual invaders replacing random extinctions of n-individual species. Further trials confirm that reduced dispersal limitation exacerbates the negative skew from the log-normal distribution, while sustaining a higher total density of individuals. The extinction-invasion imbalance sets the equilibrium species richness, *S*, as a power function of total population density, N. This can be expressed as the Arrhenius relationship: $S = cK^z$ (Figure 4 right-hand column) by virtue of the zero-sum relation of N to K (Figure 4 left-hand column). Further simulations show that reduced dispersal limitation raises c and reduces z, and a higher rate of new-species invasions raises c (though not z, in contrast to predictions from spatially explicit neutral models [29]).

The closely aligned proportionality of total individuals to habitable area for all communities illustrates emergent zero-sum dynamics for neutral and non-neutral scenarios (Figure. 4 left-hand column). Despite sharing this type of pattern, and rather similar densities of species (Figure 4 right-hand column), the non-neutral communities sustain more than double the total individuals. This difference is caused by a more than halving of their competition coefficients on average (all $\alpha_{ij} = 1$ for neutral, mean α_{ij} $(i \neq j) = 0.45$ for Lotka–Volterra, mean ratio of 0:1 values = 58:42 for dominant-fugitive). The zero-sum gradient of N against K is simply the equilibrium fraction of occupied habitat, which is $1–1/R$ for a closed neutral scenario, where R is per capita lifetime reproduction before density regulation (b/d in Materials and Methods Equation 1 [23, 24]). The closed dominant-fugitive scenario modeled in Figure 1 has a slope of $k_F/K = (1–1/R)/\alpha$, where R and α are system averages. Further simulation trials show the slope increasing with immigration, for example by a factor of 1.9 between closed and fully open (dispersal unlimited) Lotka–Volterra communities. Dispersal limitation therefore counterbalances effects of the net competitive release obtained in niche scenarios from $\alpha_{ij} < 1$ (as also seen in models of heterogeneous environments [19]).

The less crowded neutral scenario sustains a somewhat higher density of species than non-neutral scenarios (comparing Figure 4 z-values for right-hand graphs), and consequently it maximizes species packing as expressed by the power function predicting S from N in Figure 5. With no species intrinsically advantaged in the neutral scenario, its coefficient of power is higher than for pooled non-neutral scenarios (0.594 and 0.384 respectively, loglog covariate contrasts: $F_{1,42} = 122.72$, $P < 0.001$). The lower coefficients of Lotka–Volterra and dominant-fugitive scenarios are further differentiated by competitive asymmetry (0.412 and 0.355 respectively, $F_{1,42} = 7.24$, $P < 0.01$). In effect, the neutral scenario has the lowest average abundance of individuals per species, n, for a community of size K with given average R, which is also reflected in the modal values in Figure 3 histograms for $K = 1{,}000$ patches.

The lower N and n predicted for the intrinsically neutral scenario point to a detectable signal of steady-state intrinsically neutral dynamics: $\alpha = 1$ for all, because intrinsically identical species cannot experience competitive release in each others' presence (cf. $\alpha_{ij} < 1$ in niche models). These interactions may be measurable directly from field data as inter-specific impacts of equal magnitude to intra-specific impacts; alternatively, Lotka–Volterra models of the sort described here can estimate average competition coefficients at an observed equilibrium N, given an average R (a big proviso, as field data generally measure realized rather than intrinsic vital rates). This distinction of intrinsically neutral from non-neutral dynamics has been masked in previous theory by the convention for neutral models either to fix N [1, 11, 12] or to set zero interspecific impacts [13, 16]. By definition, identical species cannot be invisible to each other unless they are invisible to themselves, which would require density independent dynamics. Simulations of non-interacting species under density-dependent regulation therefore embody an extreme version of niche theory whereby each species occupies a unique niche, somehow completely differentiated by resource preferences rather than partially by trade-offs in vital rates. These models fit well to species abundance distributions in rainforests and coral reefs [13-16], though without providing any explanation for what attributes would allow each species to be invisible to all others (in contrast to the trade-off models). Indeed the condition is unrealistic at least for mature trees that partition a homogeneous environment by each making their own canopy. This so-called neutral scenario ([13, 16], more appositely a neutral-niche scenario) has no steady state outcomes in the analyses and simulations described here, because setting all $\alpha_{ij} = 0$ ($i \neq j$) allows indefinite expansion of S and hence also of N. A slightly less extreme neutral-niche community is modeled by setting all interspecific impacts to a common low value. Simulations at $\alpha_{ij} = 0.1$ for all $i \neq j$ give a zero-sum relation $N = 4.026K$, which has >4-fold steeper gradient than that for the Lotka–Volterra scenario (Figure 4) reflecting its >4-fold reduction in α and consistent with its representation of a highly niched scenario.

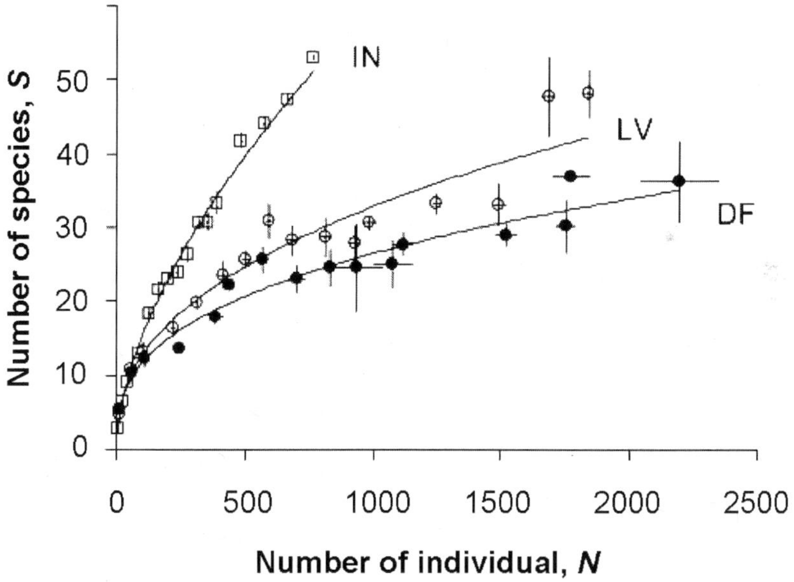

Figure 5. Simulated steady-state relationships of species to individuals. Each point shows the mean ± s.e. of the three replicate communities in Figure 4, and regression lines on the means are the power functions for intrinsically neutral (top) Lotka–Volterra (middle) and dominant-fugitive (lower) scenarios.

DISCUSSION

Although intrinsic identity is clearly not a necessary condition of ecological equivalence or of zero-sum abundances at dynamic equilibrium, only neutral models sustain these outcomes over all frequencies. It is their good fit to steady-state patterns of diversity and abundance even for communities subject to species turnover in ecological drift that has argued powerfully for niche differences having a limited role in community structure. The Figure 3 simulations reveal these types of patterns to be equally well represented by niche models, however, despite constituent individuals and species achieving fitness equivalence only at dynamic equilibrium. Non-neutral dynamics of a mature community express the community-wide average of fluctuations either side of equilibrium. Outcomes regress to the equilibrium mean for a random assembly of species undergoing stochastic extinctions of rare members, regulated by spatially autocorrelated immigration, and replacement by initially rare invaders. The predicted power of neutral theory can be taken as evidence for ecological equivalence at the co-existence equilibrium of species with more or less different intrinsic attributes.

Modeling zero-sum ecological drift as an emergent property reveals a key distinguishing feature of truly neutral communities. Their intrinsically identical species self-regulate to a lower total density as a result of inter-specific impacts equaling intra-specific impacts. Any empirical test for competitive release is therefore also a test for niche structure. For example, removing habitat is predicted to give a relative or absolute advantage to species towards the fugitive end of a dominant-fugitive spectrum, which may be picked up in correlated life-history traits for winners or losers under habitat loss or degradation [23, 24]. In contrast, neutral dynamics lead to sudden biodiversity collapse at a system-wide extinction threshold of habitat [17]. The extinction threshold of habitat for a resource-limited metapopulation is set by the fraction *1*/*R* [30, 31]. The value of *R* is thus an important yardstick of resilience in conservation planning. A neutral model fitted to empirical zero-sum abundances will overestimate their community-wide *R*, and hence overestimate community resilience, if α_{ij} are over-valued by setting all to unity. Likewise, a neutral model that sets all $\alpha_{ij} = 0$ $(i \neq j)$ will underestimate *R*, and hence resilience, if the α_{ij} are undervalued by setting all to zero.

Ecological equivalence is a much more permissive requirement for neutrality than is currently acknowledged in theoretical developments on HNT. Co-existence equilibria largely achieve the neutrality-defining mission, to eliminate all of the forces competing for a place in explanations of pattern. It remains an open question whether they do so best among species with most or least competitive release in each others' presence (e.g., Figure 1 versus Figure 2 respectively, and Figure 3 dominant-fugitive versus Lotka–Volterra respectively; [7, 10, 32]). Models need to incorporate the ecologically realistic dynamics of interspecific interactions simulated here in order to explore the true nature of competitive release between extreme scenarios of niches that are all intrinsically identical (HNT [1]) and intrinsically unique [13], [16]. Simulations of niches distributed along environmental gradients have found emerging groups of intrinsically similar species over evolutionary timescales [33]. For the spatially homogeneous environments modeled here, competition-recruitment trade-offs will always sustain species differences. In their absence, however, homogenous environments will tend to favor fast-recruiting competitive dominants. This species type may eventually prevail, with runaway selection checked by other forces such as predation, disease, mutation accumulation, and environmental variability. These systems would merit further study because many of their attributes could be those of intrinsically neutral dynamics.

MATERIALS AND METHODS

The following protocols apply to simulations of single and multi-niche communities with density-dependent recruitment and density-independent loss of individuals. They produce the outcomes illustrated in Figures 3–5 from input parameters specified at the end of this section. The general model has species-specific vital rates; the intrinsically neutral and dominant-fugitive scenarios are special cases of this model, with constrained parameter values.

The community occupies a homogenous environment represented by a matrix of K equally accessible habitat patches within a wider meta-community of K_m patches. The dynamics of individual births and deaths are modeled at each time step by species-specific probability b of each resident, immigrant, and individual of new invading species producing a propagule, and species-specific probability d of death for each patch resident. Recruitment to a patch is more or less suppressed from intrinsic rate b by the presence there of other species according to the value of α_{ij}, the impact of species j on species i relative to i on itself, where the intraspecific impact $\alpha_{ii} = 1$ always. A patch can be occupied by only one individual of a species, and by only one species unless all its resident $\alpha_{ij} < 1$. Conventional Lotka–Volterra competition is thus set in a metapopulation context by equating individual births and deaths to local colonizations and extinctions (following [24], consistent with [20]). A large closed metapopulation comprising S species has rates of change for each species i in its abundance n_i of individuals (or equally of occupied patches) over time t approximated by:

$$\frac{dn_i}{dt} = b_i n_i \left(1 - \sum_{j=1}^{S} \alpha_{ij}\, n_j / K\right) - d_i n_i. \qquad (1)$$

This is the rate equation that also drives the dynamics of Figures 1 and 2, where $k_i = (1-d_i/b_i)K$. Co-existence of any two species to positive equilibrium n_1, n_2 requires them to have intrinsic differences such that $k_1 > \alpha_{12}\, k_2$ and $k_2 > \alpha_{21}\, k_1$.

Each time-step in the simulation offers an opportunity for one individual of each of two new species to attempt invasion (regardless of the size of the meta-community). Each new species i has randomly set competitive impacts with respect to each other resident species j, of α_{ij} received and α_{ji} imposed. It has randomly set b_i, and an intrinsic lifetime reproduction $R_i = b_i/d_i$ that is stratified in direct proportion to its dominance rank among residents, obtained from its ranked mean α-received minus mean α-imposed. For example, an invader with higher dominance than all of three resident species will have random R_i stratified in the bottom quartile of set limits R_{min} to R_{max}. Communities are thereby structured on a stochastic life-history trade-off between competitive dominance and population growth capacity. This competition-growth trade-off is a well-established feature of many real communities, which captures the fundamental life-history principle of costly adaptations [11, 17, 21]. Its effect on the community is to prevent escalations of growth capacity or competitive dominance among the invading species. Neutral communities are a special case, with identical values of b and R for all species and $\alpha = 1$ for all.

At each time step, new invaders and every resident each have species-specific probability b of producing a propagule. Each propagule has small probability ν of speciation (following [1]). The sample community additionally receives immigrant propagules of its resident species that arrive from the wider meta-community in proportion to their expected numbers out there ($[K_m/K\text{–}1]n_i$), assuming the same density n_i/K of each species i as in the sample community, and in proportion to their probability (K/K_m) of landing within the sample community, and modified by a dispersal limitation parameter ω. In effect, for each resident species in the community, $[(1\text{–}K/K_m)\,n_i]^{1\text{-}\omega}$ external residents each produce an immigrating propagule with probability b_i. Thus if $K_m >> K$ and $\omega = 0$, a colonist is just as likely to be an immigrant from outside as produced from within the sample community (no dispersal limitation, following [1]). This likelihood reduces for $\omega > 0$, and also for smaller K_m. None of the propagules generated within the sample community emigrate out into the meta-community, making K a sink if smaller than K_m (sensu [34]), or a closed community if equal to K_m. The simulation is thus conceptually equivalent to randomly assembled S-species systems previously studied (e.g., [35]), except that it additionally accommodates a random drift of invasions to sustain the dynamics of recruitment following deaths and extinctions.

Each propagule lands on a random patch within the sample community and establishes there only if (a) its species is not already present, and (b) it beats each probability α_{ij} of repulsion by each other resident species j, and (c) it either beats the odds on repulsion by all other propagules simultaneously attempting to colonize the patch, or benefits from the random chance of being the first arrival among them. Each preestablished resident risks death with species-specific probability $d_i = b_i/R_i$ at each time step. Each patch has probability X of a catastrophic hazard at each time step that extirpates all its occupants. The model thus captures the principles of stochastic niche theory [21, 22] and pre-emptive advantage [20].

Each of the replicate communities contributing to distributions and relationships in Figures 3–5 is represented by values averaged over time-steps 401–500, long after the asymptote of species richness. For all graphs in Figures 3–5, meta-community carrying capacity $K_m = 10^6$, dispersal limitation parameter $\omega = 0.5$, speciation probability per resident propagation event $\nu = 10^{-12}$, two invasion attempts per time-step (setting Hubbell's [1] fundamental biodiversity number $\theta \sim 4$ independently of K_m), probability of catastrophe per patch $X = 0.01$. For neutral communities, all species take competition coefficients $\alpha = 1$, individual intrinsic propagation probability $b = 0.5$, individual intrinsic lifetime reproduction $R = 1.5$ (so lifespan $R/b = 3$); for Lotka–Volterra communities, each species i takes random $0 \leq \alpha_{ij} \leq 1$, random $0 \leq b_i \leq 1$, R_i between 1.2 and 1.8 and proportional to dominance rank; dominant-fugitive communities are as Lotka–Volterra except for random binary $\alpha_{ij} = 0$ or 1. All scenarios are thereby sampled from a large meta-community with moderate dispersal limitation, low extrinsic mortality, and sufficient invasions to sustain a reasonably high asymptote of species richness from the starting point of two species each occupying five patches. Skew in the lognormal distribution of species abundances (Figure 3) was measured for each replicate in its dimensionless third moment about the mean, g_1 [36], and confidence limits for the sample of six values were tested against $H_0{:}g_1 = 0$.

KEYWORDS

- **Hubbell's 2001 neutral theory**
- **Lotka–Volterra analysis**
- **Neutral scenario point**
- **Propagule**
- **Species-abundance distributions**
- **Zero-sum patterns**

AUTHORS' CONTRIBUTIONS

Conceived and designed the experiments: C. Patrick Doncaster. Performed the experiments: C. Patrick Doncaster. Analyzed the data: C. Patrick Doncaster. Contributed reagents/materials/analysis tools: C. Patrick Doncaster. Wrote the chapter: C. Patrick Doncaster.

ACKNOWLEDGMENTS

I thank Simon J. Cox for valuable discussions, and David Alonso and an anonymous referee for helping to sharpen my focus.

Chapter 2

Energy Sprawl or Energy Efficiency: Climate Policy Impacts on Natural Habitat

Robert I. McDonald, Joseph Fargione, Joe Kiesecker, William M. Miller, and Jimmie Powell

INTRODUCTION

Concern over climate change has led the US to consider a cap-and-trade system to regulate emissions. Here we illustrate the land-use impact to US habitat types of new energy development resulting from different US energy policies. We estimated the total new land area needed by 2030 to produce energy, under current law and under various cap-and-trade policies, and then partitioned the area impacted among habitat types with geospatial data on the feasibility of production. The land-use intensity of different energy production techniques varies over three orders of magnitude, from 1.9–2.8 km^2/TW hr/yr for nuclear power to 788–1,000 km^2/TW hr/yr for biodiesel from soy. In all scenarios, temperate deciduous forests and temperate grasslands will be most impacted by future energy development, although the magnitude of impact by wind, biomass, and coal to different habitat types is policy-specific. Regardless of the existence or structure of a cap-and-trade bill, at least 206,000 km^2 will be impacted without substantial increases in energy efficiency, which saves at least 7.6 km^2/TW hr of electricity conserved annually and 27.5 km^2/TW hr of liquid fuels conserved annually. Climate policy that reduces carbon dioxide emissions may increase the areal impact of energy, although the magnitude of this potential side effect may be substantially mitigated by increases in energy efficiency. The possibility of widespread energy sprawl increases the need for energy conservation, appropriate siting, sustainable production practices, and compensatory mitigation offsets.

Climate change is now acknowledged as a potential threat to biodiversity and human well-being, and many countries are seeking to reduce their emissions by shifting from fossil fuels to other energy sources. One potential side effect with this switch is the increase in area required by some renewable energy production techniques [1-5]. Energy production techniques vary in the spatial extent in which production activities occur, which we refer to as their energy sprawl [2, 3], defined as the product of the total quantity of energy produced annually (e.g., TW hr/yr) and the land-use intensity of production (e.g. km^2 of habitat per TW hr/yr). While many studies have quantified the likely effect of climate change on the Earth's biodiversity due to climate-driven habitat loss, concluding that a large proportion of species could be driven extinct [6-8], relatively few studies have evaluated the habitat impact of future energy sprawl. It is important to understand the potential habitat effects of energy sprawl, especially in reference to the loss of specific habitat types, since habitats vary markedly in the species and ecosystem processes they support.

Within the US, the world's largest cumulative polluter of greenhouse gases, concern over climate change has led to the consideration of a Cap-and-Trade system to regulate emissions, such as the previously proposed Lieberman–Warner Climate Security Act (S. 2191) [9] and the Low Carbon Economy Act (S. 1766) [10]. Major points of contention in structuring a cap-and-trade system are the feasibility and desirability of carbon capture and storage (CCS) at coal plants, the creation of new nuclear plants, and whether to allow international offset programs that permit US companies to meet obligations abroad [11]. The rules of a cap-and-trade system, as well as technological advances in energy production and changes in the price of fossil fuels, will affect how the US generates energy. In this study we take scenarios of a Cap-and-Trade system's effect on US energy production and evaluate each scenario's impact on habitat due to energy sprawl. Our scenarios (Figure 1A) are based on the Energy Information Administration (EIA) forecast of energy production in 2030 [12] under current law

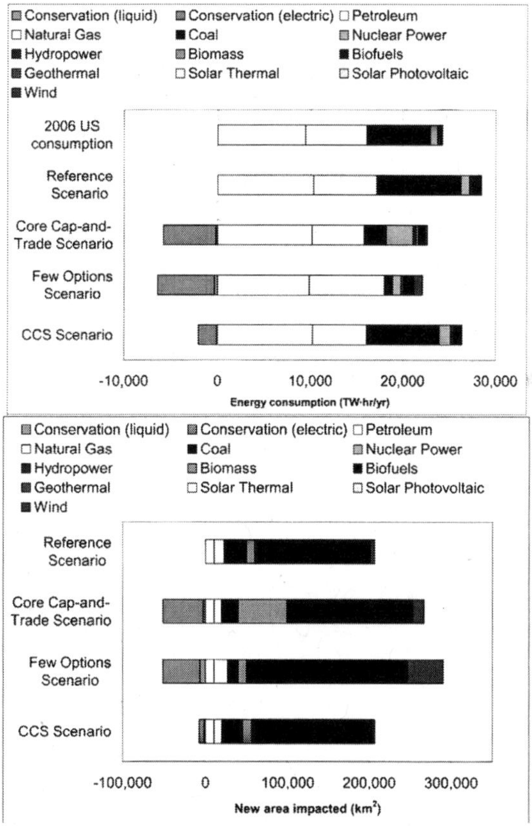

Figure 1. The US energy consumption and total new area impacted. (A) The US energy consumption in 2006 and under four EIA scenarios. Energy conservation of liquid fuels and electricity, calculated relative to the Reference Scenario, are shown as negative since they reduce consumption. (B) The total new area impacted because of development between 2006 and 2030. The new area impacted, or energy sprawl, is a product of consumption and the land-use intensity values in Figure 3. Energy conservation is calculated based on a scenario-specific weighted-average of the energy mix.

(the "Reference Scenario"), including the renewable fuel standard of the Energy Independence and Security Act of 2007, and under three Cap-and-Trade Scenarios: the "Core Cap-and-Trade Scenario", where the full Lieberman–Warner Climate Change Act is implemented; the "Few Options Scenario", where international offsets are not allowed and where new nuclear production and coal production with CCS are not possible; and the "CCS Scenario", where Congress enacts the Low Carbon Economy Act, a Cap-and-Trade system more favorable to coal with CCS.

Under each scenario, we first estimate the total new land area in the US needed to produce energy for each production technique as a function of the amount of energy needed and the land-use intensity of production. We examine the effect of US climate policy on future energy sprawl using energy scenarios based on proposed legislation, building on a body of literature on this topic [1, 2, 13-15]. Note that our analysis focuses only on US land-use implications, ignoring other, potentially significant international land-use implications of US climate policy. Second, we use available information on where new energy production facilities would be located to partition this area among major habitat types (Figure 2). We calculate the new area directly impacted by energy development within each major habitat type, but do not attempt to predict where within each major habitat type energy development will take place, nor possible indirect effects on land-use regionally or globally due to altered land markets. Our analysis provides a broad overview of what change in the energy sector will mean for area impacted in different natural habitat types, recognizing that such a broad analysis will inevitably have to simplify parts of a complex world.

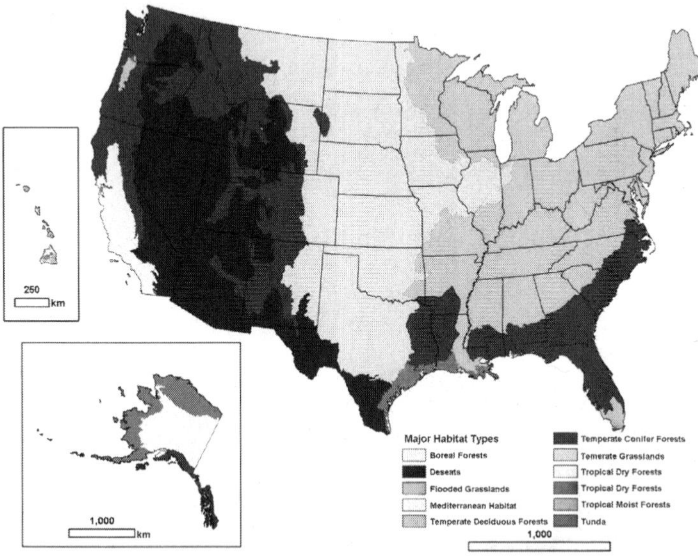

Figure 2. Major habitat types used to analyze the land-use implications of EIA scenarios. Within each major habitat type, there are a variety of land-uses, from relatively wild places to agricultural and urban systems. Our analysis estimates the new area needed for energy development within each major habitat type, without specifying where within each major habitat type this energy development might occur.

Land-use Intensity of Energy Production

The land-use intensity of different energy production techniques (i.e., the inverse of power density [16, 17]), as measured in km^2 of impacted land in 2030 per terawatt-hour per year, varies over three orders of magnitude (Figure 3). Nuclear power (1.9–2.8 km^2/TW hr/yr), coal (2.5–17.0 km^2/TW hr/yr), and geothermal (1.0–13.9 km^2/TW hr/yr) are the most compact by this metric. Conversely, biofuels (e.g., for corn ethanol 320–375 km^2/TW hr/yr) and biomass burning of energy crops for electricity (433–654 km^2/TW hr/yr) take the most space per unit power. Most renewable energy production techniques, like wind and solar power, have intermediate values of this metric.

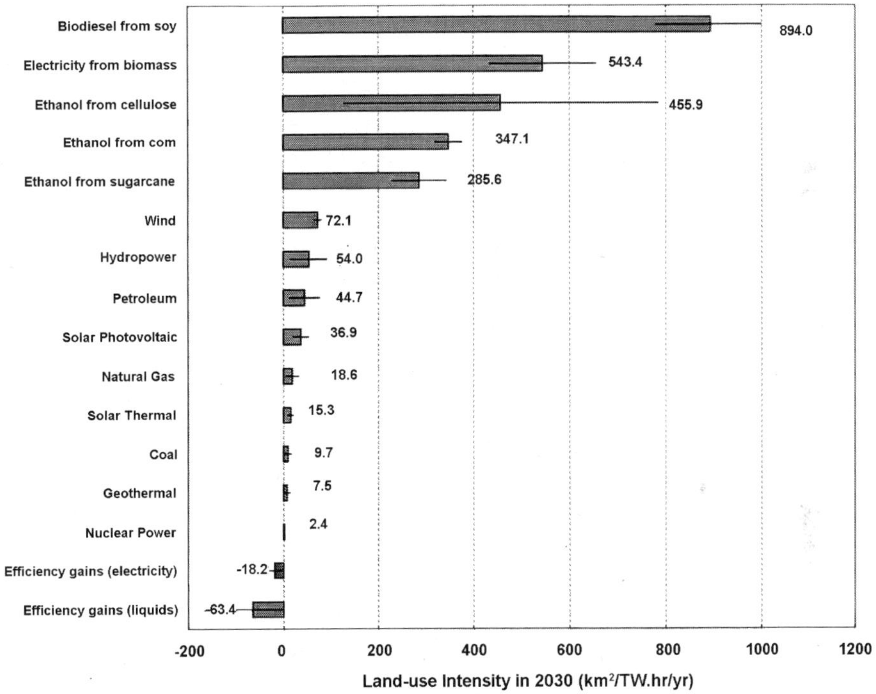

Figure 3. Land-use intensity for energy production/conservation techniques. Value shown is for 2030, as measured in km^2 of impacted area in 2030 per terawatt-hour produced/conserved in that year. Error bars show the most-compact and least-compact estimates of plausible current and future levels of land-use intensity. Numbers provided are the midpoint between the high and low estimates for different techniques. For liquid fuels, energy loss from internal combustion engines is not included in this calculation.

Energy conservation can reduce overall energy consumption thus reducing the area impacted by energy development. For every TW hour decrease in annual electric power consumption, a weighted-average of electricity use under the Reference Scenario suggests 7.6–28.7 km^2 of avoided impact. The corresponding figure for liquid fuels (27.5–99.3 km^2 of avoided impact per TW hr/yr) is higher because of the relatively large land-use intensity of biofuels.

Our definition of impact varies among energy production techniques, so a less compact way of generating energy does not necessarily mean that an energy production technique is more damaging to biodiversity, but simply that it has a larger spatial area impacted to some degree. Moreover, many energy production techniques actually have multiple effects on biodiversity, which operate at different spatial and temporal scales. Biodiversity impacts that are likely to scale with areal impact include habitat replacement and habitat fragmentation. Energy production impacts on biodiversity not related to land use intensity include impacts on air quality (e.g., acid rain, particulates), water quality (e.g., mercury, eutrophication), water consumption (e.g., irrigation water, evaporation from hydroelectric reservoirs), and water flows (e.g., dam-based hydroelectric). Further, the longevity of the impacts described here varies. For example, radioactive nuclear waste will last for millennia, some mine tailings will be toxic for centuries, and other mines may be reclaimed for agriculture within decades.

A full discussion of the impacts on biodiversity of energy production is beyond the scope of this chapter, but one fundamental distinction is worth making. Some energy production techniques clear essentially all natural habitat within their area of impact. A review of the literature found this to be true for coal, nuclear, solar, and hydropower, as well as for the growth of energy crops for biofuels or for burning for electricity. Energy crop production is a particularly complex situation because even if new energy crop production occurs on land that was previously in agricultural production, remaining global demand for agricultural commodities may spur indirect effects on land-use elsewhere, potentially causing an agricultural expansion in areas far from the location of energy crop production [18]. Other energy production techniques have a relatively small infrastructure footprint and a larger area impacted by habitat fragmentation and other secondary effects on wildlife. A review of the literature found that production techniques that involve wells like geothermal, natural gas, and petroleum have about 5% of their impact area affected by direct clearing while 95% of their impact area is from fragmenting habitats and species avoidance behavior. Wind turbines have a similar figure of about 3–5% of their impact area affected by direct clearing while 95–97% of their impact area is from fragmenting habitats, species avoidance behavior, and issues of bird and bat mortality.

Energy Sprawl in 2030

Regardless of climate change policy, the total new area affected by energy production techniques by 2030 exceeds 206,000 km^2 in all scenarios (Figure 1B), an area larger than the state of Nebraska. Biofuels have the greatest cumulative areal impact of any energy production technique, despite providing less than 5% of the US total energy under all scenarios. Biofuel production, and hence new area impacted, is similar among scenarios because EIA's economic model suggests that, under current law, incentives for biofuel production cause expansion of this energy production technique regardless of climate policy.

Nevertheless, in the scenarios we considered there is a tendency for greater reductions in greenhouse gas emissions to be associated with a greater total new area affected by energy development, particularly under the Core Cap-and-Trade and Few Options Scenario (Figure 4). A decrease in US emissions increases the new area impacted, although the magnitude of the effect is policy specific. Under the Core Cap-and-Trade Scenario, the burning of energy crops for electricity becomes profitable after the price of electricity rises due to the cap-and-trade system, resulting in a large new areal impact. Similarly, wind power is very important in the Few Options Scenario, where new electric production from coal and nuclear is not an option, and causes a large new areal impact. Conversely, in scenarios where there is not control on carbon emissions (Reference Scenario) or in cases where CCS is viable (e.g., CCS Scenario), coal production has a large new areal impact. The infrastructure for CCS is actually a small fraction of the area impacted by coal mining itself, so the major land-use change implication of the viability of coal with CCS is the continuation of coal mining.

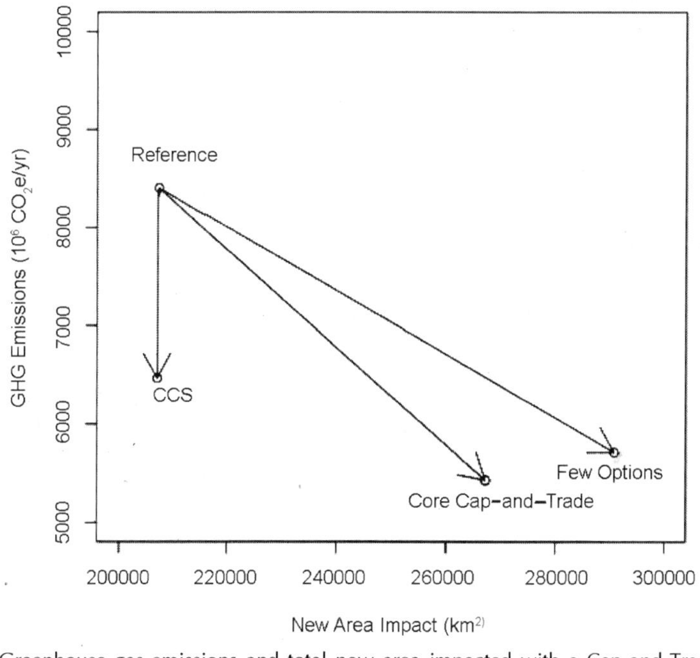

Figure 4. Greenhouse gas emissions and total new area impacted with a Cap-and-Trade system. Arrows depict the difference between the Reference Scenario, with no Cap-and-Trade system, and three other scenarios where a Cap-and-Trade system is implemented. Greenhouse gas emissions measured in million tons of carbon dioxide equivalent.

Our results stress the importance of energy conservation for reducing energy sprawl. Relative to the Reference Scenario, all Cap-and-Trade Scenarios involve a reduction in energy consumed (Figure 1B), because of energy efficiency and foregone consumption due to higher energy prices. This energy conservation is primarily in the electricity market, which is more elastic than demand for liquid fuels. Electricity conservation avoids impacts on at least 49,600 km^2 in the Core Cap-and-Trade Scenario, while at least 2,500 km^2 will be saved due to liquid fuel conservation, compared to the Reference Scenario. The EIA assumptions about the potential for energy conservation are relatively modest [19] and some groups argue that energy conservation has greater potential [20].

Habitat Impacts

The major terrestrial habitat types (Figure 2) impacted domestically by energy development varied among energy production technique (Table 1). Regardless of scenario, the major habitat types with the most new area affected, summing over all energy production techniques, are Temperate Deciduous Forests and Temperate Grasslands. In the Reference Scenario, Temperate Deciduous Forests have between 95,000 km^2 (most compact estimate) and 229,000 km^2 (least compact estimate) impacted, while Temperate Grasslands have 65,000–168,000 km^2 impacted. In the Core Cap-and-Trade Scenario these types have 119,000–254,000 km^2 and 88,000–191,000 km^2 impacted, respectively. Patterns of total new areal impacts are driven by biofuel production, which peaks in these two habitat types. Biomass burning for electricity and coal mining are also concentrated in Temperate Deciduous Forests and Temperate Grasslands. Wind production onshore is likely to affect Temperate Conifer Forests and Temperate Grasslands in the western US disproportionately. The least impacted habitats are: Tundra; Boreal Forest; Tropical Dry Forests; Flooded Grasslands; and Tropical Moist Forests. All of these habitat types have less than 150 km^2 impacted by energy development in the Reference Scenario and less than 600 km^2 impacted by energy development in the Core Cap-and-Trade Scenario, using the minimal sprawl estimates from Figure 3.

Table 1. Minimum new area (km^2) of habitat types impacted in the US.

Habitat Type	Coal	Biomass	Biofuels	Wind
Boreal Forests	94 (- 27)	2 (+12)	3 (+0)	6 (+9)
Deserts	2,310 (- 662)	257 (+1244)	372 (+14)	884 (+ 1 ,300)
Flooded Grasslands	0 (0)	30 (+143)	41 (+1)	0 (0)
Mediterranean Habitat	5 (- 1)	123 (+596)	1,699 (+37)	54 (+79)
Temperate Conifer Forests	4,936 (- 1,413)	1,883 (+9,106)	12,977 (+739)	2,835 (+4, 169)
Temperate Deciduous Forests	10,297 (- 2,945)	4,014 (+19,415)	76,841 (+6,751)	428 (+630)
Temperate Grasslands	7,508 (-2,147)	3,760 (+18,185)	46,821 (+4,136)	1,392 (+2,047)
Tropical Dry Forests	0 (0)	4 (+18)	5 (0)	34 (+50)
Tropical Grasslands	1 ,304 (- 373)	59 (+284)	1,583 (+65)	3 (+5)
Tropical Moist Forests	0 (0)	7 (+32)	9 (0)	78 (+115)
Tundra	0 (0)	0 (0)	0 (0)	0 (0)

Data are from the Reference Scenario for four major types of energy development: coal production, biomass burning for electricity, biofuel production, and wind power. Energy development is partitioned among habitat types, as depicted in Figure 2. Numbers in parentheses are the change in value under the Core Cap-and-Trade Scenario. For example, boreal forests have 94 km^2 affected under the Reference Scenario by coal, but have 27 fewer km^2 affected by coal under the Core Cap-and-Trade Scenario (i.e., 67 km^2).

doi: 1 0.1371 /journa l.pone.0006802.t001

The major habitats impacted by new energy development also varied among scenarios for certain energy production techniques. For example, scenarios where continued coal power generation is viable, either because of no restrictions on carbon emissions (Reference scenario) or because CCS is viable (e.g., CCS Scenario) have greater impacts on those major habitat types with major coal seams (e.g., Temperate Deciduous Forests). Conversely, scenarios where coal power generation is less viable have greater production from wind, affecting specific habitat types where those production techniques are favorable (e.g., Temperate Conifer Forests). Climate policy thus controls the extent to which specific habitat types are at risk from new energy development.

DISCUSSION

Our analyses show that, regardless of scenario, at least 206,000 km^2 of new land will be required to meet US energy demand by 2030. Further, implementing a cap-and-trade system may increase the total new area impacted by energy development and change its distribution among habitat types, relative to the Reference Scenario. Energy production will shift from fossil fuels to energy production techniques that draw more diffuse energy from a broader spatial area. Note that because the EIA analysis assumes that the energy market responds to price signals and does not explicitly attempt to minimize land-use per se, it is theoretically possible that there are other, more expensive mixes of energy production that would satisfy US energy needs in 2030 but would take less space. Although policies that reduce carbon emissions with minimal new land use are possible, none of the different policy EIA scenarios we considered were designed with that goal in mind. As shown by Wise et al. [21], if there were financial incentives to minimize land-use in energy production like a tax on greenhouse gas emissions from land-use change, the energy market response to a Cap-and-Trade might be very different from the response depicted in the EIA scenarios.

There are at least four ways to achieve emissions reduction but avoid the potential side effect of energy sprawl. First, energy conservation can help reduce the new energy needed by the US, reducing the area impacted by new energy development. Second, because end-use generation of electricity often occurs on already developed sites, it has minimal habitat impacts, and policy instruments that encourage end-use generation can also decrease the total area impacted. Third, our results suggest that energy sprawl is less severe when the cap-and-trade bill is more flexible, allowing for CCS, new nuclear plants, and international offsets. Fourth, many areal impacts can be mitigated or eliminated with appropriate site selection and planning for energy development. The new area affected by energy development within each major habitat type might, for example, have minimal biodiversity effects if sited in already disturbed places.

The areal impacts on habitat types will vary among scenarios, along with the potential biodiversity impacts of US climate policy. While not all impacts on biodiversity are strictly related to the areal impact, it is likely that energy production techniques with a large areal impact will have a relatively large biodiversity impact. Thus, the details of climate change policy, by favoring particular energy production techniques, pick biodiversity winners and losers. For instance, the Few Options Scenario assumes that international offsets, actions taken abroad to prevent carbon emissions or sequester more carbon, are not allowed under a cap-and-trade regime. The major response forecasted by EIA is an increase in wind production domestically relative to the Reference Scenario, affecting especially Temperate Conifer Forests and Temperate Grasslands. This increase in wind production may be compatible with biodiversity if properly sited, but certainly will pose a challenge for conservationists, because of the large area impacted and the threat of bird and bat mortality [22]. On the other hand, the biodiversity impacts of international offsets are beyond the scope of this chapter, but could conceivably be negligible (e.g., scrubber on Chinese coal plant smokestacks), negative (e.g., replacing natural grasslands with plantation forests), or positive (e.g., reducing emissions from degradation or deforestation).

Regardless of whether or not a cap-and-trade system is implemented, the EIA analysis forecasts that biofuels will increase dramatically in importance, with large areal impacts. In the Reference Scenario 141,000–247,000 km^2 will be impacted by biofuels. Within the US much energy crop production will occur on grassland or forest sites already in use for agricultural production, although increased aggregate demand for agricultural commodities may still spur agricultural expansion domestically or internationally (i.e., indirect land-use effects). In part, the large increase in biofuels forecasted by EIA simply reflects their assumption that current law, including the renewable fuel standard defined by the Energy Independence and Security Act of 2007, is maintained to 2030. In part it also reflects the inelasticity of the market for liquid fuels, relative to the electricity market, and the likely high cost of petroleum over the long-term [23]. It seems likely that under current law there will be a large areal impact from biofuels regardless of the Cap-and-Trade system put in place. Given the high land-use intensity of biofuels, techniques for either increasing the efficiency of biofuel production or making sites of energy crop production more biodiversity-friendly should be a high priority for research.

Our results demonstrate that, under certain policy scenarios, one potential side effect of reducing emissions is an increase in habitat impacts. The impact of a cap-and-trade system will be less, however, than from biofuel production already mandated by current law. Aggressive energy conservation, appropriate siting, sustainable production practices, and reduction of greenhouse gas emissions will all be necessary to minimize the impact of future energy use on habitat and wildlife. Energy sprawl deserves to be one of the metrics by which energy production is assessed.

MATERIALS AND METHODS

Our analysis proceeded in two phases. First, we calculated the new land area of energy development necessary to meet the EIA scenarios. Not all biodiversity impacts are directly related to the amount of land taken up by a technology, but it is likely that an energy production technique that takes a lot of land will have a relatively large biodiversity impact, so the total new area impact is a useful quantity to measure. Second, we partitioned this new land area among different geographic regions. We focused on domestic impacts for this analysis, ignoring the future habitat impacts of foreign-produced energy, principally future oil imports from Canada, Latin America, and the Middle East, as well as future ethanol imports from Brazilian sugarcane plantations. Similarly, we focused on terrestrial impacts for this analysis, ignoring potential freshwater or marine impacts by hydropower, wind, oil, and natural gas development in US waters. Finally, we are calculating direct land-use (how much land will we need to produce energy?), and did not attempt to estimate secondary changes in the land market in response to the direct land-use.

Scenarios

Our analysis is based on the EIA's 2008 scenarios of energy markets and the economy [12]. These scenarios were calculated by EIA's National Energy Modeling System, a comprehensive econometric model of US energy production, imports, and consumption. The scenarios supply information on the amount of additional energy consumed in 2030 in a particular sector (e.g., million barrels per day of petroleum, billion KW hr of new generation capacity by solar power). We use four scenarios in our analysis. A Reference Scenario describes what will likely happen in US energy markets under laws in force as of April, 2008. The Core Cap-and-Trade Scenario forecasts the effect of the full Lieberman–Warner Climate Security Act (S. 2191), which regulates emissions of greenhouse gases through a cap-and-trade system and provides economic incentives for increased energy efficiency [24]. One variant of S. 2191 is considered, the Few Options Scenario, where the use of international offsets in greenhouse gas emissions is either not economically feasible or is severely limited by regulation and where there is no increase in nuclear, coal with CCS, and imports of liquefied natural gas over current levels (the EIA called this the "Limited Alternatives/No International Offsets case"). Finally, the CCS Scenario forecasts the effect of the Low Carbon Economy Act (S. 1766), which also sets up a cap-and-trade system for greenhouse gases but offers strong incentives for the development and deployment of CCS [25]. Overall energy consumption by each sector for each scenario is shown in Figure 1A.

Four things about the scenarios are worth noting. First, other scenarios of the likely effect of S. 2191 [11, 24] are available from other groups, although they are broadly similar to the EIA analysis. Second, all scenarios of the effect of a Cap-and-Trade bill are tentative due to uncertainty about the pace of technological change, among other things. Thus, the EIA scenarios we use in this analysis must be taken as indicative of future trends, but not definitive [24]. Third, the EIA scenarios model the likely response of the US energy sector in response to a set of policy assumptions, and do not consider land-use per se. The EIA scenarios predict the most likely response, and do not attempt to find a more expensive energy mix that would minimize the total land-use. Finally, US energy policy has changed rapidly since the EIA's 2008 analysis, which still is the most current available full analysis of a cap-and-trade bill. The Warner–Lieberman bill is no longer considered an active bill, and most activity in Congress has focused on the Waxman–Markey bill (H.R. 2454), which proposes a very similar Cap-and-Trade system. Additionally, the passage of the American Recovery and Reinvestment Bill of 2009 (i.e., the stimulus bill) has provided significant support to renewable energy producers, particularly wind producers. Thus, the Reference Case discussed in this manuscript may underestimate the amount of renewable energy production in that case.

Despite these changes in US climate policy since the EIA's 2008 analysis, the broad results of our analysis will likely apply to any similar Cap-and-Trade system.

Calculating Area Requirements

Our general strategy was to estimate reasonable most-compact and least-compact values of the amount of area needed to produce a certain amount of energy in a year, the land-use intensity of production. We then multiplied the needed energy by our measurement of land-use intensity of production (km^2/energy/year) to obtain the "energy sprawl," the total new area needed for new energy production in that sector. Calculation of land-use intensity in this manner is useful for the goals of our analysis, allowing calculation of the total new area impacted by energy development. It is similar to the measurement of "area efficiency" [26], in that it does not attempt to account for site preparation prior to energy production nor potential site reclamation after production has ceased, which are sometimes considered in a full life-cycle analysis [3]. Ecologically, over the time period of our study few sites will be reclaimed to vegetation approaching their original habitat value [27]. Numerically, our land-use intensity values represent a lower estimate because they do not include land-use during site preparation or reclamation. While our methodology does not allow a full statistical analysis of the uncertainty of our estimates, the variation between the least compact and most compact estimates of land-use intensity captures most of the uncertainty.

Estimating areal impacts of new methods of electricity generation is relatively straightforward. The EIA forecasts new power generating capacity needed (billion KW), after accounting for likely retirement of existing generation capacity and the nameplate capacity factors of different energy production techniques. We used values from the literature to calculate the km^2 of impact per GW of new generating capacity (see Tables 1 and 2). Our approach ignores the importation of electricity from Canada or Mexico, on the grounds that this is predicted by the EIA to remain a minor component of total electricity generation. End-use generation of electricity, which is tracked separately for the EIA for several sectors, is considered to have negligible area requirements, since it by definition occurs on previously developed sites. Similarly, the energy efficiency increases in the EIA scenarios are considered to have negligible area requirements, since they occur through upgrades in existing building and infrastructure. We have not attempted to estimate the new area impacted by new long-distance transmission lines needed to carry new capacity, as the length and location of new lines is very uncertain and depends on future energy production mixes as well as federal and state policy.

Our large-area estimates of land-use intensity of production are generally from current operating plants, whereas our small-area estimates represent expert opinion about expected future technological advances in efficiency. The definition of impact implied by these estimates was designed to match the majority of published studies of the severity of impact on biodiversity, and thus varies slightly among energy technologies. For example, for hydropower we have assumed that new dam construction inundates an area of terrestrial habitat, removing it of most of its native biodiversity. Note also that the different categories of electricity generation derive from the EIA report. For instance, the EIA chose to recognize solar photovoltaic and electricity from solar thermal as two different energy production techniques, and we follow their convention in our analysis. The EIA partitioned solar photovoltaic use between end-users, assumed to have negligible land-use implications in our analysis, and large-scale generation, which does have significant areal implications. We track end-use and large-scale generation separately in this and other cases, following EIA's methodology.

Table 2. Land-use intensity of production for coal mining.

Geographic region	Proportion surface mining	Most compact ha/mmt	Least Compact ha/mmt	Notes
Appalachia U.S.	0.352	11.6 pit, 79.1 surface	115.7 pit, 791.4 surface	Proportion of surface mining from EIA's 2006 Coal production in the United states fad sheet (uses 2003 data). Surface coal yields most compact figure based on Spitzley and Keoleian [3], least compact figure on Flattop mine has 2.3 million tons of coal on 600 acres. Pit mining assumed 10% of area as surface mining.
Interior U.S.	0.643	11.6 pit, 79.1 surface	115.7 pit, 791.4 surface	As above
Western U.S.	0.898	11.6 pit, 79.1 surface	115.7 pit, 791.4 surface	As above
Import	0.671	11.6 pit, 79.1 surface	115.7 pit, 791.4 surface	Proportion surface mining assumed to be same as overall U.S. average.

For each geographic region of coal, we show the proportion of the coal that is from surface mines versus pit mines, as well as least-compact and most-compact estimates of the area requirements of coal mining, in hectares per million metric tonnes of coal. Impact for coal mines is defined as the area directly surrounding the mine site.
doi:1 0.1 371/journal.pone.0006802.t002

The EIA forecasts biofuel (ethanol, biodiesel, and other liquids from biomass) production and total use, with the vast majority of use being ethanol in transportation. Moreover, they estimate the proportion of domestic ethanol production from corn, cellulose, and other feedstocks, as well as net imports of ethanol and biodiesel. For each type of biofuel and its feedstock, we estimated least compact and most compact estimates of the number of m^2 of feedstock cropland per liter of biofuel. In general, least compact estimates are for current agricultural yields (kg/m^2) and biofuel production efficiencies (l/kg), while most compact estimates are a product of future agricultural yields and biofuel production efficiencies. For some crops like soy, the difference in least compact and most compact estimates is primarily due to a predicted increase in yield, while for other biofuels like cellulosic ethanol the difference is largely due to differences in biofuel production efficiency.

For many biofuels, farmers also make a portion of their income from coproducts, portions of the crop that are not used to make biofuels but have another economic market. We use a market-value allocation approach, defining the actual increase in production area of a crop as a function of the fraction of the economic value of the crop that is embodied in the biofuel [28]. Note that our methodology tracks the direct land-use needs of biofuel production, and does not consider indirect effects on land-use via agricultural commodity markets. For example, if a soy field in the US is switched to corn to make ethanol, than soy production will likely expand elsewhere either domestically or internationally. A full accounting of the demand and supply curves of the various agricultural crops, biofuels, and their coproducts is beyond the scope of this project, but is an active area of research [18, 29-31].

Estimating the areal impact of fossil fuels was done in an analogous manner. The EIA analyses divided domestic coal production into three geographic regions (Appalachia, Interior, and West). We separated the coal produced in each region into the proportion mined underground and the proportion mined at the surface, using EIA's factsheet on coal production in the US. Then, using data on the amount of coal removed per unit area, we calculated area impacted (Table 2). For this analysis, we ignore the relatively small areal impact of coal burning power plants, which comprise a small fraction of the areal impact of coal mining.

For oil production, the EIA estimated imports as well as domestic production from three geographic regions: the land surface of the contiguous 48 US (lower 48 onshore); water bodies in or close to the contiguous 48 US (lower 48 offshore) and the state of Alaska. Note that because EIA assumes existing law will continue, including the ban on new oil drilling in the Arctic National Wildlife Refuge, Alaska oil production actually falls slightly under all scenarios. Based on historical data from the United States Geological Survey (USGS), we estimated the proportion of oil production that is from oil wells (as opposed to incidental production from gas wells) and the average number of barrels per day per development well (Table 3). We also estimated the number of development wells that are abandoned per year. Using these data, we calculated the number of new development wells needed to maintain current production and the number of wells needed to achieve any production increase forecasted by the EIA. By using this approach we are accounting for the tendency of older wells to fall in production over time and be abandoned, necessitating new wells just to maintain current production levels.

Table 3. Land-use intensity of oil and natural gas production.

Natural Resource	Proportion from well type	Average production	Most compact (ha/well	Least Compact (ha/well)	Notes
Oil	0.862 from oil wells for onshore production in lower 48, 0.636 for Alaska.	1.14 m3/day of crude from lower 48 onshore, 56.13 m3/day of crude from Alaska onshore.	5.67	32.38	See text for estimation of trends in oil production. Pinedale Anticline spacing is taken as most-compact estimate and Jonah field spacing as least-compact estimate.
Natural Gas	0.948 from natural gas wells for U.S. onshore.	2,821 m3/day of dry natural gas from U.S. onshore.	5.67	32.38	As above.

For each geographic region (lower 48 onshore, lower 48 offshore, and Alaska), we used estimates of the proportion of the resource that is withdrawn from each type of well and average well productivity, as well as least-compact and most-compact estimates of the area affected by each well, in hectares per well. Impact for wells is defined as both the well area and the surrounding habitat fragmented by wells, access roads, and other structures. See text for details on impact calculations of oil and gas pipelines.

doi: 1 0.1371 /journal.pone.0006802.t003

For natural gas production the EIA provides one aggregate domestic production figure, but does differentiate between pipeline natural gas imports, which are predicted to decline, and liquefied natural gas imports, which are predicted to increase. Following a similar approach to the petroleum case, we estimated the proportion of gas production that is from gas wells (as opposed to incidental production from oil wells) and the average annual thousand cubic feet per well (Table 3). Methodology generally followed that used for oil wells.

The EIA provides explicit estimates of the emissions avoided (million metric tons of CO_2 equivalent) by the use of CCS technology, relative to a 2006 baseline, for three sectors of electricity generation (petroleum, natural gas, and coal). Because of the EIA assumptions about the cost-effectiveness of implementation of CCS and the incentive structures in S. 2191 and S. 1766, power plants burning petroleum for electricity do not generally implement CCS, whereas power plants burning natural gas or coal do whenever there is a carbon cap in place (i.e., not the Reference Scenario) and when the CCS technology is available (i.e., not the Few Options case).

For new petroleum and natural gas production, we estimated the amount of new pipeline needed, based on current ratios of kilometers of pipeline to wells, assuming that these ratios held constant into the future (0.9 km/well for oil production, 1.3 km/well for gas production). For CCS, we also estimated the new pipelines needed to move CCS (0.5 km/well): the length of new pipeline per CCS injection site is likely to be more limited than in the petroleum or natural gas case because CO2 has little economic value [32]. For all pipelines, we then estimated the area impacted on either side of the pipe (most-compact estimate 0.3 ha/km of pipe, least-compact estimate 1.8 ha/km of pipe, based on common right of ways of pipelines). By estimating the area impacted by pipelines in this way, we are assuming that the process of pipeline construction removes most native biodiversity, and that any revegetation after pipeline construction will have minimal biodiversity value.

A literature review revealed that many energy production techniques actually have multiple effects on biodiversity, which operate at different spatial and temporal scales. A full discussion of the impacts on biodiversity of energy production is beyond the scope of this chapter, but we recorded quantitative data on the proportion of our defined impact zone that was directly affected by land clearing, as opposed to more diffuse processes such as habitat fragmentation and organism avoidance behavior. Studies with useful quantitative or semi-quantitative data on this topic include: Coal [33], Nuclear [34-37], Solar [38, 39], Hydroelectric [40, 41], Biofuels [5, 18, 28-31, 42], Geothermal [43], Natural Gas and Petroleum drilling [44], [45], and Wind [22, 46-51].

Where Energy Development Occurs

The goal of this phase of the analysis was to partition the total area of new energy development among geographic regions. We ignored energy production techniques that had no significant cumulative areal impact as calculated above (i.e., end-use power generation, energy efficiency gains). For our regionalization analysis, we chose definitions of geographic regions that have maximal relevance to biodiversity yet are coarse-scaled enough to average over errors and uncertainty in more fine-scaled input data on energy resource availability and demand. For terrestrial impacts we used the 11 major habitat types of the US, as defined by Nature Conservancy ecoregions [52-54]. For each major habitat type, we estimate the total area of new energy development, without attempting to specify where within each major habitat type development will take place. Within each major habitat type, there are a variety of land-uses, from relatively wild places to agricultural and urban systems. Thus specific siting decisions, while outside the scope of our analysis, will be important in determining actual biodiversity impact.

Throughout our analysis, we excluded certain areas as being protected or restricted from development, modeling our decision rules on those used in the Department of Energy's report "20% Wind Energy by 2030 [47]." We excluded areas that were protected areas with a Gap Analysis Program code of 1 or 2 (i.e., permanent protection excluding development), based on the Protected Area Database of the United States, version 4 [55]. We also excluded airports, urban areas, and wetlands/water bodies from development, based on vector layers included with Environmental Systems Research Institute's (ESRI) ArcGIS package. Areas with an average slope greater than 20% were also excluded, based on a surface analysis of the GTOPO global digital elevation model [56]. Finally, for wind power we assumed that areas within 3 km of an airfield or urban area were not developable.

For each energy production technique, we partitioned its land use among regions in one of two methods. For some energy production techniques, continuous (i.e., interval or ratio scale) estimates of the supply of that resource were available for different geographic regions. For example, the Department of Energy publishes a continuous estimate of the water power potential in MW of the different hydrologic regions of the US [57]. In these cases with continuous estimates of resource supply, we assumed that the area of energy development in each geographic region was proportional to the total supply in that region. For some resources, the geographic units in which data was available did not match those of our analysis units, and we used geographic information system (GIS) analyses to partition the resource among habitat types, making the simplifying assumption that the resource was evenly distributed within the original geographical units of the data. To give an example from one particularly dataset, potential biomass estimates were available from the National Renewable Energy Laboratories (NREL), summarized per county [58]. We calculated tons/km^2 for each county, digitized the data to a 1 km raster resolution of the US, and then used ESRI ArcGIS ZonalStatistics commands to sum up the total available biomass in each of our major habitat types.

Other energy production technologies had data on the supply of the resource that were categorical (i.e., ordinal scale). For example, NREL wind power maps rank sites on a scale of 1–7, based on the quantity of wind available as well as its consistency. In these cases with categorical data, we reclassified the US into excellent, good, and poor regions for development of that energy resource. In some cases a continuous estimate of a proxy for a resource was available rather than a direct estimate of power availability, and in these cases we classified the resource into categorical categories based on published opinion about what sites were developable. While our decision rules are admittedly arbitrary, they are derived from common GIS analysis for site selection in the energy industry, and we believe any reasonable set of decision rules would provide qualitatively similar results. In general, we looked at both the supply of a particular resource (e.g., how much sunlight is there?) and the demand (e.g., how far away is the nearest electric transmission line to carry the power to market?). We then calculated how much of each geographic region was in the three categories (excellent, good, and poor) using ESRI ArcGIS ZonalStatistics commands. Next, we assumed that the area of energy development in each geographic region was proportional to the area classified as excellent in that region. If all areas categorized as excellent were developed without meeting the total areal target, the remaining development was assumed to "spill-over" to the good category, where it was similarly divided among geographic regions.

KEYWORDS

- **Climate policy**
- **Core Cap-and-Trade Scenario**
- **Energy conservation**
- **Energy Information Administration**
- **Land-use intensity**

AUTHORS' CONTRIBUTIONS

Designed the project, completed portions of the analysis, and wrote the chapter: Robert I. McDonal. Analyzed biofuel and biomass trends: Joseph Fargione. Analyzed trends in oil and natural gas exploration: Joe Kiesecker. Analyzed trends in solar power generation: William M. Miller. Designed the project: Jimmie Powell.

ACKNOWLEDGMENTS

We thank M. Wolosin, P. Kareiva, J. Hoekstra, and B. Thomas for reviewing earlier drafts of this manuscript. J. Slaats provided GIS support for this project. We thank all of the organizations that created data that made this analysis possible, including: NETL, DOE NREL and EERE, NASA, USGS, and Ventyx Corporation.

Chapter 3

Mediterranean Ecosystems Worldwide: Climate Adaptation

Kirk R. Klausmeyer and M. Rebecca Shaw

INTRODUCTION

Mediterranean climate is found on five continents and supports five global biodiversity hotspots. Based on combined downscaled results from 23 atmosphere-ocean general circulation models (AOGCMs) for three emissions scenarios, we determined the projected spatial shifts in the mediterranean climate extent (MCE) over the next century. Although, most AOGCMs project a moderate expansion in the global MCE, regional impacts are large and uneven. The median AOGCM simulation output for the three emissions scenarios project the MCE at the end of the 21st century in Chile will range from 129 to 153% of its current size, while in Australia, it will contract to only 77–49% of its current size losing an area equivalent to over twice the size of Portugal. Only 4% of the land area within the current MCE worldwide is in protected status (compared to a global average of 12% for all biome types), and, depending on the emissions scenario, only 50–60% of these protected areas are likely to be in the future MCE. To exacerbate the climate impact, nearly one-third (29–31%) of the land where the MCE is projected to remain stable has already been converted to human use, limiting the size of the potential climate refuges and diminishing the adaptation potential of native biota. High conversion and low protection in projected stable areas make Australia the highest priority region for investment in climate-adaptation strategies to reduce the threat of climate change to the rich biodiversity of the mediterranean biome.

The mediterranean biome is a global conservation priority [1, 2] owing to high plant species diversity and density that rivals that of tropical rainforests [3, 4]. The biome's mild climate and proximity to the ocean also makes it attractive to humans, resulting in disproportionately high conversion for agriculture, development, and other human uses [5, 6]. Found on five continents, the mediterranean biome includes the Mediterranean Basin, the western US (California) and Mexico (northwest Baja), central Chile, the cape region of South Africa, and south and southwestern Australia [4]. These five areas cover just 2% of the Earth's land area, but support 20% of the Earth's known vascular plant diversity [3, 7]. Despite this biome's relative biological wealth, formal land management for biodiversity conservation is lagging, as it has the second lowest level of land protection of all the 13 terrestrial biomes [5]. By 2,100, the mediterranean biome is projected to experience the largest proportional loss of biodiversity of all terrestrial biomes due to its significant sensitivity to multiple biodiversity threats and interactions among these threats [8].

The mediterranean biome's extraordinary plant diversity and endemism are a result of the evolutionary processes induced by the characteristically unique annual cycles of extended summer drought and cool wet winter, high topographic variation, and low soil fertility [9]. Climate change resulting from increases in atmospheric concentrations of greenhouse gases will impact the extent and distribution of the mediterranean climate, posing a threat to the survival of many species. While biome level analyses are rare, there has been a recent proliferation of climate change impacts studies specific to species and habitats in each of the five mediterranean regions [10-14]. These studies generally project significant reductions in endemic species range sizes. For example, in California, 66% of the endemic plant taxa will experience >80% range reductions within a century [14]. Midgley et al. projected a 51–65% reduction in the mediterranean biome in South Africa by 2050, and that only 5% of the endemic Proteaceae species modeled would retain more than two-thirds of their current range [15]. However, each of these studies is limited to one of the five mediterranean regions and generally focuses on the results from one to a few of the 23 AOGCMs. In this analysis, we focus on projected shifts in the mediterranean climate using a consistent methodology worldwide. This allows for comparisons between regions and highlights areas that are in most urgent need for climate change adaptation action. We present the first biome-level analysis of global climate change using all AOGCMs analyzed in the Intergovernmental Panel on Climate Change's (IPCC) Fourth Assessment Report [16]. Finally, we estimate the potential for facilitation of species adaptation within a region via the climatic stability of protected areas or via the migration pathways to optimal climatic conditions, based on current distribution of areas managed for biodiversity conservation, current patterns of land conversion, and magnitude of future impacts of climate change. We refer to this measure as extrinsic adaptation potential which illuminates characteristics of the landscape that facilitate species adaptation, in contrast to species-specific characteristics that determine intrinsic adaptation potential such as dispersal ability or genetic diversity. Intrinsic and extrinsic adaptation potential together defines the adaptation potential of a species.

MATERIALS AND METHODS

The mediterranean biome is typically mapped using a combination of climate characteristics and plant assemblages that vary by region. One widely-used delineation of this biome is a collection of ecoregions mapped by the World Wildlife Fund that covers 2.2% of the earth and is based upon climate and plant associations [2]. As the climate changes, the impacts on the climate characteristics across all five mediterranean regions will be mechanistically similar, but the impacts on the plant assemblages will differ as the definition and composition of mediterranean vegetation differs among regions. For this reason, this analysis focuses on the climatic impacts and maps the MCE across all five regions based solely on climatic factors and not plant assemblages.

Although there are varying definitions for the mediterranean climate, we chose one definition that can be readily mapped with available climate data and has minimal over prediction into areas that are not part of the mediterranean biome. According to this definition, published by Aschmann [17], an area is within the MCE if it meets five conditions; (1) The winter must be wetter than the summers (>65% of the precipitation falls in the winter half of the year), (2) the annual precipitation must not be too low (>275 millimeters (mm)), (3) nor too high (<900 mm), (4) the winter must be cool (<15°Celsius (C) mean temperature for the coldest month of the year), but (5) it cannot have too much frost (<3% of the annual hours are below freezing). We used the WorldClim [18] high resolution (2.5 arc-min or ~5 kilometer (km) horizontal resolution at the equator) grids of global climate data summaries from 1960 to 1990 to map the current MCE where all five Aschmann conditions are met.

Data for projections of future climate conditions were derived from the results of the AOGCMs run to support IPCC's Fourth Assessment Report. The data [World Climate Research Programme's (WCRP) Coupled Model Intercomparison Project (CMIP) phase 3 multi-model dataset] include seven future emissions scenarios. Three of these scenarios are used most often by modeling groups and are considered representative of low (B1 or stabilization at 550 ppm atmospheric CO_2), moderate (A1b or stabilization at 720 ppm atmospheric CO_2) and high (A2 or no stabilization) emission trajectories [19]. We compiled the AOGCM output data for monthly surface air temperature and precipitation flux for the 20th century and the 21st century for three future emissions scenarios. While some modeling groups have generated multiple simulations for a given scenario and others have done no simulations for a given scenario, we analyzed all available AOGCM simulations in the CMIP multi-model dataset, including 48 low emission simulations, 52 moderate emission simulations, and 36 high emission simulations, for a total of 136 simulations of future climate. By doing so, we treat each AOGCM simulation for a given emissions scenario as a unique and probable experimental outcome and average the results, thereby elucidating a more robust set of potential climate outcomes.

To reduce the variability associated with annual climate projections, we averaged the monthly data in the AOGCM simulation results to two 30-year periods; one "current" and one "future." The WorldClim data is primarily derived from 1960 to 1990 weather records, so we averaged the monthly data from January, 1960 to December, 1989 for each of the modeled 20th century simulations to generate the current time period. The majority of the model simulations end in 2100, so we averaged the monthly data from January, 2070 to December, 2099 for each of the modeled 21st century simulations to generate the future time period. We then subtracted the modeled current data from the modeled future data to reduce modeling biases and generate projected climate anomalies. For example, an AOGCM simulation may have modeled the average July temperature for a specific area to be 24°C for the current time period (1960–1989) and 27°C for the future time period (2070–2099), so the projected climate anomaly for that area would be 3°C. We used the change factor approach to downscale the projected climate anomalies from the coarse resolution of the AOGCMs to the finer resolution of the WorldClim data. This method involves interpolating the projected climate anomalies and adding the interpolated data to the current climatology.

We applied Aschmann's [17] conditions to generate binary maps of the projected future MCE for each AOGCM simulation. The size of the projected future MCE in each region was compared to the size of the current MCE for each AOGCM simulation, and the average and 5, 10, 25, 50, 75, 90, and 95 percentiles of the projected percent change for all simulations in a given emissions scenario were calculated. This provides both a measure of the range and the central tendency of the ensemble of projected changes [20]. We considered using a weighted-average, but since we do not have a testing dataset for the future locations of the MCE, and other studies have found little increases in predictive power with a weighted average compared to a non-weighted average [21], we used only the average.

We spatially combined the current MCE and all of the future MCE projections to calculate the percent of AOGCM simulations predicting an expansion or contraction of the MCE for each grid cell on the globe. We defined seven categories based on whether areas were in the current MCE or not, and the number of AOGCM simulations that project the area will be in the future MCE (Table 1). These categories were mapped using the suite of AOGCM simulations for each emissions scenario.

Table 1. Mapped categories for the MCE future projections.

Category	Area incurrent MCE?	Percent of AOGCM simulations projecting area will be in future MCE
Confident Stable	Yes	90-100%
Likely Stable	Yes	66-90%
Uncertain	Yes	33-66%
Likely Contraction	Yes	10-33%
Confident Contraction	Yes	0-10%
Confident Expansion	No	90-100%
Likely Expansion	No	66-90%

doi: 1 0.1371 / journal.pone.0006392.t001

We determined the amount of land protected and modified through development and land conversion within the current and projected future MCE [5]. We used all World Conservation Union categories (I–VI) in the 2006 World Database on Protected Areas to map areas that are protected in the current MCE and in the areas where the MCE is projected to expand (www.unep-wcmc.org/wdpa) [22]. Marine protected areas were not included, and protected areas with only point location were mapped as circles with the correct area. We converted the polygon data to a binary 2.5 min resolution grid by assigning a grid cell a value of 1 if the center of the cell falls within a protected area polygon, and 0 if not. For spatial data on modified or converted areas, we used the areas classified as "cultivated and managed areas" and "artificial surfaces" in the Global Land Cover 2000 (www-gvm.jrc.it/glc2000) [23]. We converted these data to a binary 2.5 min resolution grid where 1 indicated a grid cell is converted and 0 indicated it is not. We performed a spatial combination of these two binary grids with the current and projected future MCE grids. Areas that were classified as both protected and converted were considered converted. From this combination of grids, we could determine the percent of the MCE that is protected and converted to human land uses.

The MCE at the end of the 20th century covered just over 1.5 million km^2, according to Aschmann's [17] definition. This is a conservative definition of the MCE and reflects the core areas of the mediterranean biome. For comparison, on a commonly used map the mediterranean biome covers 3.2 million km^2 [2], or over twice the area in the current MCE. Approximately 60% of the current MCE occurs in the Mediterranean Basin, and covers a portion or all of the following countries: Algeria, Cyprus, Greece, Iran, Iraq, Israel, Italy, Jordan, Lebanon, Libya, Malta, Morocco, Portugal, Spain, Syria, Tunisia, and Turkey. The remaining MCE occurs in Australia (25%), the US/Mexico (9%), Chile/Argentina (4%), and South Africa (2%).

The majority of AOGCM and emissions scenario projections of the MCE at the end of the 21st century (or future MCE) are larger than the MCE at the end of the 20th century (or the current MCE). The median future MCE increases to 106, 107, or 111% of its current size, for the low, medium, and high emissions scenarios, respectively (Figure 1). However, this pattern is not consistent within each region. Instead, there is a disparity between the regions with some projected to experience an increase in the MCE in the future and some projected to experience a decrease in the MCE. Almost all of the AOGCM simulations project an increase in the MCE in the Mediterranean Basin with the median future MCE increasing to 115, 126, or 132% of its current size for the low, medium, or high emissions scenarios, respectively. The median projected increases are greater in Chile/Argentina, ranging from 129% for the low to 153% for the high emissions scenario. In the US/Mexico region, the projected change in the MCE is less dramatic, with the median future MCE decreasing to 96, 95, and 94% of its current size, for the low, medium, and high emissions scenarios, respectively. In South Africa, greater than 90% of the AOGCM simulations project a decrease in the future MCE, with the median estimates ranging from 83% of the current MCE for the low to 60% for the high emissions scenario. In Australia, the projected area reduction is more extreme, with median estimates ranging from 77% of the current MCE for the low to 49% for the high emissions scenario.

By overlaying all of the future MCE projections, we were able to map all of the grid cells on the globe where the MCE is likely to expand, contract, and remain stable with varying levels of confidence (Figure 2). We show that there are areas of contraction even within regions that are projected to have a net increase in the MCE. For example, the median projection of the future MCE in the Mediterranean Basin is larger than the current MCE, but most AOGCM simulations project contractions in Morocco and in the Middle East. The geographic separation between the areas of contraction and expansion within each region highlighted in Figure 2 will have important implications for adaptation of native plants and animals with limited dispersal or migration capabilities.

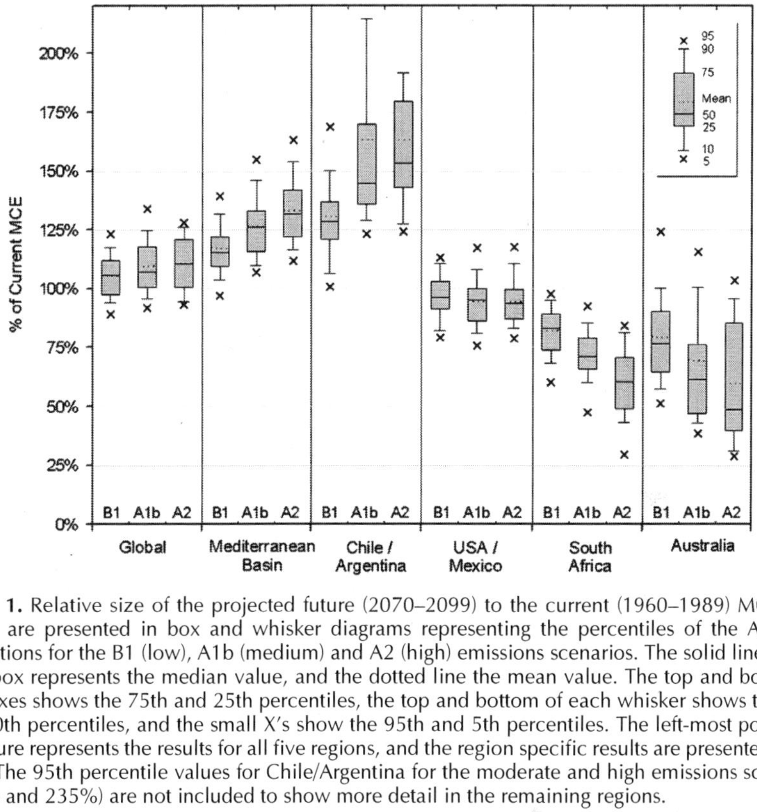

Figure 1. Relative size of the projected future (2070–2099) to the current (1960–1989) MCE. The results are presented in box and whisker diagrams representing the percentiles of the AOGCM simulations for the B1 (low), A1b (medium) and A2 (high) emissions scenarios. The solid line within each box represents the median value, and the dotted line the mean value. The top and bottom of the boxes shows the 75th and 25th percentiles, the top and bottom of each whisker shows the 90th and 10th percentiles, and the small X's show the 95th and 5th percentiles. The left-most portion of the figure represents the results for all five regions, and the region specific results are presented to the right. The 95th percentile values for Chile/Argentina for the moderate and high emissions scenarios (320% and 235%) are not included to show more detail in the remaining regions.

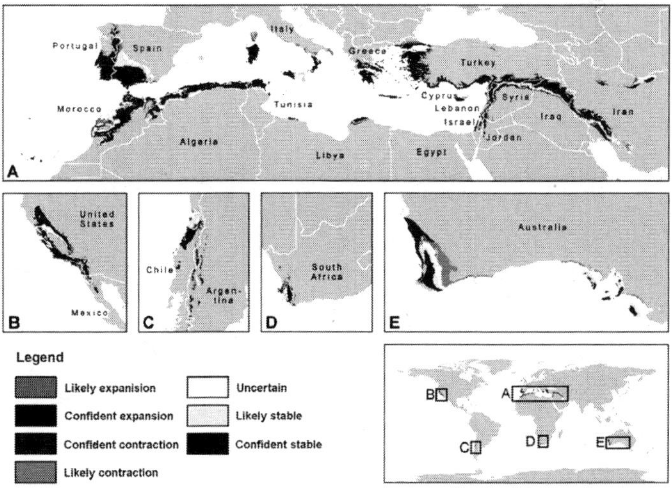

Figure 2. Projected status of the MCE in 2070–2099 relative to 1960–1989 under high (A2) emissions scenario. The projected status is considered likely if at least 66% of the AOGCM simulations agree, confident if at least 90% agree. Maps A. through E. are un-projected at 1:50,000,000 scale.

In addition to showing where the MCE is projected to contract, we also wanted to determine if there was AOGCM agreement on why it will contract (e.g., a warmer winter temperature, or less annual precipitation). These changes will have important implications for the persistence of native plants and animals in the mediterranean biome. To do this, we determined the level of agreement among the AOGCM simulations on which Aschmann's conditions were no longer met for the areas with a projected contraction. We report the results for areas where at least 90% of the AOGCM simulations agree that the MCE will contract and agree on the changing climatic condition causing the contraction under the high emissions scenario. Across all five regions, we can project with the most confidence that 7.2% of the current MCE will contract. Over half of this projected contraction, or 4.0% of the current MCE, results from warming in the winter. Almost a quarter of the projected contraction, or 1.7% of the current MCE, results from a drop in total annual precipitation below the 275 mm threshold. For one fifth of the projected contraction, or 1.5% of the current MCE, the AOGCM simulations agree that an area will contract, but they do not agree on which condition will change. By country, most of the loss in Australia, the US, Iran, Israel, and Libya is attributable to a warming winter, while the majority of the loss in Argentina, South Africa, Morocco, and Syria is due to a drop in annual precipitation.

Current land conversion and protection status and configuration relative to these climatic changes will play an important role in determining the extrinsic adaptation potential for the species of the mediterranean biome. Approximately one third of the area in the current MCE has already been converted to agricultural and urban land uses. If most of the converted land is in areas where the MCE is projected to contract, extrinsic adaptation potential will not be significantly reduced because these areas are poor habitat for native species. However, if the areas projected to have a stable or expanded MCE are disproportionately converted, this will exacerbate the negative impacts of climate change on biodiversity in the mediterranean biome. When looking across all five regions, we found the land conversion patterns are similar in areas where the MCE is projected to contract with confidence (23%) and in areas projected to remain stable with confidence (29%), but the regional patterns were more variable (Figure 3). In California and Mexico, extrinsic adaptation potential is conserved in the future because most of the conversion lies in areas that are projected to contract, and there is little conversion in the areas of stability. Similarly, Chile/Argentina and South Africa also have low levels of conversion in the confident stable areas. The opposite is true in Australia, where 64% of the likely stable area and 49% of the confident stable area is already converted, greatly diminishing the extrinsic adaptation potential of native biota.

At 4%, the level of protection for biodiversity in the current MCE is below that of the more expansive mediterranean biome (5%) and well below the global average (12%) for all terrestrial biome types [5]. We wanted to determine if the level of protection is higher or lower in areas with high likelihood of retaining a mediterranean climate. For the entire biome, just over half of the existing protected areas are projected to retain the mediterranean climate with high confidence, even under the high emissions scenario (Figure 4). The projected status of protected lands in some regions is much better, as over 70% of the protected areas in California/Mexico, South Africa, and the Mediterranean Basin are in the confident stable areas. In stark contrast, the MCE is projected to contract or is uncertain in over 75% of the protected areas with mediterranean climate in Australia, and there is almost no projected expansion into other protected areas to offset this loss.

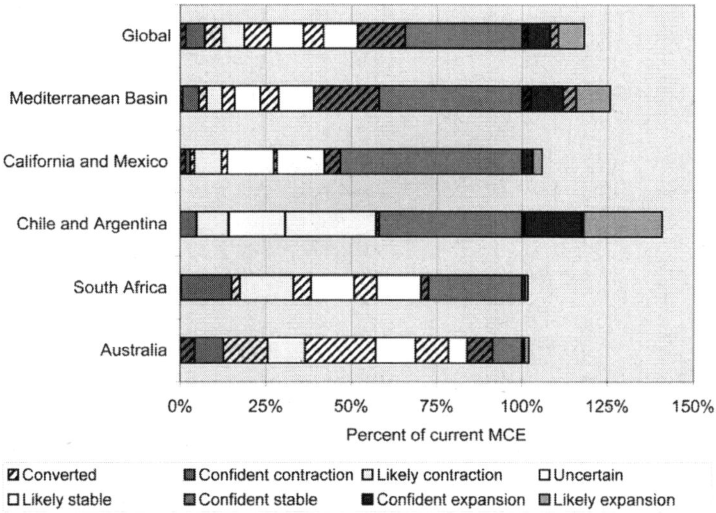

Figure 3. Projected future changes and current conversion status of the MCE under the high (A2) emissions scenario. The percentages indicate the portion of the area within the current MCE.

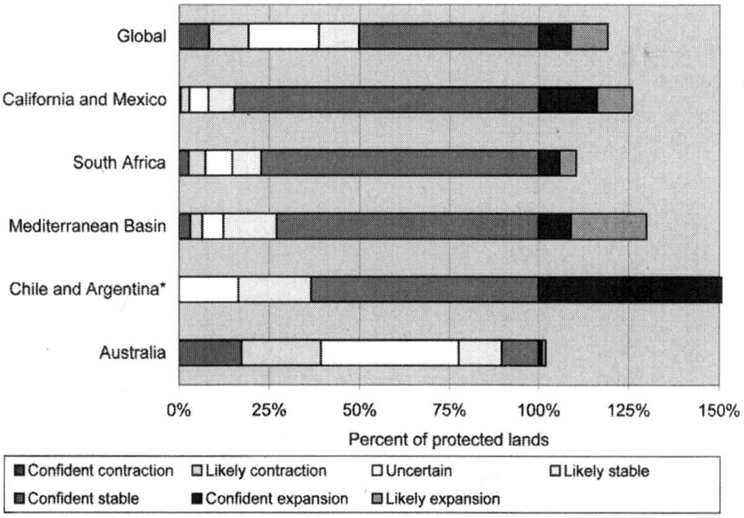

Figure 4. Projected future status of the MCE on protected lands under the high (A2) emissions scenario. The percentages indicate the portion of the current protected lands within the current MCE. There is over five times the amount of protected area in the confident and likely expansion area in Chile and Argentina relative to the amount in the current MCE, but the full extent is not shown in the chart to show more detail in the other regions.

DISCUSSION

This analysis provides the most comprehensive assessment to date of how climate change is projected to impact mediterranean climates in all five mediterranean regions across the globe and the implication of the climate change for adaptation potential of the native biota. By analyzing the full suite of multiple simulations of 23 AOGCMs under multiple emissions scenarios, we are able to present quantitative estimates of the level of agreement in the projected contractions in the MCE. Previous studies focus on one or several "bookend" AOGCM runs that attempt to encompass the range of variability in the future projections. If one model projects a wetter future and the other a drier future, the recommendations for conservation action could be vastly different with no method for determining which future is more likely. This study quantifies the level of agreement between a full suite of projected climatic futures, providing a more robust and conservative estimate of impact and thus more confidence to support conservation action.

One of the first challenges to projecting the impacts of climate change on mediterranean ecosystems worldwide is defining the MCE. The extent of the mediterranean biome is typically mapped based on a combination of climate characteristics and plant assemblages that vary from region to region [2]. Projecting how plant assemblages will shift in response to climate change is subject to significant uncertainty because it requires compounding the uncertainty with projecting climate change with the uncertainty inherent in projecting future distributions of individual species [24, 25]. In this analysis, we minimize the uncertainty and focus on mediterranean climate shifts in the future. As such, we utilize a conservative definition of mediterranean climate [17] that is consistent across all mediterranean regions and minimizes "false-positives" (areas that are not considered part of the mediterranean biome). Some areas that are traditionally considered part of the mediterranean biome are not within our current or future MCE, including the south coast of France, western Italy, northeastern Spain, portions of central Chile, and the southern coast of South Africa. Most of these areas receive less than 65% of their rain in the winter, and thus do not meet the first Aschmann condition (Figure 5). Despite the conservative nature of this definition, we do include some false-positives, including parts of Argentina and the Middle East. These commissions could be the result of a lack of climate station data in these more remote and mountainous areas. We performed a sensitivity analysis with the Köppen definition of mediterranean climate, which is less conservative. This definition identifies more of the traditional areas, but also includes large areas of false-positives. However, when focusing in on the five mediterranean regions, the results of the projected fate of the MCE are very similar using both definitions.

While there are some significant discrepancies between our map of the current MCE and commonly used maps of the mediterranean biome, such as the one mapped by Olson et al. [2], preliminary analysis indicates significant overlap between the current MCE and the mapped hotspots of plant richness and endemism within the mediterranean biome. Our current MCE corresponds well with the biogegraphical sectors with high incidences of plant endemism in the Mediterranean Basin [7], areas of high modeled endemic *Banksia* species richness in Western Australia [11], and areas of high modeled endemic plant richness in the California Floristic province [14]. In South Africa, an analysis of both modeled and observed plant richness shows the area identified in the current MCE in the Western Cape as a hotspot [26]. This correlation between areas of high plant richness and the MCE does not diminish the need for conservation action in all areas of the mediterranean biome, but it does provide support for the theory that the core mediterranean climate is an important driver of plant endemism and diversity, and that changes in climate could threaten the survival of these endemic plants.

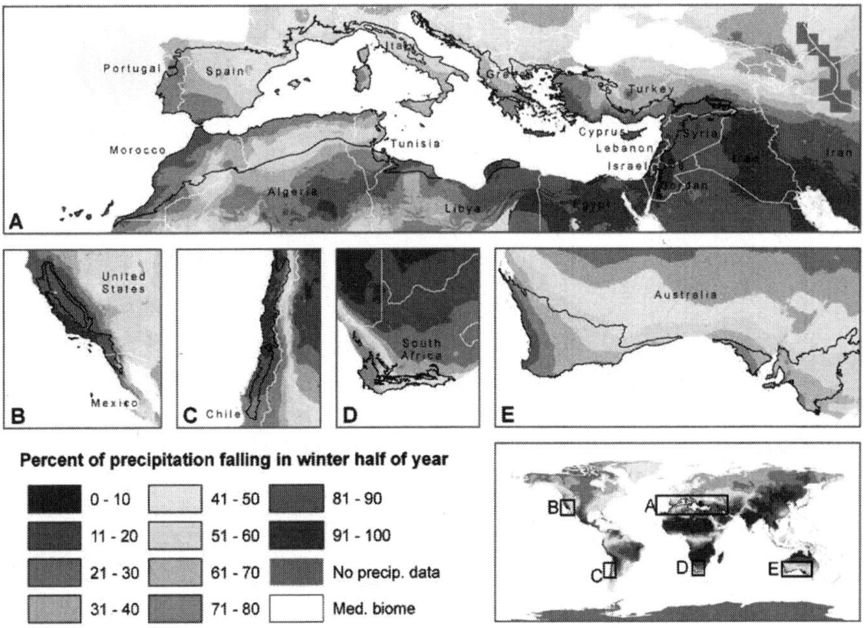

Figure 5. Percentage of precipitation falling in the winter half of the year from the WorldClim dataset. The depiction of the mediterranean (med.) biome is from Olson et al. [2] and is provided to show areas that are often considered part of the mediterranean biome but have less that 65% of the precipitation falling in the winter half of the year, and thus are not included in the current MCE.

Across the entire mediterranean biome, most of the 23 AOGCMs project a minor increase in the MCE, with significantly large increases in some regions, and significant decreases in others. Not surprisingly, the physical characteristics in each region help to explain some of the disparity between areas of contraction and expansion. High topographic diversity and contiguous land toward the nearest pole provide room for the expansion of the MCE in the US, Chile, Argentina, Greece, Turkey, Spain, and Portugal. In South Africa, there is topographic relief so the MCE can retreat upslope into the Western Cape Fold Mountains, but there is no contiguous land toward the South pole cutting off future expansion of the MCE. Similarly in Morocco, the Atlas Mountains provide topographic diversity, but the Mediterranean Sea blocks expansion toward the north. Southwestern Australia is a flat highly weathered plateau [27] and is contained by the Indian Ocean to the south, resulting in the largest projected contraction of all the regions.

While, this is the first global assessment of the impacts of climate change on the MCE, our results are consistent with other regional biome and species level analyses. The IPCC determined that the mediterranean biome as a whole is threatened by desertification from expansion of semi-arid and arid system under relatively minor warming and drying scenarios, and projects significant regional vegetation and species range shifts [28]. In South Africa, despite differences in the current spatial extent, Midgley et al. projected areas of future mediterranean biome contraction and stability in similar areas as our analysis [15]. Similarly, Williams et al. and Hannah et al. found the higher elevation areas of the Western Cape in South Africa support high levels of endemic Proteaceae species richness, and will be important habitat for dispersal by 2050, while the low-lying areas north of Cape Town are high in richness now, but are not protected to support the species in the future [29, 30]. These areas correspond well with our projected areas of higher elevation stability and lower elevation contraction in South Africa. Fitzpatrick et al. studied potential range shifts for native Banksia species in Western Australia and found the areas of greatest percent loss in richness in the arid interior, while the projected loss was less severe in the coastal areas and Fitz–Sterling Ranges [11]. These results are consistent with the results presented in Figure 2, although Fitzpatrick et al. did project species richness increases along the western coast of Australia, while our analysis found almost no projected expansion of the MCE in Australia. Loarie et al. identified areas of future refugia for species with projected range reductions in the mountainous areas along the central coast and foothills of California, USA and Baja California, Mexico [14], which correspond well with our areas of projected MCE stability. Benito-Garzón et al. studied a series of tree species in the Iberian Peninsula and found that some of the mediterranean species had the least projected range reductions and could potentially expand north and into higher elevations [13], which is consistent with our findings in Spain and Portugal. In general, the results from these biome and species level analyses that focused on one region at a time support our findings as well as the urgent need for conservation action to increase the extrinsic adaptation potential of the native biota of the mediterranean biome.

This analysis shows that climate change puts areas with the some of the highest levels of plant diversity and endemism on the Earth at risk. As shown in Figure 2, the Mediterranean Basin, Morocco, and Israel contain large areas of projected contraction with no adjacent areas of expansion. The mediterranean portions of Morocco contain almost 13 plant species per 1,000 km^2, while Israel contains 200 [7]. The cape region in South Africa and southwest Australia are the other two regions with large projected losses, and they contain 95.5 and 71 plant species per 1,000 km^2, respectively [7]. For comparison, there is one plant species/1,000 km^2 in Europe, 6.5 in Brazil, and 40 in Columbia [7]. The current MCE in these four threatened countries contain 22,400 plants species, 12,925 of which are found nowhere else [7].

The adaptation potential of the species native to the mediterranean biome will be further limited beyond the direct impacts of climate change analyzed here. Native plant species in all mediterranean regions, except perhaps Chile, are well adapted to natural fire regimes, but a hotter and drier climate has been observed to promote [31] and is projected to promote significant alterations to the fire regime beyond those created by decades of human fire management [28, 32, 33]. While rising atmospheric CO_2 levels could provide benefits to mediterranean plant species [34], the effects are altered when multiple factors of change are considered, including fire, drought, temperature increase, nitrogen deposition, and invasive species [35-38]. The rich plant diversity of mediterranean systems is explained in part by the plant adaptations to survive in low nutrient soils, such as California's serpentine soils, Australia's kwongan, and South Africa's fynbos [3, 4, 9]. The patchy nature of soils will act as a barrier and will make species migration in response to a changing climate more difficult. The intrinsic adaptation potential of some mediterranean endemics, particularly in South Africa and Australia, is limited by the relatively short seed dispersal distances and lack of colonization ability of these plants [11, 29, 39, 40]. While these indirect and interacting impacts of climate change are not explicitly considered in this analysis, they are likely to further limit the adaptation potential of mediterranean species.

Despite the significant projected contractions in the MCE, this analysis does offer some reasons for hope and some guidance to direct future conservation action. Approximately 50% of the biome is projected to remain stable with confidence, even under the high emissions scenario. Establishment and management of protected areas in these areas in all five regions represent sound investments given our current understanding of future change, and will help to secure future refugia for endemic species from other threats such as land conversion. However, given the uncertainties associated with the indirect effects of climate change, a conservative conservation approach should also include gene banking and *ex-situ* conservation for the rich floras of highly threatened regions like Israel, Morocco, South Africa, and Australia.

Since this analysis was conducted using a consistent methodology across all five regions, we can use the results to determine the highest priority regions for action. As shown in Figure 3, South Africa and Australia have the large projected contractions in the MCE. The existing protected area network covers almost 7% of the current MCE in these two regions. In South Africa, the protected areas are concentrated in the higher elevations where the MCE is projected to remain stable, so 77% of the current protected areas are projected to retain mediterranean climate and only 3% are projected to have a different climate in the future with high confidence, even under the high emissions scenario. In contrast, the protected areas in southwest Australia are concentrated in the drier inland portions of the MCE, so only 10% of the current protected areas will remain stable while 17% will likely shift to a new climate. These results suggest that some of the first strategies to enhance the extrinsic adaptation potential and to reduce the threat of climate change to the rich biodiversity of the mediterranean biome might include establishing new protected areas within areas projected to remain stable, improving land management, and restoring native habitat in southwest Australia. In particular, restoration efforts could focus on creating corridors or stepping stones of native habitat to connect isolated remnant vegetation patches in the areas with projected MCE contraction to the native habitat in the areas projected to remain stable. While this strategy has been advocated before [41, 42], and is currently underway in projects like the Gondwana Link (www.gondwanalink.org), this analysis highlights specific areas where this strategy could be implemented to improve the extrinsic adaptation potential of native species in the face of climate change.

Climate observations over the past century indicate the climate in the mediterranean biome is changing [43]. There is also a high level of agreement in the AOGCM simulations that most of the biome will continue to get hotter and drier. This study shows where there is agreement that the biome will shift, how the threats to biodiversity from these shifts will be exacerbated by current land use and land protection patterns, and highlights which regions are in need of the most urgent conservation attention. This conservation attention is required to establish protected areas and connectivity pathways to ensure the current investment in protected areas remains secure and facilitates and enhances the extrinsic adaptation potential of the species of the mediterranean biome.

KEYWORDS

- **Extrinsic adaptation potential**
- **Genetic diversity**
- **Mediterranean biome**
- **Mediterranean climate**

AUTHORS' CONTRIBUTIONS

Conceived and designed the experiments: Kirk R. Klausmeyer and M. Rebecca Shaw. Performed the experiments: Kirk R. Klausmeyer. Analyzed the data: Kirk R. Klausmeyer. Contributed reagents/materials/analysis tools: Kirk R. Klausmeyer and M. Rebecca Shaw. Wrote the chapter: Kirk R. Klausmeyer and M. Rebecca Shaw.

ACKNOWLEDGMENTS

We thank Peter Kareiva, Ken Caldeira, Dominique Bachelet, Jaymee Marty, Edwin Maurer, Stacey Solie, Emma Underwood, and William Klausmeyer for helpful discussions and reviewing drafts of this manuscript. We acknowledge the climate modeling groups, the Program for Climate Model Diagnosis and Intercomparison and the World Climate Research Programme's Working Group on Coupled Modeling for their roles in making available the Coupled Model Intercomparison Project phase 3 multi-model dataset. Support of this dataset is provided by the Office of Science, US Department of Energy.

Chapter 4

Environmental Knowledge Management

Po-Shin Huang and Li-Hsing Shih

INTRODUCTION

Owing to the green revolution, environmental problems have now become some of the most important issues worldwide. Environmental knowledge management (EKM), which combines the strengths of environmental management (EM) and knowledge management (KM), will become a popular tool for businesses in the near future. In this chapter, through interviews of staff at different levels and in different departments of the business, that is managers and engineers and using the environmental knowledge circulation process (EKCP), the authors evaluate the success of EKM when applied to China Steel Corporation (CSC) in Taiwan. From the case study, the authors found that CSC has applied the EKCP for over 30 years. The company continually improves its environmental and financial performance through Environmental knowledge creation (EKC), environmental knowledge accumulation (EKA), (EKS), environmental knowledge utilization (EKU) and environmental knowledge internalization (EKI). Water pollution and air emissions have reduced year on year and total energy consumption has reduced by 20% from 1979 to 2006. On the other hand, CSC also makes a profit and reduces cost through energy sold, by-products and recycling. Continuous improvement in EKM has rendered CSC in the most profitable steel company in Taiwan and the world's 25th largest steel producing company in 2006.

Through globalization and the Earth's natural processes, local environmental problems have transformed into international pollution issues. In order to solve these pollution problems, EM and green supply chain management (GSCM) have been widely adopted by enterprises with the aim of reducing pollution and evaluating environmental performance [1-3]. Environmental protection and performance have become two of the world's most important priorities in order to attain sustainable development [4]

Over the previous decade, EM has moved away from a regulatory-based approach to one focusing on continuous development. The KM has also rapidly developed over the last decade. A review of the literature shows that KM was first introduced in theoretical research and has since been developed for application in the identification and solving of a variety of problems. The EKM is also defined as a system to connect data, analysis and people which presents an opportunity to formalize industrial ecology in a business setting [5] It is also clear from the literature review that KM and EM have both been studied for many years; however, the key aspect of successful EM is effective KM which provides employees at all levels of an organization with a computerized and systemized EKM system enabling staff to gain environmental awareness and make environmentally responsible business decisions.

The objective of this study is to identify the extent and scope of EKM used in the case study and evaluates its performance from the viewpoint of the knowledge circulation process (KCP) [6]. This chapter begins with a discussion of KM and EM and then outlines EKM using the EKCP. The case study in which EKM principles are applied and the business responsibility for the sustainable development of the enterprise is derived may lead to the improvement of environmental and financial performance.

This study has been surveyed between May 2007 and February 2008.

ENVIRONMENTAL MANAGEMENT AND KNOWLEDGE MANAGEMENT

Environmental Management

Petak [7] defined EM as managing human affairs so as to achieve acceptable balance between the quality of the human environment and the quality of the natural environment. At that time, EM was still being used to find a balance between the environment, ecology protection and economics. According to Lorrain-Smith [8], the element of "control" should be added to EM; hence, he redefined EM as an action taken by society, a section of society and an organization to improve environmental quality by developing plans, implementing them and continuously reviewing them introducing the concept of control after the pollution problems have been caused. Traditional end of pipe pollution control mechanisms were widely adopted in this period; however, they are neither technologically efficient nor cost effective and may cause serious environmental impacts [9].

The US Environmental Protection Agency (US EPA) and the Pollution Prevention Act [10] define pollution prevention as source reduction, material substitution and more efficient processes in order to prevent all kinds of pollutants from the very beginning of the manufacturing process. In 1995, EM was used to enhance environmental sustainability development [11]. The three requirements of economic growth, social justness and sustainability in the use of natural resources should be kept in balance [12]. As indicated above, it is clear that great progress has been made in EM in recent years developing from passive pollutant control to the initiation of prevention right from the start of the process and moving forward to encompass the concept of environmental sustainable development.

Knowledge Management (KM)

In today's fast changing environment, knowledge is the highest source of power [13] and how a company creates and shares its knowledge is the main source of sustainable competitive advantage and profitability [14-17]. Hence, knowledge is a very important and powerful resource for organizations to use it in the preservation of heritage, accumulation of experience, creation of new ideas and sharing of new knowledge.

In the knowledge-based economy, KM has replaced the traditional factors of production, including land (i.e., natural resources), labor and capital [18] and has become a key factor in the success of a business. According to Malhotra [19], The KM is defined as containing organizational processes that seeks a synergistic combination of data and the information-processing capacity of information technologies and the creative and innovative capacity of human beings. Quinn et al. [20] defined KM generally as including any processes and practices that are concerned with the creation, acquisition, capture, sharing and use of knowledge, skills and expertise. The key purpose of an enterprise KM is to make information and knowledge accessible, reusable and sharable [21] and concerns the harnessing of the intellectual assets of an organization in order to improve its learning capability [22].

In general, KM is the collection, integration and classification of disordered information, technology and expertise in an enterprise and provides systematic and organized information for internal employees or management-level staff to consult when undertaking new tasks and decision-making.

The KM is a kind of enterprise asset that can help an organization to create new technology and achieve specific outcomes such as shared intelligence, improved performance or competitive advantage.

The question of how to gain control of knowledge assets has become a major organizational concern [16, 22]. Knowledge and expertise have been viewed as the most strategic and efficient inputs for an enterprise in the quest to achieve a sustainable competitive advantage [23]. The KM represents a new science and technology that provides competitive conditions and advantages for the modern business [24].

Moreover, Sarvary [24] also stated that with a good KM system, the job is much more challenging and people can concentrate on problem solving rather than on number crunching and data collection. Effective KM has a great impact throughout the enterprises from the lower levels of manufacturing to the management levels of strategic decision-making. Hence, through the efficient management of knowledge, an enterprise can create new technology and a unique business strategy that will improve its performance and competitive advantage.

ENVIRONMENTAL KNOWLEDGE MANAGEMENT

Environmental Knowledge and Environmental Knowledge Management

The ideas of the sustainable development as introduced in the Brundant report [25] have being become increasingly important. Achieving sustainable development to combine EM and KM are an indispensable ingredient. Wagner [26] found that for firms with pollution prevention-oriented corporate environmental strategies, the relationship between environmental and economic performance is more positive thus making improvements in corporate sustainability are more likely. However, it is a quite new issue pertaining to how to combine KM with EM and apply them into real situation. Nonaka et al. [27] defined tacit knowledge as a kind of personal characteristic that is too abstract to transfer or even express using words. Howells [28] thought that tacit knowledge is a kind of expertise that is not editable. This knowledge is obtained through informal learning behavior and a sequence process.

Therefore, EKM must combine the management of explicit knowledge and tacit KM in an organization and control the environmental impact via the accumulation, utilization, sharing and creation of environmental knowledge.

Environmental knowledge is a kind of general knowledge and includes the concepts of environmental protection, the natural environment and ecosystems [29].

This means that environmental knowledge involves what people know and are concerned about regarding to the natural environment, their responsibilities towards environmental protection and the relationship between the economy and sustainable development.

People with environmental knowledge will process information using this knowledge (system knowledge), know what can be done about the environmental problems (action related knowledge) and understand the benefits (effectiveness) of environmentally responsible actions [30].

A current business trend is towards the use of KM, EM, governments, and organizations are concentrating on these concepts. The concept of EKM combines the management of explicit knowledge and tacit knowledge in an organization and controls the environmental impact via the accumulation, utilization, sharing, and creation of environmental knowledge; it is essentially a fusion of KM and EM. The EKM also integrates environmental problems into an organization's routine operations in order to reduce environmental pollution and increase business responsibility and concern towards the natural environment.

Tatsuki and Masahisa [31] stated that KM can effectively solve environmental issues. They found that a large quantity of data and information collected from EM projects can be systemized and organized into a KM system and can be used to solve environmental problems. The EKM combines tacit knowledge from employees' experience and explicit knowledge from environmental tasks in order to improve teamwork efficiency and solve environmental problems. This can not only reduce pollutant emissions from production processes, but also enhance the precontrol and prevention of environmental pollution.

THE DIFFERENCE BETWEEN EKM AND THE ENVIRONMENTAL MANAGEMENT INFORMATION SYSTEM (EMIS)

The EMIS is an important component of EM that can assist both environmental and non environmental managers who fulfill their daily tasks. The EMISs have been broadly defined as computer based technologies that support environmental management systems (EMS). However, Finster et al. [32] define EKM system as the consistent of the tools, mechanisms, processes, structures, people, policies, strategies, data, and information that enable the creation, capture, accumulation, storage, retrieval, use, and transfer of knowledge that improve an organization's overall impact on the environment. Therefore, EMIS is one specific part of EKM and could not represent whole of it.

ENVIRONMENTAL KNOWLEDGE CIRCULATION PROCESS (EKCP)

Lee et al. [6] measured KM performance using the KCP which has five components: Knowledge creation (KC), knowledge accumulation, knowledge sharing, knowledge utilization, and knowledge internalization. Based upon KCP theory, this study will also evaluate the environmental performance of CSC via the EKCP which also has five components: EKC, EKA, EKS, EKU, and EKI.

Enterprises adopt the EKCP to convert external environmental knowledge into internal enterprise values with regards to environmental responsibility and sustainable development. In order to transform environmental issues into enterprise values and culture [33], companies should develop a training strategy designed to motivate and improve employees' environmental knowledge and awareness [34, 35]. Under increasing pressure from the trend of green innovation, enterprises can only remain competitive through the use of EKM to enhance their profit and ensure their international competitiveness in the field of sustainable development.

The Components of the EKCP

The EKCP is composed of continuous environmental knowledge activities distributed among EKC, EKA, EKS, EKU, and EKI. The KC is further an understanding of the need to develop newer technologies and systems in order to enhance creativity and competitiveness [20]. Nonaka et al. [36] indicated that there are four modes of knowledge conversion:

- Socialization (from tacit knowledgetotacit knowledge).
- Externalization (from tacit knowledge to explicit knowledge).
- Combination (from explicit knowledge to explicit knowledge).
- Internalization (from explicit knowledge to tacit knowledge).

The same concept is applied for KC with the transition mechanism, the tacit knowledge and explicit knowledge of EKC expanding in both quality and quantity throughout the above conversion processes.

The two essential components of EKA are experience and experimentation. From a resource based perspective, technological knowledge, skills and employees' experience are valuable assets that are accumulated over time in an enterprise and are even embedded in the organizational culture. Dierickx and Cool [37] indicated that an organization's strategic asset stocks accumulates integrated resources over a period of time; however, a stock of knowledge usually resides in a particular person in the organization [38] and hence, EKA requires continuing efforts to acquire employees' experiences and the considerable allocation of organizational resources. Enterprises engage in environmental creation and R & D activities by EKA and increase their market competitiveness and advantage.

The EKS is a kind of knowledge transaction between individuals and groups. Andersen and APQC, [39] proposed the following Equation:

$$KM: K = (I+P)^s.$$

The EKI is the process of learning by converting explicit knowledge into tacit knowledge [36]. Employees can access the environmental information and knowledge needed to complete tasks through a series of environmental training sessions and exercises.

The EKCP is a never ending process that continuously improves ensuring the obtaining of competitive advantage by creating, accumulating, sharing, utilizing and internalizing EM. Moreover, the production of cleaner products may enable companies to be continuously improved by identifying ways to maximize profits through creating and improving eco friendly products, reducing waste and pollution and increasing invisible corporate assets–an awareness and sense of environmental protection and eco-efficiency.

MATERIALS AND METHODS

The study is based on semistructured interviews carried out in 2007 with managers and engineers of CSC (Tables 1 and 2) and information collected from the Internet, thematic papers, case studies and CSC environmental reports.

The objectives of this research are:

- The implementation of EKCP in CSC.
- The transition of environmental knowledge in CSC (from the knowledge spiral theory viewpoint).
- Discussion of whether EKM is able to effectively improve the performance of EM in CSC.

The authors are interested in how the EKM system works and why implementation is so successful in CSC.

This study will evaluate the environmental performance of CSC using the EKCP as described by Lee et al. [6]. The evaluation components are as follows: EKC; EKA; EKS; EKU, and EKI which include the organization capacity to internalize task related to environmental knowledge, the environmental education opportunities and the level of organizational learning [6]. The concept of this chapter is to study whether the performance of EKM will improve and whether environmental targets will be achieved when the EKCP is effectively improved.

Company Background

The CSC established in 1971 was the first integrated steel mill in Taiwan. After privatization in 1955, CSC concentrated on developing its steel business, as well as expanding its business into the CSC Group which operated and invested in diversified business areas. With maximally efficient resource integration among the subsidiaries and affiliates and an output of 10.7 mmt of crude steel of CSC had become the world's 25th largest steel producing company by 2006 [40].

To enhance their international competitive advantage, CSC set up a KM Committee in 2003 which was responsible for implementing KM in order to pass down the accumulated experience, technology and expertise of over 30 year of operation. This is not only a corporate asset for future development and forward innovation, but also the best approach to sustainable development. According to Tony Chao, the manager of the Manpower Development Section, HR development of CSC indicates that employees' experiences are the most valuable company asset of CSC (n.d.). The CSC invested capital in an e-learning program and combined this program with KM systematically transforming accumulated skills into a digital system. Overcoming the limitations of location, time and budget, e-learning provides an organized and systematic KM system for all staff levels in CSC.

In addition to the development of the KM system, CSC has also demonstrated outstanding achievement in EM. Over the past two decades, CSC has contributed significantly to environmental protection initiatives; it integrated environmental protection activities and functions into the EMS and obtained ISO14001 certification in 1997. Moreover, a multi-examination system was successfully developed into a total quality standard control and management system (TP00) which combines the benefits of a Quality Management System (ISO9001), EMS (ISO14001) and occupation health and safety management (OHSAS18001). All levels of staff in CSC are able to search and analyze all kinds of information and data more efficiently through the TP00 system. Furthermore, an EM meeting is held every 6 months and EM review is held every year by the EM Committee to ensure that the EMS is working effectively and appropriately [41]. Hence, CSC has received different types of awards from the government for industrial safety, hygiene and environmental protection since 1999.

Table 1. Details of interviews conducted on 2007/08/30.

Name	Position and Department
K. C. Liu	General manager, Environmental Protection Department
I. Y. Chen	Environmental Planning Section, Environmental Protection Department

Table 2. Details of interviews conducted on 2007/05/02.

Name	Position and Department
W. S. Tsai	President of the Labor Union of China Steel Corporation
T. Chao	Manager of the Manpower Development Section, Human Resource Department.
C. P. Chang	Manager of the Employee Relations Section, Human Resource Department.
C. F. Lee	Engineer in the Labor Safety and Health Section, Industrial Safety and Environmental Protection Department.
C. L. Wu	Engineer in the Labor Safety and Health Section, Industrial Safety and Environmental Protection Department.

Steel Production Flow Chart and Environmental Problems

The environmental problems in steel production processes are complicated; therefore, the solutions are also complicated. Despite the complexities, it is possible to highlight the key emissions problems, including air pollution, water pollution and solid waste. Air pollution presents the main problems due to CO_2, SO_2, NO_2 and particulate emissions. Gas scrubbing water, the water used in the rolling mills which can be of high oil content, electrolytic cleaning water, the water used for cooling, and the rust removal and lubrication processes are the main causes of the water pollution problems. Volumes of slag and the other by-products emanate from the production processes causing solid waste.

From the steel production flow chart (Figure 1), the input sources of materials and resources and the output results of remnants and waste can be clearly identified. To solve these problems, CSC combines its accumulated production experience and EMS to continuously develop new technologies to reduce pollution during the production processes which is an example of the performance of EKM in CSC (Kuo–Chung Liu, general manager of the Environmental Protection Department (EPD), Production Division of CSC). In addition, CSC has implemented an environmentally-friendly production strategy and has invested more than nine hundred thousand USD on various environmental equipments in order to improve environmental quality through pollution prevention and waste reduction [41].

Environmental Knowledge Management in CSC

What has CSC done to promote environmental protection?

In order to show the special emphasis and the business responsibility that CSC places on environmental protection, the EPD, a new and independent department, was separated out from the Industry Safety and EPD in 2007. It is a professional department strongly supported by its high-level manager who is responsible for EKM and environmental issues. The change of the original organizational structure gives the EPD more power in resource integration and energy saving; it also increases the awareness of the EKCP in CSC.

Environmental Knowledge Creation

The EKM system is a complete and user friendly information system containing vast amounts of environmental knowledge and employees can search for related information and E-learning courses on the CSC internal database. The official environmental website related to environmental policies and foreign environmental information can also be searched through the CSC website and intranet. Outside the EKM system, entry-level engineers can also learn directly from senior engineers gaining environmental knowledge and experience (tacit knowledge) that cannot be obtained by a search of the network. The CSC also offers staff the opportunity to investigate new environmental protection knowledge and technology abroad.

Under the EKM system, predecessors usually give different challenges to their subordinates according to their ability and job attributes, the task specialty and complexity. Environmental engineers who are in charge of environmental tasks in different positions need a full understanding of the core environmental knowledge.

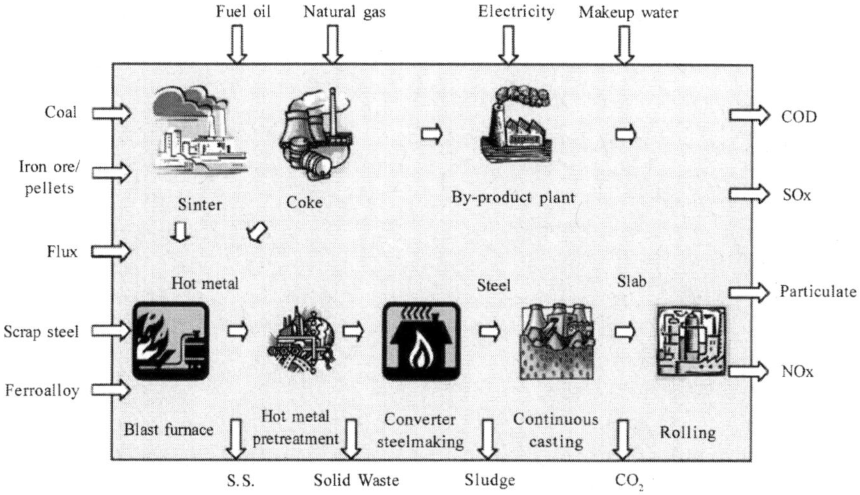

Figure 1. Steel production flow chart.

For example, engineers who are responsible for air pollution prevention must:

- Read the air pollution prevention policy;
- Undergo capability building and training programs for 1 month;
- Obtain related certification and become a professional in air pollution prevention management;
- Fully understanding of the concept of EM and apply it to air pollution prevention and control.

The EPD usually convenes brainstorming meetings to obtain useful environmental information, knowledge and suggestions. For instance, the EPD can communicate directly with the different departments, subsidiaries and affiliates of CSC during the weekly meeting to obtain information valuable for decision-making and the EM strategy. Due to this continuous brainstorming, engineers in the EPD are able to take up new challenges upon new EKC in order to create new technologies to reduce pollution and apply them to new tasks.

Environmental engineers will also search for suggestions and solutions through communities when they encounter thorny environmental problems. Therefore, professional environmental knowledge can be circulated freely through the systematic knowledge communities and the e-learning systems that were built up by the professional senior engineers.

Environmental Knowledge Accumulation

Wu, Engineer in the Labor Safety and Health Section, Industry Safety and EPD, usually searches for the necessary information or experts experience from various databases, communities and environmental knowledge bases while facing a difficult problem or a case similar to one that has occurred before. The environmental knowledge system of CSC is being continuously improved moving towards full computerization and a paperless system.

Moreover, Wu uses the safety auditing system as an example to explain how the necessary information and knowledge are searched through the environment-related databases to reach the environmental goals. According to WU, CSC usually makes routine checks and audits and then uploads the final results on the internal information system. This forms the training material for future use creating knowledge and increasing the benefits of KM. For environmental protection and true environmental benefits, CSC observes environmental policies and continually strives to make the factory greener and improve the ecological environment and environmental quality. Environmental knowledge will be accumulated by learning stored in the environmental manual and shared via the intranet after completion of the pollution prevention projects or EM projects. Moreover, tacit knowledge will also be transformed into explicit knowledge and stored as electronic files or in the manual such as the transformation of presentations into media files or presentation articles. In the accumulation of environmental knowledge and environmental education, CSC conducts environmental training and provides abundant information on the courses which employees can also download from the internet.

Environmental Knowledge Sharing

According to the attributes and features of environmental knowledge and information, knowledge sharing can be classified as:

- Latest information: CSC usually shares the latest information in meetings and in conferences;
- Special information for a specific person/position: Different environmental information and knowledge is passed on via documents or Emails to a specific person or particular positions;
- General information: General information is posted on the electronic bulletin board to share with all employees of CSC.

The CSC shares environmental knowledge and information with external and internal teams through the regular meetings and conferences which encourages employees to be more interested in the environmental work of the company. With regards to external teams, CSC usually exchanges environmental opinions and new technologies with the CPC Corporation, Taiwan Power Company and the Formosa Plastics Group as for internal teams, environmental protection and pollution control information is widely circulated through conferences in order to improve environmental knowledge sharing and achieve the organization environmental goals. Wu, President of the Labor Union of CSC, emphasizes that there is no secret in the skills and techniques of environmental protection and industrial safety issues. Moreover, a successful and responsible enterprise should be concerned about its social and environmental responsibility, contribute to environmental infrastructure development, integrate resources and take action towards attaining global sustainability. The CSC has developed metadata, a file system defined as the data about the data in the knowledge database system. It includes that description of content, file size, author, publication date, and physical location is an effective data management system that varies with the type of data, the context of use and the content of files. Using metadata, it is easier to increase the system speed, enrich the searchable resources and share environmental knowledge. Of course, document security has been seriously considered and ensured during knowledge sharing.

Environmental Knowledge Utilization

An electronic exchange system is used in EKM. A systematic EKM system enables environmental tasks to be implemented smoothly. In addition, participation in environmental activities leads to better environmental performance relying on the utilization of environmental knowledge in CSC. According to Wu, CSC has a high quality environmental education program and has taken advantage of the existing knowledge creation and sharing in the organizational culture to improve the EKM system. The CSC provides incentives such as product awards, individual awards and innovation awards for the employees who have demonstrated outstanding performance in EKM, knowledge conservation and environmental protection.

Environmental Knowledge Internalization

The CSC has compiled handbooks and manuals with a view to improve employees' occupational knowledge. This enables the employees to better understand their job and the work required skills. In addition, some departments download the environmental course information from the internet and some also use the environment related courses to improve employees' capability and develop their sense of environmental responsibility.

Under EPD, the environmental resource and information standard system is set up to assist employees to implement environmental task. The professional information system is reviewed and updated continuously which allows employees to search the environmental information and accumulate knowledge through the system that can be used to complete new tasks. Through this process, employees can internalize explicit environmental knowledge into their tacit knowledge and gain environmental awareness. The use of an environmental information system speeds up the sharing and internalization of environmental knowledge.

RESULTS AND DISCUSSION

Based on their existing environmental knowledge experience, CSC continuously creates new environmental knowledge through brainstorming sessions and meetings. It also implements master and apprentice schemes and reward systems to inspire employees to pass down their accumulated knowledge and experience. According to the attributes of the knowledge and the level of the employees, the sharing of environmental knowledge enables employees to have a greater understanding of the environmental information which means that the environmental knowledge is passed down to the right person at the right time making the EKCP more active and effective. The mechanism of the conversion activity enables the environmental knowledge to be internalized into the employee personal values and improve the awareness of environmental protection issues. Environmental knowledge is also internalized knowledge improving employees' personal potential capability of creating new environmental knowledge, technologies, and skills. Circulation activities maximize the value of environmental knowledge which is also the purpose of EKM. According to IISI [42], CSC had become the world's seventeenth largest (and Asia's eighth-largest) steel making group by 2000. In addition, CSC was listed in a list on Forbes [43] as one of the 400 best large companies of the world. Moreover, CSC was designated by IISI as 25th out of 80 top steel producing companies in 2006 [42] which demonstrates the performance and implementation of EKM in CSC.

From the interviews conducted in this study, it is clear that CSC uses environmental knowledge activities to its advantage and continuously creates, accumulates, shares, utilizes, and internalizes environmental knowledge enabling EKM to mesh with CSC's ethos, mission and vision (Figure 2). External stresses such as environmental policies, WEEE, RoHS, awareness of the environment, green innovation, and social responsibility all speed up the improvement of EKM in CSC. Moreover, in consistently improving with EKM activities, environmental knowledge is internalized CSC business spirit, organizational values, competitive advantage, and organizational culture which drive CSC to use its environmental knowledge and ideas in product design and the production process. Therefore, CSC has built up a complete and high quality EMS through the EKCP providing a good example for other enterprises to follow.

Environmental Performance

According to the environmental load of Chinese Steel Corporation, this report will be referred to the environmental load generated in 2004, including input of resources and energy, as well as gas, liquid and solid output. This report indicates the current status along with improvement measures taken in the past years.

Indicators are calculated based on each ton of steel products, including:

- Unit consumption of energy and discharge of CO_2 gas.
- Unit discharge of NO_2, SO_2 and particulates from waste gases.
- Unit consumption of water and unit discharge of chemical oxygen demand (COD) and suspended solids (SS) in effluent.
- Generation of the by-products process and the recycled or disposed percentages. Therefore, environmental performance really can be improved through the performance of EKM.

In response to environmental regulations and increasing of environmental awareness, CSC implemented environmental technology and managed solutions to reduce emissions and natural resource demands and control waste. For instance, water treatment has been a major concern; it includes physical, chemical and biological treatment to remove or change the composition of coke effluent before releasing or recycling. The water recirculation system ensures water efficiency and enables reuse of water to reduce production costs. The CSC is still searching for new environmental knowledge to reduce the environmental pollution, energy demand and manufacturing cost via the newest recycling technologies.

Water Pollution Control and Water Consumption

The CSC method of water pollution control and consumption is an in-plant process that reduces and eliminates the generation of pollutants and waste to reduce risk to human health and the environment. The CSC has taken effective measures to reduce the environmental risk and preserve natural resources through wastewater treatment systems, the secondary water reuse system and the rain recycling system.

The CSC plantwide water recycling and reuse plan maximize the efficiency of water consumption and the recycling of 97.6% of water is normally achievable. A significant development of CSC in the area of water consumption is the reduction and prevention of water pollution through water recycling and saving.

Figure 3 shows the trend of COD and SS over the past 7 years for CSC. The CSC waste water treatment system has reduced the content of COD from 71 ppm in 2000 to 41 ppm in 2006 and that of SS from 9.7 ppm in 2000 to 4.4 ppm in 2006. This not only reduces the total operating costs, but also achieves CSC environmental objectives.

Air Emission Control

The Environmental Surveillance Center of CSC established in 1995 continuously monitors, analyzes and controls the emissions and effluents from the manufacturing process. Three monitoring systems are used in pollution prevention and EM: Periphery air monitors and displays boards, continuous monitors for off gases and year round visual monitors. With the aim of increasing environmental protection, the high performance of the air control system is shown in Figure 4; from the relative index, the emissions of particulates, SO_2 and NO_2 in 2006 were 91%, 79%, and 67%, respectively which are much lower than the same values measured in 1992 [44].

The important outcomes of the resource reducing project and the energy saving program in CSC are shown in Table 3. The CSC reduces greenhouse gases through internal and external improvement programs such as waste heat recovery, the application of energy-saving in the producing process, improving the yield rate, regional energy integration, the zero-waste project, etc.

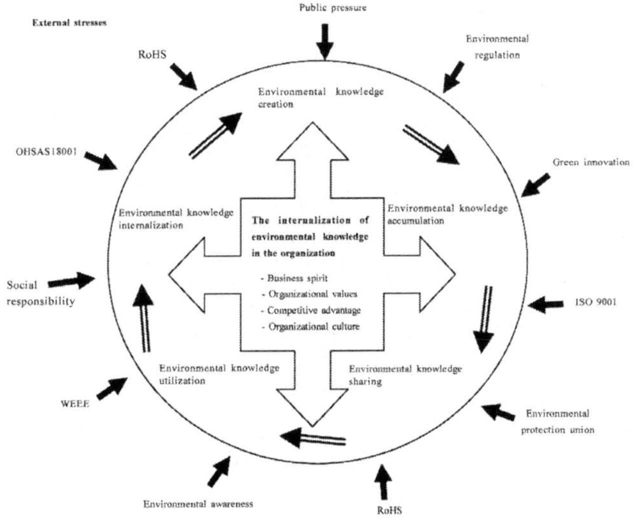

Figure 2. The circulation of environmental knowledge in CSC.

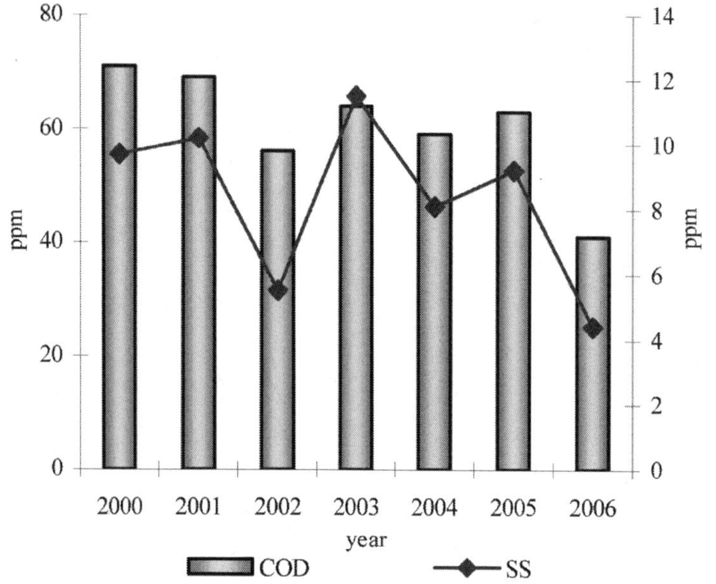

Figure 3. COD and SS of the effluent [44].

Table 3. The performance of CO_2-reducing projects [44].

Project	Contribution (1,000 ton-CO_2/ year)	
Waste heat recovery Project	1,300	Total energy saving from 1997 to 2006
2010 energy-saving Project	160	Total energy saving in 2006
Eco-industrial park*	210	Ton-CO_2/KLOE
Water-quenched slag replacing the use of Concrete	2,170	CO_2 reduction: 0. 79 per Ton of steel

* Eco-industrial park means CSC sold energy which it did not need and gases such as N_2 , O_2 , coal breeze and steam to increase income and reduce pollution.

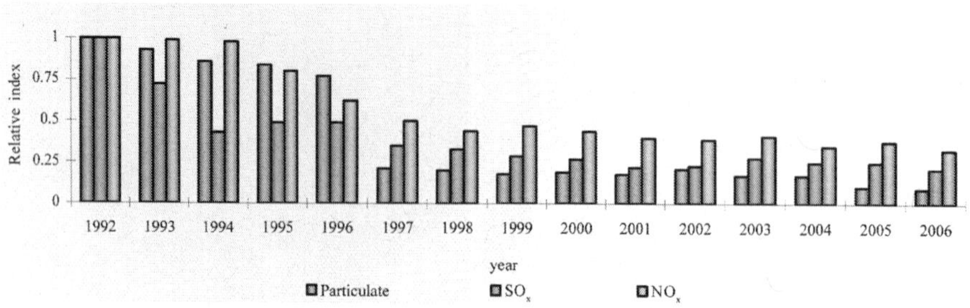

Figure 4. Air pollution per ton of steel production (relative index) [44].

Energy Consumption

As the steel making industry is an energy-intensive industry and this energy constitutes up to 20% of the cost of steel making, energy consumption has become a major factor affecting global competitiveness. As shown in Figure 5, the major input of energy consumption is coking coal (73.8%); others are BF injection coal (16.1%), electricity (5.1%), steaming coal (3.8%), etc. Moreover, considering the various changes in the international market and the shortage of natural resources, CSC felt it necessary to refocus on energy efficiency and reducing the consumption rate. Hence, an advanced steel making technology was created and is used in CSC. Compared with the energy consumption in 1979, the relative index shows that it has been reduced by 20% in 2006 as previously reported by CSC (Figure 6).

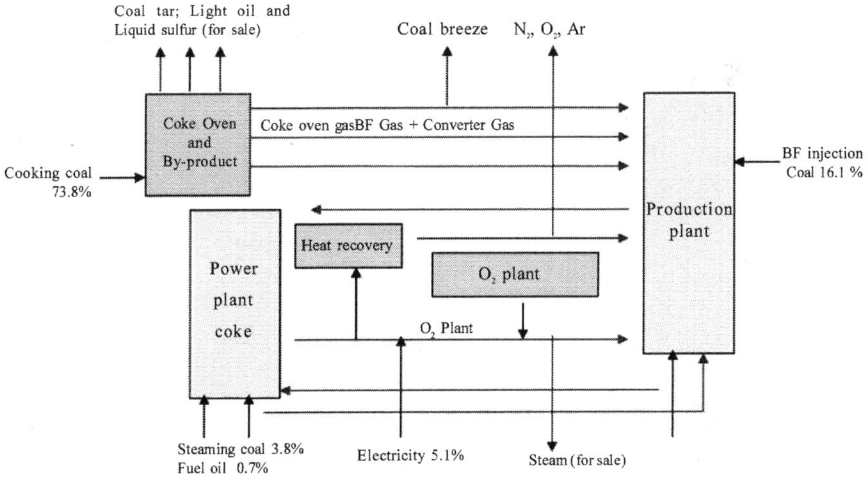

Figure 5. Major inputs and outputs and flowchart (dry basis, for each ton of steel produced) [41].

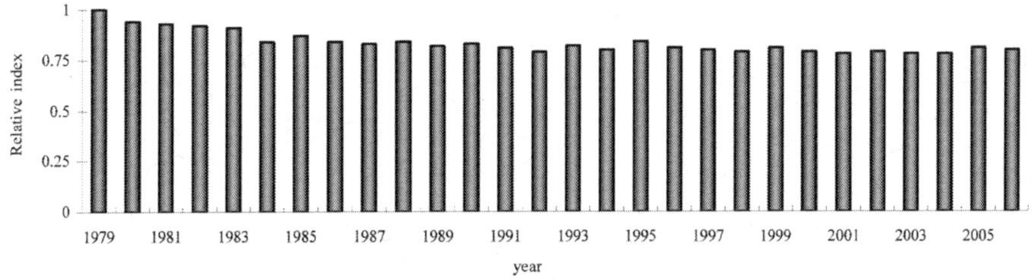

Figure 6. Energy consumption of CSC, 1979–2006 (relative index) [44].

Reduction of Process Residues and Recycling Toward Zero Waste

The CSC zero waste philosophy in environmental matters leads to the recycling, reuse, conservation and by-products production of useful mineral products, materials, and resources. The benefits of a zero waste policy are that it reduces operation costs, increases profit and complies with environmental regulations. The CSC continuously creates new technology to reduce waste or transfer it into useful resources through continuous improvement in the manufacturing process, consistently moving towards a zero waste system. With the improvements in technology and investment in the R & D Department over the past two decades, CSC has accomplished its missions and goals gradually.

Its performance in reducing residues and recycling will be shown in the Table 4 and Figure 7.

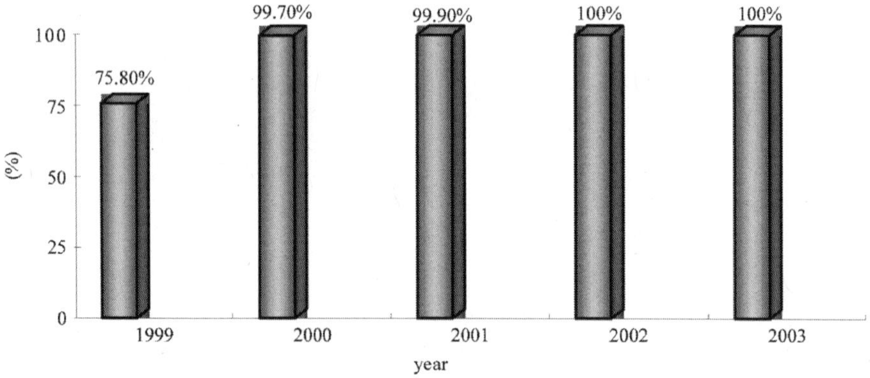

Figure 7. Percent of sludge recycled at CSC since 1999 [42].

Table 4. By-products of process residues in 2004 [44].

Type of residue	Description	Generation (10,000 tons)	%oftotal residues
BF slag	By-products of BF after smelting or raw materials	303.9	49.0
BOF slag	By-products of the basic oxygen furnace (BOF) after refming of steel	128.2	20.6
De-S slag	By-products of the hot metal desulfurizing process	36.8	5.9
Dust	Particulates collected from de-dusting system	34.1	5.5
Sludge	Solid cakes or mud from wastewater treatment after condensing and dehydrating	44.8	7.2
Mill scale	Rust of steel products or semi-products	29.4	4.7
Spent refractory	Used refractory from high temperature process	4.0	0.6
Civil refractory	Residual soil, concrete, etc.	19.8	3.2
Limestone cake	Filter cake from limestone washing water	5.0	0.8
Other	Waste oil, waste packaging materials, refuse, etc.	15.5	2.5
Total		621.5	100

Figure 7 presents the percentage of sludge recycled at CSC. The recycling of sludge has improved over the years and the goal of 100% of recycling was achieved in 2000.

Financial Performance

The strategy of CSC eco-industrial park is not only to maximize energy efficiency and financial performance, but also reduce the environmental burden. Steam, N_2, O_2 and Ar are generated during the steel making process and are sold to external plants such as China Petrochemical Development Corporation at KHH, Tang Eng Stainless Steel Plant, China Steel Chemical Corporation, etc. The financial gain from sold energy is shown in Figure 8 from which it can be seen that profit has increased from 4.12 million USD in 1994 to 33.66 million USD in 2006.

In addition to the increase in revenue gained from the energy sold to nearby plants, CSC also reduces its costs by increasing the pulverized coal injection (PCI) rate and reducing the fuel rate which can save 6.59 million USD each year. Also, the rain recycling system can collect 150,000 tons of water which can save around 66,000 and USD each year; the waste heat recovery system can save 72.46 million USD each year and the acid regeneration plant (ARP) will reduce the cost of acid purchasing by 3.95 million USD. Hence, to eliminate energy waste and increase financial performance, CSC also continuously improves its management strategies and technology in order to improve the overall energy efficiency.

Application of EKM in Strategy Planning and Performance Improvement

The change in the organizational structure of CSC has enabled the EPD to concentrate on strategy and planning related to environmental issues, integration of natural resources and prevention of pollution in production. Strong support from executives drives the success of CSC in EKM and environmental performance. In addition, managers usually apply the newest techniques to reduce pollution due to the environmental concerns and social responsibility. The organizational change of CSC was also significant in establishing dynamic strategy planning, executing EKM and achieving continuous environmental improvement through EKM activities. The implementation of CSC environmental protection strategy enables engineers to prevent problems and prepare possible solutions in advance. According to the manager of environmental protection, CSC used to invest in new mills without considering the environmental loading until problems began to appear. To avoid the same pollution problems, the EPD integrates all the related departments and production resources, including upstream, downstream, cross departments, the EMS, and so forth to reduce the total environmental loading and pollution. The main concept is to implement pollution prevention and core process management strategies in advance instead of having to find a solution to the pollution problem. Using an iterative four step problem solving process, the PDCA (Plan-Do-Check-Action), the EPD collects related production data and develops evaluation forms for pre-evaluation, in process-evaluation and post-evaluation during new plant construction. The EKM system enables the EPD to develop new technical knowledge through KC, estimate the results from the accumulated experience and utilize pollution prevention technologies in advance.

Another example is the system improvement project which helped the EPD to estimate the pollution results during the project planning stages. The system improvement project integrated the EPD, Energy Department and R & D Department and focused on both comprehensive and objective pollution prevention in order to achieve the company's pollution reduction goals. Moreover, the basic aim of the system improvement project was to achieve user-friendliness, simplification and efficiency; it was well-accepted and effectively implemented by all departments because of its standard processes and simple design.

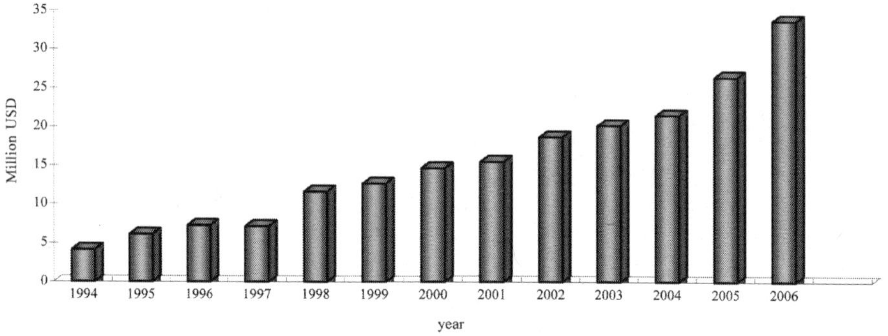

Figure 8. Energy sold from 1994 to 2006 (million USD) [44].

Through a combination of strategy planning and EKM, EPD can calculate the environmental loading exactly according to the accumulated data and experience which helps to manage pollution prevention, green design, energy saving and resource integration. The EPD calculates not only environmental loading in CSC, but also the secondary environmental loading upstream and downstream. Therefore, EKM is a continuous environmental activity involving the creation, accumulation, sharing, utilization and internalization of knowledge which helps CSC to take action before environmental accidents occurence, increase R & D investment, create green technology and develop green products; it also helps managers to make eco-efficient and strategy efficient decisions and identify the best investment projects in order to reach the goal of sustainable development.

CONCLUSION

From the perspective of the knowledge spiral theory [27], it has been demonstrated that the conversion of EKM in CSC results in:

1. Socialization: Tacit knowledge is gained from accumulated experience;
2. Externalization: Accumulated experience and knowledge are converted into systematic explicit knowledge;
3. Combination: Opinions are exchanged between the knowledge community and community of practice, both internally and externally;

4. Internalization: The environmental knowledge is internalized into CSC company ethos and social responsibility.

The CSC is one of the obvious organizational learning example which can develop "The Five Learning Cycle of Competent Organization" [45] because it has improved their environmental performance and has benefited greatly from EKM and the five learning circle such as in financial savings, value-added product design, energy savings and green competitiveness. It has improved its EKM through the EKCP: The creation, accumulation, sharing, utilization, and internalization of environmental knowledge. The sharing of information is the foundation for creating and implementing a method of improving and developing new technology and designing environmentally friendly products. New technology creation also enables CSC to play a leading role in green product and steel making technology development. With continuous improvement of the EKCP, energy efficiency can often be achieved through integrated strategy planning. Installing water saving technology, introducing treatment techniques and using recirculating systems can lead to the reduction of water consumption. Moreover, the elimination of air emissions, water pollution and waste ensure the conservation of a better environment for future generations. Hence, CSC consistently enhances its international competitiveness and protects the quality of the environment which also helps the company to create a win–win situation corporate and environmental sustainable development.

KEYWORDS

- **Environmental performance**
- **Explicit knowledge**
- **Green supply chain management**
- **Tacit knowledge**

ACKNOWLEDGMENTS

The authors would like to thank the CSC for providing internal information. The authors especially wish to thank Kuo-Chung Liu, I-Yueh Chen, Wu Shun Tsai, Tony Chao, Chio-Po Chang, C. F. Lee, and C. L. Wu for participating in the interviews and providing helpful assistance. The authors also especially wish to thank the financial support of Meiho Institute of Technology.

Chapter 5

Electromagnetic Radiation Effect on Foraging Bats

Barry Nicholls and Paul A. Racey

INTRODUCTION

Large numbers of bats are killed by collisions with wind turbines and there is at present no accepted method of reducing or preventing this mortality. Following our demonstration that bat activity is reduced in the vicinity of large air traffic control and weather radars, we tested the hypothesis that an electromagnetic signal from a small portable radar can act as a deterrent to foraging bats. From June to September, 2007 bat activity was compared at 20 foraging sites in northeast Scotland during experimental trials (radar switched on) and control trials (no radar signal). Starting 45 min after sunset, bat activity was recorded for a period of 30 min during each trial and the order of trials were alternated between nights. From July to September, 2008 aerial insects at 16 of these sites were sampled using two miniature light-suction traps. At each site one of the traps was exposed to a radar signal and the other functioned as a control. Bat activity and foraging effort per unit time were significantly reduced during experimental trials when the radar antenna was fixed to produce a unidirectional signal therefore maximizing exposure of foraging bats to the radar beam. However, although bat activity was significantly reduced during such trials, the radar had no significant effect on the abundance of insects captured by the traps.

The UK government is committed to ensure that 10% of the country's electricity will be generated from renewable sources by 2010 with an aspiration to double this figure by 2020. Unfortunately, the drive to ameliorate the indirect impact of energy production on the environment has led to a more immediate impact on local fauna. The exploitation of wind as a renewable and pollution-free source of energy has led to the proliferation of wind farms across the UK where 206 are currently operational, comprising 2,381 turbines and with an estimated 444 sites proposed for future development [1]. Several studies have highlighted the problem of birds colliding with turbines placed along traditional migratory routes [2-6] but until recently the impact of wind turbines on bats has received little attention.

The scale of the problem became apparent in 2004 when, during a 6-week period, an estimated 1,764 and 2,900 bat fatalities were recorded at two wind farms in Pennsylvania and West Virginia respectively [7]. The number of collision mortalities reported in America are greater than in Europe, where surveys have begun more recently. However, 15 of the 35 species of European bat have been recorded as regular victims of turbine collisions, and an Intersessional Working Group of Eurobats listed 20 species thought to be at risk of collision due to their foraging and commuting behavior [8]. Currently, research in Europe is concentrated on arriving at scientifically credible mortality estimates to assess the extent of the problem. Although this is clearly important, the rapid proliferation of wind turbines requires a more urgent response. Research has to be focused on the underlying reasons behind these collisions and potential methods of mitigation to prevent what is undoubtedly an increasing threat to bat populations.

Attempts at mitigating bird collisions with wind turbines have typically involved the application of visual stimuli to increase the conspicuousness of the turbine blades [9, 10], but for bats, where audition is the primary sensory modality, this is clearly not appropriate. The design of an acoustic deterrent for bats, as used to mitigate cetacean entanglement in drift nets [11-13], is complicated by the intrinsic properties of ultrasound, which attenuates rapidly in air [14]. Despite this inherent problem, a recent study [15] revealed a significant aversive response by big brown bats (*Eptesicus fuscus*) following exposure to broadband white noise in a laboratory. However, when an acoustic deterrent was deployed at a wind farm in New York State, USA, results were more equivocal, and researchers concluded that the acoustic envelope of the deterrent system was probably not large enough to consistently deter the activity of bats within the large volume of the rotor-swept zone [16].

A more promising solution is offered by curtailing the operations of wind turbines during high-risk periods. A substantial portion of bat fatalities at operating wind farms occurs during relatively low-wind conditions during the bat migration period [17]. Some curtailment of turbine operations during these conditions, and during this period, has been proposed as a possible means of reducing impacts to bats [17, 18]. Recent results from studies in Canada [19] and North America [20] indicate that changing turbine "cut-in speed" (i.e., the wind speed at which wind generated electricity enters the power grid) from the customary 3.5–4.0 m/s, on modern turbines, to 5.5 m/s, resulted in at least a 50% reduction in bat fatalities. This requires considerable cooperation on behalf of the project operators as curtailing turbine operations, even on a limited basis, clearly poses operational and economic restrictions resulting in some loss of revenue. This method does however offer a promising solution, particularly in areas where it has been proven that bat mortalities occur over a clearly defined and restricted time period. It is not yet clear whether this method of mitigation will prove sufficiently feasible and effective at reducing impacts to bats at costs that are acceptable to companies that operate wind energy facilities. Therefore, given the problems associated with the existing proposed methods of mitigation it is essential to investigate all other alternatives.

It has been suggested that the radio frequency (RF) radiation associated with radar installations could potentially exert an aversive behavioral response in foraging bats [21]. In 2006 Nicholls and Racey recorded bat activity along an electromagnetic gradient at 10 radar installations throughout Scotland. Their results revealed that bat activity and foraging effort per unit time were significantly reduced in habitats exposed to an electromagnetic field (EMF) strength of greater than 2v/m when compared to matched sites registering EMF levels of zero. Even at sites with lower levels of EMF exposure (<2v/m), bat activity and foraging effort was significantly reduced in comparison to control sites.

Ahlén et al. [22] also reported anecdotal evidence that bats foraging offshore in Sweden avoided an area around Utgrunden lighthouse where a powerful radar was in permanent operation. However, although it has been demonstrated that large air traffic control and weather radars appear to exert an aversive response on foraging bats [21], this has little practical application in preventing bats from colliding with turbine blades. It is therefore necessary to establish whether a deterrent effect can be replicated with a small, portable radar system. It is also possible that the electromagnetic radiation from the radar may not be affecting bats directly but rather the insects upon which they feed. Bat activity within an area is strongly correlated with insect density [23, 24] therefore any reduction in insect density would result in a concurrent reduction in bat activity. In order to provide an efficient deterrent it is necessary to determine whether any observed reduction in bat activity is a direct result of exposure to electromagnetic radiation or an indirect result of a localized reduction in insect density.

Therefore the aims of the present study were to test the following hypotheses:

1. Bat activity will be reduced following exposure to a pulsed electromagnetic signal from a small portable radar unit.
2. The abundance of aerial insects will be reduced following exposure to a pulsed electromagnetic signal from a small portable radar unit.

MATERIALS AND METHODS

Study Sites and Sampling Protocol

In Britain, foraging bats are predominantly associated with areas where insect density is high: broadleaved woodland, particularly woodland edge, linear vegetation (tree lines and hedgerows) and riparian habitat. More open and intensively managed areas are avoided. In order to assess the impact of radar on foraging bats it was important to locate foraging sites with a high level of bat activity. Using existing knowledge obtained from detailed radio telemetry projects [25] in conjunction with extensive acoustic surveys, 20 foraging sites, with a high and consistent level of bat activity, were selected. All foraging sites were located within a 100 km radius of Aberdeen in northeast Scotland (latitude 57°23'N, longitude 02°45'W) and were separated by a minimum straight-line distance of >1 km to ensure independence. Twelve of these sites were located within riparian habitats (small ponds, rivers and streams) and the remainder along the edge of woodland where the radar signal would not be attenuated by any obstruction.

The radar used throughout the study was a Furuno FR—7062 X-band marine radar (peak power 6 kW, beamwidth: horizontal −1.9°, vertical −22°, rotation 24 rpm, or 48 rpm) with a slotted waveguide array antenna (1.2 m) capable of transmitting at pulse lengths of 0.08–0.8 μs depending on the range selected. At each site the radar antenna was placed on a platform 2 m above ground level, such that the core area of bat activity was directly in line with the radar beam. At each foraging site a control (no radar signal) and experimental trial (radar switched on) were carried out. Starting 45 min after sunset, bat activity was recorded for a period of 30 min during each trial and the order of trials were alternated between nights. To avoid pseudoreplication, recordings were carried out only once at each of the 20 sites.

As in most radar systems, the antenna of the radar usually swept through 360 degrees. For the current experiment this would reduce the extent of exposure along any radius. Therefore the experiment was repeated with the antenna of the radar fixed such that the radar signal was orientated directly towards the area of highest bat activity. Similarly the duration of exposure to the radar signal is dependent on the duty cycle of the radar transmitter (pulse length×pulse repetition frequency). Therefore the experiment was repeated at each site using two different pulse length/pulse repetition rates (0.08 μs/2100 Hz, 0.3 μs/1200 Hz,) with the radar antenna fixed to maximize exposure. A portable electromagnetic field meter (PMM 8053-Accelonix Ltd.) and isotropic field probe (EP-330 Isotropic E-Field probe-Accelonix Ltd.) were used to measure the maximum value (peak hold) of the electromagnetic field strength (EMF) of the radar in volts per meter (v/m) at three distances from the radar antenna (10, 20, 30 m) for each of the two radar settings implemented throughout the study.

Bat Activity Recording

At each foraging site bat activity was recorded at three distances from the radar antenna (10, 20, 30 m) using automatic bat-recording stations [26]. Each automatic station consisted of a Batbox III heterodyne bat detector (Stag Electronics, Sussex, UK) linked to a count data logger (Gemini Data Loggers, UK Ltd, Chichester, UK) via an analogue to digital signal converter (Skye instruments, Ltd). The signal converter converts analogue signals from the bat detector into digital signals that can be recorded by the data logger. Every 0.5 sec a positive or negative signal is sent to the data logger indicating the presence or absence of ultrasound respectively. Therefore, the recorded number of bat active half seconds referred to as "bat counts" over a 30-minute trial provides a quantitative index of bat activity during that period. Most narrowband detectors will detect a range of frequencies centered on the value shown on the tuning dial. For the Batbox III this window is ±8 kHz of the tuned frequency, therefore the frequency was set to 50 khz in order to effectively detect each of the five breeding species of bat in Scotland (*Pipistrellus pipistrellus, Pipistrellus pygmaeus, Myotis daubentonii, Myotis nattereri,* and *Plecotus auritus*). The component parts of the system were housed in large plastic boxes with a hole cut for the bat detector microphones. Automatic recording stations were positioned on platforms 1.5 m above the ground and orientated perpendicular to the radar signal (Figure 1).

In conjunction with the automatic recording stations bat activity was recorded continuously during each trial using a frequency division bat detector (S-25, Ultrasound Advice, London). This method of ultrasound transformation allows calls to be recorded in real time on audiocassettes and the number of recorded passes provides a quantitative assessment of bat activity Bat detectors were linked to a tape recorder (Sony Professional Walkman, Tokyo, WMD6C) containing metal-tape cassettes. At each site the bat detector was placed at a distance of 20 m from the radar antenna and the height and direction remained constant at 70 cm. The 60 min recording at each site were analyzed using BatSound software (BatSound Pro, Pettersson Elektronic AB, Uppsala Sweden). In addition to the total number of bat passes, terminal feeding buzzes at each site were counted. These characteristic sounds are produced by aerial hunting and trawling vespertilionid bats when prey capture is attempted [27] and can be used to quantify foraging activity within a site. Foraging rate is expressed as the ratio of terminal buzzes to bat passes; this feeding buzz ratio (FBR) provides a measure of foraging intensity per unit of flight activity [28]. The use of frequency division detectors also allowed accurate species identification at each site.

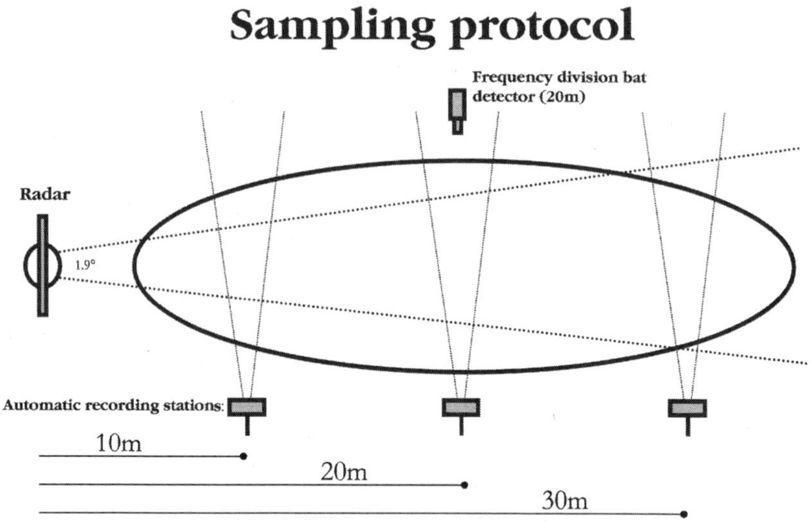

Figure 1. Sampling protocol of experimental trials carried out at 20 independent sites from July to September, 2007. At each site bat activity was recorded for 1 hr at three distances from the radar antenna (10, 20, 30 m) using three automatic bat recording stations orientated perpendicular to the radar beam. A frequency division bat detector was positioned 20 m from the radar antenna to provide further information on bat foraging activity during this period.

Insect Abundance

From July to September, 2008 aerial insects were sampled using two identical Pirbright-Miniature light-suction traps (PMLT) [29] equipped with 8 W UV light bulbs. Each trap operated at 220 V transformed to 12 V to run from a car battery. At the base of each trap was a water-filled collecting vessel containing 2–3 drops of detergent. Most large insects were excluded by a large-mesh screen immediately above the fan and below the light bulb. The traps were deployed at 16 of the 20 foraging sites described above and were switched on for 1 hr prior to sunset. At each site the traps were positioned approximately 40 m apart with their trap inlets 2 m above ground level. On each sampling night the radar antenna was positioned on a platform 2 m above ground level and 10 m from one of the traps such that the antenna was orientated directly towards the trap inlet and fixed to produce a unidirectional signal. The second trap was positioned perpendicular to the radar beam to prevent any potential exposure to electromagnetic radiation and left to function as a control. To avoid any potential bias the selection of traps used as the control was alternated each night. The parameters of the radar tested were identical to those described above (Pulse length/pulse repetition rate: 0.08 μs/2100 Hz; 0.3 μs/1200 Hz,) no test was carried out with the antenna rotating.

Immediately following sampling, the insect catch was transferred from the collecting column into a 70% ethanol in water solution using a fine brush. Insects were then counted using a dissecting microscope (×30). Any insects with wingspans exceeding 20 mm were removed from the catch, as they would exceed the range of insect sizes captured by the species recorded throughout the study [30]. Following counting and sorting, the dry mass of insects was recorded by drying the samples in an oven until a constant mass was achieved (21 hr).

Statistical Analysis

Differences in bat activity (bat counts and bat passes), bat foraging activity (FBRs) and insect abundance between experimental and control trials were analyzed using paired t tests. To account for multiple comparisons in paired t tests, we applied a manual Bonferroni correction (P-values×number of comparisons). However, since the application of the Bonferroni correction increases the risk of making more type II errors, that is not recognizing a true effect as significant [31] we report both corrected P Bonferroni and uncorrected P-values. The effect of distance from the radar antenna was analyzed using one-way ANOVA. Analyses were carried out using Minitab version 14 [32].

Ethics Statement

The authors' work on bats is licensed by the statutory nature conservation organization in Scotland (Scottish Natural Heritage).

RESULTS

Bat Activity

Experimental trials were carried out during 58 nights from July, 2007 till September, 2007 representing a total of 58 hr of recording data within the following parameters:

1. Rotating antennapulse length/pulse repetition rate (0.08 μs/2100 Hz)–20 hr
2. Fixed antennapulse length/pulse repetition rate (0.08 μs/2100 Hz)–20 hr
3. Fixed antennapulse length/pulse repetition rate (0.3 μs/1200 Hz)–18 hr

The maximum value (peak hold) of the electromagnetic field strength within these parameters is shown in Table 1. Field strength diminished slightly with increasing distance from the antenna under all radar parameters. However, when the radar antenna was fixed to emit a unidirectional signal a fourfold increase in field strength was observed at all distances (Table 1).

Table 1. The maximum value (peak hold) of the electromagnetic field strength (v/m) at three distances from the radar antenna.

Antenna position	Pulse length (μs)	Pulse Repetition rate (Hz)	Duty Cycle (%)	EMF (v/m) Peak hold (10m)	EMF (v/m) Peak hold (20m)	EMF (v/m) Peak hold (30m)
Rotating	0.08	2100	0.0168	5.58	5.11	3.79
Fixed	0.08	2100	0.0168	26.24	22.99	20.25
Fixed	0.3	1200	0.036	25.52	18.68	17.67

doi:1 0.1371 /journal.pone.0006246.t001

The three automatic stations recorded a total of 102,810 bat counts during 58 hr of recording (Table 2). No significant difference was observed in the number of bat counts recorded between automatic stations positioned at 10, 20, and 30 m from the radar antenna (ANOVA, rotating antenna with pulse length 0.08 μs: P = 0.57; fixed antenna with pulse length 0.08 μs: P = 0.64; fixed antenna with pulse length 0.3 μs P = 0.68) therefore all further tests were carried out on the average of these three values. A further 53,731 bat passes were recorded with the frequency division detector (Table 2). As expected, the majority of passes (84%) were attributed to the two cryptic pipistrelle species: *Pipistrellus pygmaeus* and *P. pipistrellus* (51% and 33% respectively) which are the most common and abundant bats in Scotland. A further 16% of bat passes were attributed to *Myotis daubentonii*.

Table 2. Total numbers of bat counts, bat passes and feeding buzzes recorded within treatment and control trials during 58 hr of recording.

Index of bat activity	Rotating antenna (0.08 μs/2100 Hz)		Fixed antenna (0.08 μs/2100 Hz)		Fixed antenna (0.3 μs/1200 Hz)	
	Treatment	Control	Treatment	Control	Treatment	Control
Bat passes	11160	11599	8065	9305	5367	8235
Feeding buzzes	3711	4015	2386	3300	1563	2720
Bat counts (10 m)	6052	6275	4998	5974	3241	5517
Bat counts (20 m)	6364	6820	5261	6183	3494	5525
Bat counts (30 m)	7066	7386	5744	6792	3879	6239

doi:1 0.1371 /journal.pone.0006246.t002

Total bat activity was invariably higher during the control trials when compared to experimental trials (Table 2). However paired t tests carried out on all indices of bat activity (bat counts, bat passes, FBRs) revealed no significant difference in bat activity between control and experimental trials when exposed to a short pulse length (0.08 μs) radar signal from a rotating antenna (bat counts: $t = 1.50$; $P = 0.151$; $P_{Bonferroni} = 0.453$; Figure 2a. Bat passes: $t = 1.89$; $P = 0.074$; $P_{Bonferroni} = 0.222$; Figure 3a. The FBR: $t = 1.80$; $P = 0.088$; $P_{Bonferroni} = 0.264$; Figure 4a). Paired t tests carried out on all indices of bat activity (bat counts, bat passes, FBRs) showed that bats were significantly less active during experimental trials than during control trials when exposed to a short pulse length (0.08 μs) radar signal from a fixed antenna (bat counts: $t = 2.87$; $P = 0.010$; $P_{Bonferroni} = 0.030$; Figure 2b. Bat passes: $t = 2.54$; $P = 0.020$; $P_{Bonferroni} = 0.060$; Figure 3b. The FBR: $t = 3.82$; $P = 0.001$; $P_{Bonferroni} = 0.003$; Figure 4b). However, following Bonferroni correction the difference in the number of bat passes between experimental and control trials was no longer significant. Bats were also significantly less active during experimental trials than during control trials when exposed to a medium pulse length (0.3 μs) radar signal from a fixed antenna (bat counts: $t = 3.95$; $P = 0.001$; $P_{Bonferroni} = 0.003$; Figure 2c. Bat passes: $t = 3.69$; $P = 0.002$; $P_{Bonferroni} = 0.006$; Figure 3c. The FBR: $t = 6.78$; $P<0.001$; $P_{Bonferroni} = 0.003$; Figure 4c). A summary of these results is presented in Table 3.

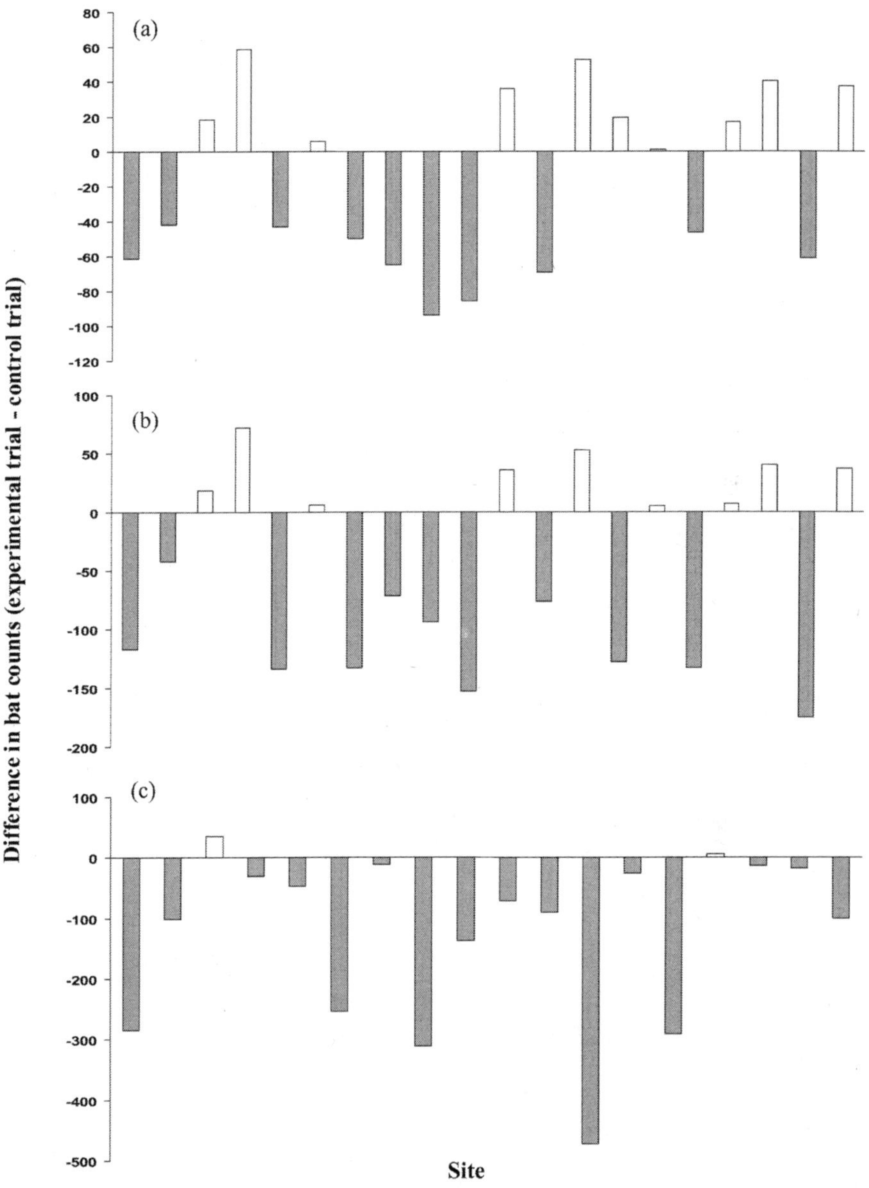

Figure 2. The response of bats to: (a) short pulse length (0.08 μs) radar signal from a rotating antenna. (b) short pulse length (0.08 μs) signal from a fixed antenna. (c) medium pulse length (0.3 μs) signal from a fixed antenna. Each bar represents the difference in bat counts (the number of times that ultrasound was detected by the automatic bat recording stations) between control and experimental trials. A negative value indicates that bat activity was higher during the control trial than during the experimental trial when the radar was switched on.

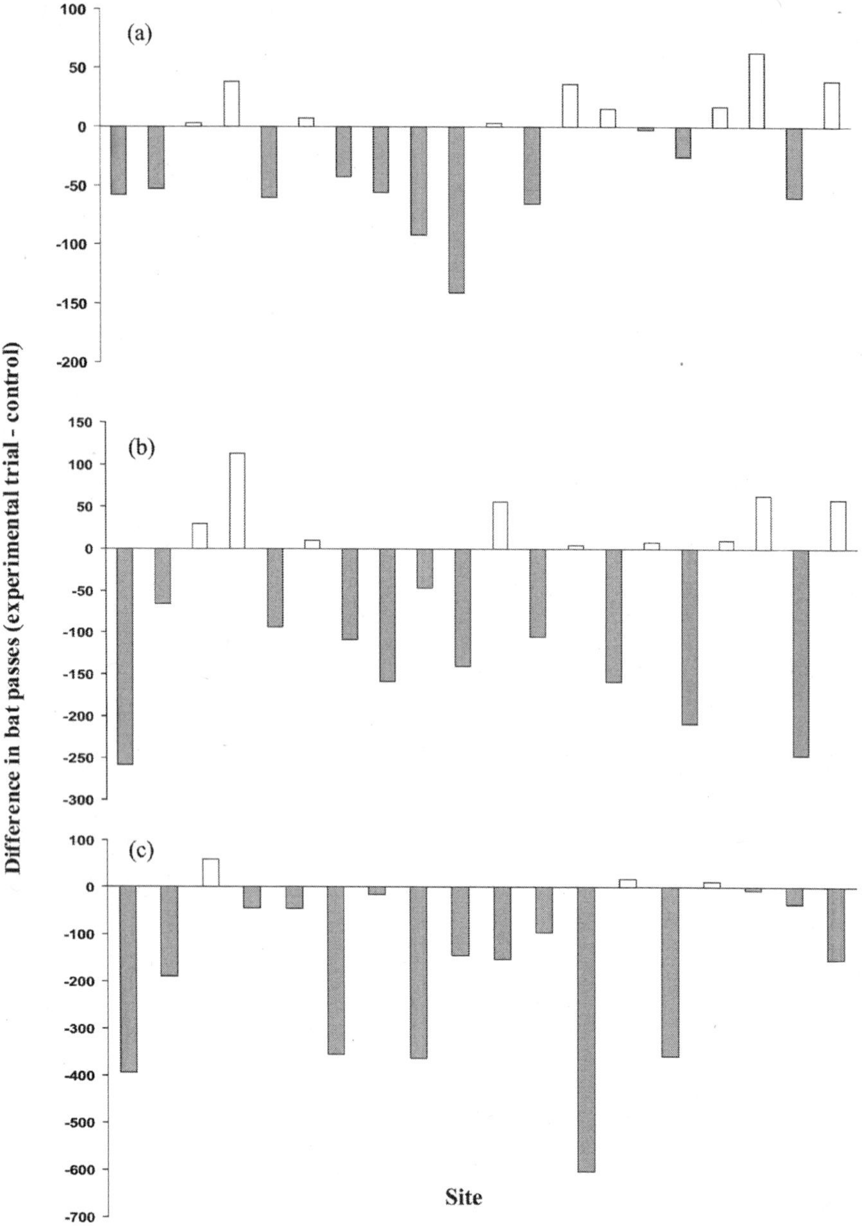

Figure 3. The response of bats to: (a) short pulse length (0.08 μs) radar signal from a rotating antenna. (b) short pulse length (0.08 μs) signal from a fixed antenna. (c) medium pulse length (0.3 μs) signal from a fixed antenna. Each bar represents the difference in bat passes (recorded using a frequency division bat detector) between control and experimental trials. A negative value indicates that bat activity was higher during the control trial than during the experimental trial when the radar was switched on.

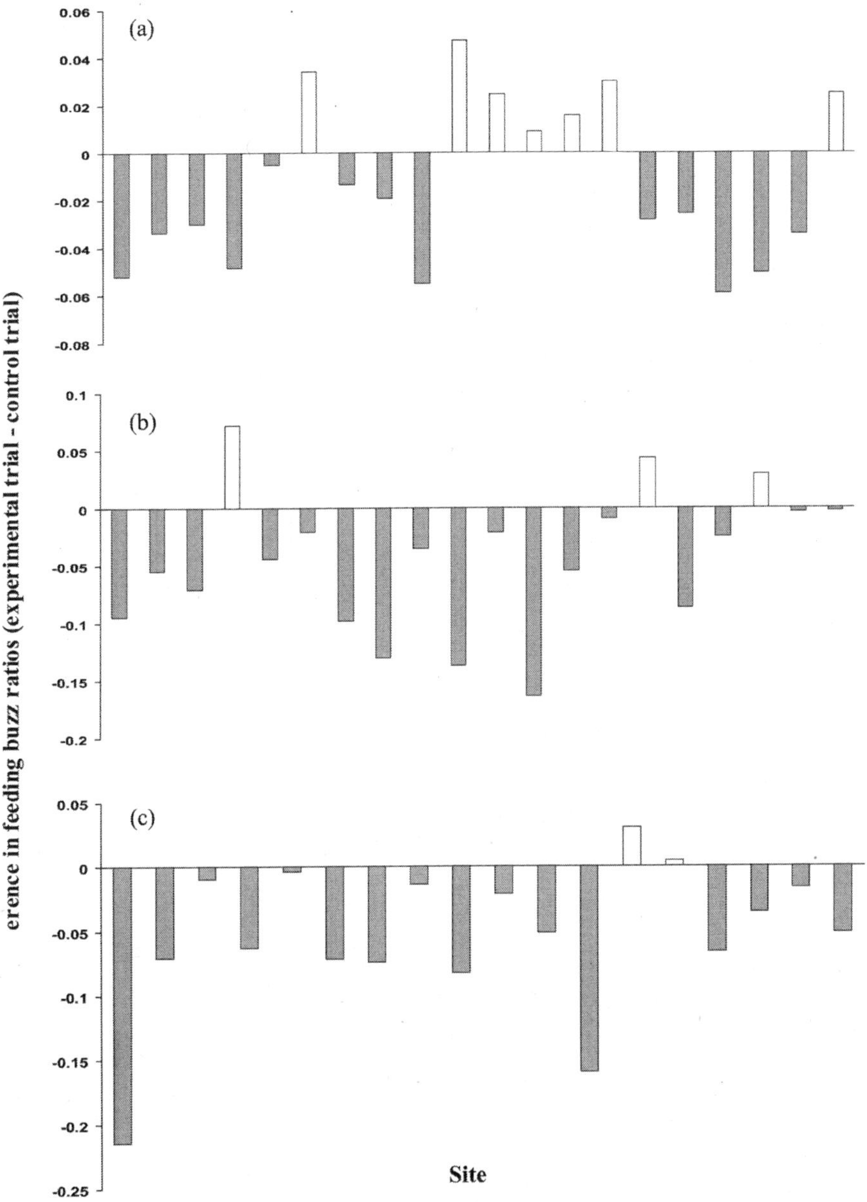

Figure 4. The response of bats to: (a) short pulse length (0.08 μs) radar signal from a rotating antenna. (b) short pulse length (0.08 μs) signal from a fixed antenna. (c) medium pulse length (0.3 μs) signal from a fixed antenna. Each bar represents the difference in foraging rate per unit time as reflected by the difference in FBR between control and experimental trials. A negative value indicates that bat activity was higher during the control trial than during the experimental trial when the radar was switched on.

Table 3. Statistical significance of differences in bat activity between control and experimental trials (*) denotes a significant result for both corrected P Bonferroni (P-values×number of comparisons) and uncorrected P-values.

Antenna Position	Index of activity	Pulse Length (μs)	Pulse Repetition Rate (Hz)	Duty Cycle (%)	n	t	p	$P_{Bonferroni}$
Rotating	Bat passes	0.08	2100	0.0168	20	1.89	0.07	0.22
	Bat counts	0.08	2100	0.0168	20	1.50	0.15	0.45
	FBR	0.08	2100	0.0168	20	1.80	0.08	0.26
Fixed	Bat passes	0.08	2100	0.0168	20	2.54	0.02*	0.06
	Bat counts	0.08	2100	0.0168	20	2.87	0.01*	0.03*
	FBR	0.08	2100	0.0168	20	3.82	0.001*	0.003*
Fixed	Bat passes	0.3	1200	0.036	18	3.69	0.002*	0.006*
	Bat counts	0.3	1200	0.036	18	3.95	0.001*	0.003*
	FBR	0.3	1200	0.036	18	6.78	< 0.001*	0.003*

doi: 1 0.1371 /journa l.pone.0006246.t003

Insect Abundance

Experimental trials were carried out during 32 nights from July, 2008 till September, 2008 representing a total of 32 hr of recording data within the following parameters:

1. Fixed antenna–pulse length/pulse repetition rate (0.08 μs/2100 Hz)–16 hr
2. Fixed antenna–pulse length/pulse repetition rate (0.3 μs/1200 Hz)–16 hr

A total of 10,430 insects were caught during 32 hr of sampling per trap. Ninety-five percent of the insects caught had wingspans <20 mm and were dried and included in further analyses. Paired t tests revealed no significant difference in insect abundance between control and experimental traps when exposed to either a short (0.08 μs) or medium pulse length (0.3 μs) radar signal (short pulse: $n = 16$, $t = 1.50$; $P = 0.151$; $P_{Bonferroni} = 0.453$; long pulse: $n = 16$, $t = 1.89$; $P = 0.074$; $P_{Bonferroni} = 0.222$).

DISCUSSION

Currently there is no accepted method of successfully mitigating bat collisions with wind turbines and attempts at deterring bats by the use of ultrasound have, as yet, been unsuccessful. Therefore, the identification of alternative methods capable of inducing an aversive response in bats approaching turbine blades is of paramount importance. Very few field experiments have been carried out to ascertain the possible effects of high frequency electromagnetic radiation on populations of wild animals. However, studies have shown that electromagnetic radiation can influence the development, reproduction, and physiology of insects [33], mammals [34], and birds [35]. Our results demonstrate that an electromagnetic signal from a small radar unit with a fixed antenna invariably reduced the foraging activity of bats within 30 m of the unit. However, no significant decrease in activity was observed when the radar antenna was rotating. This is not surprising; the length of time a bat would be exposed to the radar signal is a function of the duty cycle of the radar signal (pulse length×pulse repetition rate) and the dwell time (the duration of time that a target remains in the radar beam during each rotation). The rotation of the radar antenna would reduce the time that bats were exposed to pulse-modulated microwave radiation and would therefore attenuate any potential deterrent effect. When the radar antenna was fixed to emit a unidirectional signal a fourfold increase in field strength was observed at all distances.

When foraging sites were exposed to a short pulse length signal from a fixed antenna there was a significant reduction in bat activity during experimental trials (bat counts and bat passes dropped by 15.5% and 13.3% respectively). Although, once the Bonferroni correction had been applied, the difference in bat passes between control and experimental trials was no longer significant. An even greater level of significance was however observed when foraging sites were exposed to a medium pulse length signal from a fixed antenna (bat counts and bat passes dropped by 38.6% and 30.8% respectively). Clearly, this represents a substantial reduction in bat activity. However, bats continued to forage at each site during experimental trials, and on no occasion were bats observed behaving abnormally or actively avoiding the beam of the radar. However, temporal and spatial fluctuations in bat foraging behavior are common [23, 24] and therefore results have to be treated with caution. Despite this caveat the significant reduction in bat activity during all experimental trials with a fixed antenna supports our hypothesis that electromagnetic radiation exerts a deterrent effect on foraging bats. This raises questions regarding the mechanisms through which bats could perceive electromagnetic fields and why they would seek to avoid them.

Nicholls and Racey [21] suggest that the aversive behavioral response of foraging bats to electromagnetic radiation may be a result of thermal induction. Studies investigating the behavioral response of laboratory animals to the presence of electromagnetic fields have shown that even short-term exposure can produce a thermal burden in an organism that can result in significant behavioral and physiological changes, some of which may be harmful [36]. Behavioral effects of such exposure include perception, aversion, work perturbation, work stoppage and convulsions [37]. The wing membranes of bats present a large surface area over which radiation might be absorbed, increasing heat load on the animal. This, combined with the heat energy produced during flight, makes bats particularly susceptible to overheating [38, 39]. Furthermore, observations of captive bats have noted their aversion to even a moderate infra-red heat source [40].

However, the pulsed microwave radiation characteristic of radars is a rather inefficient source of energy. The energy produced by a radar signal can reach very high values of peak power density, at relatively low levels of power density averaged over time. This is because the pulse length of the radar signal is hundreds of times shorter than the pulse repetition rate, therefore the average value of power density is hundreds of times lower than the peak value of the radiation. Therefore, it would seem unlikely that the energy in the radar signal would be sufficient to induce a thermal burden in bats foraging within the beam. However, several studies have reported significant behavioral and physiological effects resulting from exposure to pulsed microwave radiation even when the average power density of the signal was relatively low [41-43]. The mechanism through which pulsed microwave radiation could affect behavior in this manner is unclear although one possibility is an auditory response commonly referred to as the auditory microwave hypothesis.

The auditory perception of pulsed microwaves is now widely accepted. The effect is generally attributed to the thermoelastic expansion of brain tissue following the small but rapid increase in temperature due to the absorption of the incident energy. This generates a sound wave in the head that subsequently stimulates the cochlea. Repeated or prolonged exposure to these auditory effects is considered stressful [44].

Laboratory experiments have shown that the frequency of the induced sound is a function of head size and of the acoustic properties of the brain tissue. The estimated fundamental frequency of vibration in guinea pigs, cats, and adult humans are 45, 38, and 13 kHz respectively [45, 46]. It is therefore not only plausible but probable that bats exposed to an RF pulse of sufficient power would effectively hear this pulse and the frequency detected would lie within the range of frequencies used for orientation, prey detection, and capture for the majority of bat species. It is possible that, as reported in other studies, exposure to these auditory effects may be stressful for bats or indeed it may interfere with their echolocation, inhibiting prey detection, or capture. During the present study, foraging rate per unit time was significantly reduced during experimental trials indicating that bats foraging within the exposed area were feeding at a reduced rate in comparison to those foraging during the control trials. This is particularly surprising given that exposure to the radar had no significant impact on the abundance of aerial insects, and the observed reduction in foraging rate is therefore unlikely to be linked to a decline in insect abundance. It is therefore possible that the auditory perception of the radar signal during experimental trials could have interfered with the bats ability to detect or capture prey. However, further experimentation would be required to accurately identify the causal relationship between exposure to electromagnetic radiation and the observed reduction in both bat activity and foraging rate.

Although, we have demonstrated a clear biological effect, one of the limitations of the present study was the use of a commercial marine radar that was not specifically designed for the task. With only a limited control over the parameters of the radar signal, it is difficult to determine which parameters are most effective in deterring bats. To better understand the response of bats to electromagnetic radiation, and to identify an optimum signal capable of deterring bats, will require radar engineers to work with bat biologists to develop a portable radar which can be manipulated to produce a wider range of electromagnetic outputs. The parameters most likely to be important are the frequency, pulse length/pulse repetition rate and power output of the signal. Similarly, the radar used in the present study was only effective when the antenna was fixed to produce a unidirectional signal with a horizontal beamwidth of 1.9°. A narrow unidirectional signal is clearly not appropriate to deter bats from approaching wind turbines. In order to provide an effective deterrent it would be necessary to emit a multidirectional electromagnetic signal capable of encapsulating the large volume of the rotor-swept zone.

CONCLUSION

We have demonstrated that pulsed electromagnetic radiation from a small, affordable and portable radar system can reduce bat activity within a given area. Results were most effective when the radar antenna was fixed to produce a unidirectional signal therefore maximizing dwell time within the beam of the radar. However, although bat activity was significantly reduced during experimental trials substantial numbers of bats continued to forage within the beam. It is possible that only a particular combination of wavelength, pulse repetition rate, power output and target size or orientation may provoke a reaction and further work is necessary to elucidate this relationship further.

KEYWORDS

- **Aerial hunting**
- **Electromagnetic radiation**
- **Electromagnetic signal**
- **Radar antenna**

AUTHORS' CONTRIBUTIONS

Conceived and designed the experiments: Paul A. Racey. Performed the experiments: Barry Nicholls. Analyzed the data: Barry Nicholls. Wrote the chapter: Barry Nicholls.

ACKNOWLEDGMENTS

We would like to thank J. French, J. D. Pye, and C. D. MacLeod for their comments on an earlier draft of this manuscript, and Jennifer Skinner for assisting in the fieldwork.

Chapter 6

Phylogeographic Analysis of Tick-borne Pathogen

Agustнn Estrada-Peca, Victoria Naranjo, Karina Acevedo-Whitehouse, Atilio J. Mangold, Katherine M. Kocan, and Josй de la Fuente

INTRODUCTION

The tick-borne pathogen *Anaplasma marginale*, which is endemic worldwide, is the type species of the genus *Anaplasma* (Rickettsiales: Anaplasmataceae). *Rhipicephalus* (*Boophilus*) *microplus* is the most important tick vector of *A. marginale* in tropical and subtropical regions of the world. Despite extensive characterization of the genetic diversity in *A. marginale* geographic strains using major surface protein (MSP) sequences, little is known about the biogeography and evolution of *A. marginale* and other *Anaplasma* species. For *A. marginale*, MSP1a was shown to be involved in vector-pathogen and host-pathogen interactions and to have evolved under positive selection pressure. The MSP1a of *A. marginale* strains differs in molecular weight because of a variable number of tandem 23–31 amino acid repeats and has proven to be a stable marker of strain identity. While, phylogenetic studies of MSP1a repeat sequences have shown evidence of *A. marginale*-tick co-evolution, these studies have not provided phylogeographic information on a global scale because of the high level of MSP1a genetic diversity among geographic strains.

In this study we showed that the phylogeography of *A. marginale* MSP1a sequences is associated with world ecological regions (ecoregions) resulting in different evolutionary pressures and thence MSP1a sequences. The results demonstrated that the MSP1a first (R1) and last (RL) repeats and microsatellite sequences were associated with world ecoregion clusters with specific and different environmental envelopes. The evolution of R1 repeat sequences was found to be under positive selection. It is hypothesized that the driving environmental factors regulating tick populations could act on the selection of different *A. marginale* MSP1a sequence lineages, associated to each ecoregion.

The results reported herein provided the first evidence that the evolution of *A. marginale* was linked to ecological traits affecting tick vector performance. These results suggested that some *A. marginale* strains have evolved under conditions that support pathogen biological transmission by *R. microplus*, under different ecological traits which affect performance of *R. microplus* populations. The evolution of other *A. marginale* strains may be linked to transmission by other tick species or to mechanical transmission in regions where *R. microplus* is currently eradicated. The information derived from this study is fundamental toward understanding the evolution of other vector-borne pathogens.

The genus *Anaplasma* (Rickettsiales: Anaplasmataceae) contains obligate intracellular organisms found exclusively within membrane-bound inclusions or parasitophorous vacuoles in the cytoplasm of both vertebrate and invertebrate host cells [1, 2]. The genus *Anaplasma* includes pathogens of ruminants, *A. marginale, A. centrale, A. bovis*, and *A. ovis*. Also included in this genus are *A. phagocytophilum*, which infects a wide range of hosts including humans and wild and domesticated animals, and *A. platys* that infects dogs.

To date, most research has been reported for *A. marginale*, the type species for the genus *Anaplasma* [3]. Both cattle and ticks develop persistent infections with *A. marginale* and therefore can serve as reservoirs of infection. *A. marginale* is transmitted horizontally by ixodid ticks including *Rhipicephalus spp.* and *Demacentor spp. Rhipicephalus (Boophilus) microplus* is considered the most important biological vector in tropical and subtropical regions of the world [4]. Transfer of infected blood by biting flies or blood-contaminated fomites effects mechanical transmission of *A. marginale*. The complex developmental cycle of *A. marginale* has been described and shown to be coordinated with the tick feeding cycle [3]. The midgut is the first site of infection, where membrane-bound vacuoles or colonies initially contain reticulated forms that divide by binary fission and subsequently transform into dense forms. Infection of salivary glands and other tissues then occurs which completes the developmental cycle and allows for transmission to susceptible hosts during tick feeding.

Vector-pathogen interactions involve traits from both the vector and the pathogen [5]. Several MSPs have been identified and characterized in *A. marginale* [3, 5]. The MSPs are involved in interactions with both vertebrate and invertebrate hosts [2, 3, 6-9], and therefore are likely to evolve more rapidly than other genes because they are subjected to selective pressures exerted by host immune systems. The MSP1a of *A. marginale* geographic strains differs in molecular weight due to a variable number of tandem 23–31 amino acid repeats, and the sequence of MSP1a has been shown to be a stable marker for identification of geographic strains [10, 11]. Functionally, MSP1a was shown to be an adhesin for bovine erythrocytes and tick cells [12-14]. Tick molecules involved in vector-*A. marginale* interactions were recently identified and functionally characterized [15].

The geographic strains of *A. marginale* are highly variable, as demonstrated by the analysis of MSP1a sequences [7, 11]. Such genetic heterogeneity observed among *A. marginale* strains in endemic regions could be explained by cattle movement and maintenance of different genotypes by independent transmission events, due to infection exclusion of *A. marginale* in cattle and ticks which commonly results in the establishment of only one genotype per animal [16-18]. Due to the high degree of sequence variation within most endemic areas, MSP1a sequences have failed to provide phylogeographic information on a global scale [7]. These studies also suggested that multiple introductions of *A. marginale* strains from different geographic locations had occurred in many regions.

The evolutionary history of vector-pathogen interactions can be reflected in the sequence variation of *Anaplasma* MSPs. Previous studies demonstrated that *A. marginale* MSP1a evolved under positive selection [19]. Analysis of *A. marginale* MSP1a repeats provided evidence of tick-pathogen co-evolution [5, 6, 20], a result that is consistent with the biological function of MSP1a in pathogen transmission by ticks [14]. However, the study of *A. marginale* evolutionary history and tick-pathogen co-evolution has remained elusive because of the extensive genetic diversity of MSP1a sequences.

In this study we analyzed MSP1a repeat and microsatellite sequences in order to provide information on the evolution of *A. marginale* strains by determining their phylogeographic association with world ecological regions (ecoregions) and by testing the effect of different evolutionary pressures associated with the tick vector, mainly *R. microplus*, ecology in these ecoregions.

MATERIALS AND METHODS

Anaplasma marginale MSP1a Repeat Sequences

The MSP1a sequences included in this study were obtained from *A. marginale* strains collected worldwide from infected cattle and recently reviewed by de la Fuente et al. [11]. The data on MSP1a sequences was updated by searching the Genbank sequence database [21]. The amino acid sequence of the first (R1) and last (RL) MSP1a repeats of 111 *A. marginale* strains were used in this study, from which 39 and 28 unique R1 and RL sequences, respectively, were obtained. For comparison, the sequences of Rn MSP1a repeats, located between R1 and RL, were included in some analyses. MSP1a sequences not included in these studies were from *A. marginale* strains that were not adequately geo-referenced.

Anaplasma marginale MSP1a Microsatellite Sequences

A microsatellite was located in the MSP1a 5'UTR between the putative Shine-Dalgarno sequence (GTAGG; [6]) and the translation initiation codon (ATG). The structure of the microsatellite (bold) was GTAGG (G/A TTT)m (GT)n T ATG (Table 1). The microsatellite was sequenced in 115 *A. marginale* strains collected from infected cattle in the USA, Canada, Mexico, Puerto Rico, Argentina, Brazil, Italy, Israel, South Africa, and China [11].

Table 1. Structure and ecoregion cluster distribution of the *A. marginale* MSP1a microsatellites.

Genotype	Number of strains	m	n	SD-ATG distance (nucleotides)	Genotype frequency per ecoregion cluster			
					1	2	3	4
A	2	1	7	19	0.00	1.00	0.00	0.00
B	5	1	9	23	1.00	0.00	0.00	0.00
C	7	2	5	19	0.57	0.14	0.29	0.00
D	3	2	6	21	0.33	0.67	0.00	0.00
E	12	2	7	23	0.75	0.00	0.25	0.00
F	3	3	4	21	0.00	0.00	1.00	0.00
G	78	3	5	23	0. 15	0.14	0.56	0.14
H	3	3	6	25	0.00	0.67	0.33	0.00
I	NI	4	6	29	...	...	...	...

The MSPI a microsatellite sequences were analyzed in 115 A. marginale strains. The microsatellite (sequence in bold) was located between the Shine-Dalgarno (SD; sequence in brackets) and the translation initiation codon (ATG) with the structure: GTAGG (G/A TTT)m (GT)n T ATG. The SD-ATG distance was calculated in nucleotides as (4 x m) + (2 x n) + I. Abbreviation: NI, not included in the study because the A. marginale strain was not adequately gee-referenced.

World Ecoregions and Association with *A. marginale* Strains

Ecoregions are used herein to classify the world across dynamic environmental factors. We assumed that (i) ecoregions could be delineated using quantitative abiotic characters based on well-recognized and repeatable attributes and (ii) *A. marginale* strains are associated with each ecoregion and subjected to different environmental conditions that could be analyzed by multivariate geographic clustering [22]. Multivariate geographic clustering involves the use of standardized values for selected environmental conditions in a set of raster maps. Those values serve as coordinates in the environmental data space, in which environmental conditions are further clustered according to their similarities. The feature selected to put together the clusters was the monthly Normalized Difference Vegetation Index (NDVI). The NDVI is a variable that reflects vegetation stress, a feature that summarizes information about the ecological background for tick populations [23]. We obtained a 0.1° resolution series of monthly NDVI data for the period 1986–2006. The 12 averaged monthly images were subjected to Principal Components Analysis (PCA) to obtain decomposition into the main axes representing the most significant, non-redundant information. The strongest principal axes were chosen using Cattell's Scree Test [22]. It has been found that the first principal component derived from NDVI typically represents the greenness of the surveyed area [24]. Component 2 is interpreted as a change component, taken to represent a winter/summer seasonality effect. Components 3 and 4 are also essentially seasonal, but represent areas where the timing of green-up is different from that in component 2. Our PCA analysis retained three principal axes, explaining the 92% of total variance. These three axes were related to the mean NDVI values, annual amplitude, and NDVI values in the period May to August, respectively. We then used a hierarchical agglomerative clustering on PCA values to classify multiple geographical areas into a single common set of discrete regions. Mahalanobis distance was used as a measure of dissimilarity and the weighted pair-group average was used as the amalgamation method. A value of 0.05 was used as the cut-off probability for assignment to a given ecoregion. All the procedures adhered to methods previously described [25].

The decision about the number of ecoregions to retain without any prior detail about the information they contain is a problem to which a solution has not yet been found. The main goal is to define unambiguously the *A. marginale* strains recorded mostly in a single ecoregion cluster and present in the highest number of geographical sites belonging to that cluster. The result is to refine the degree of clustering that gives the optimal degree of association between *A. marginale* strains ("species") and ecoregions ("sites"). This analysis was done using the "indicator species" method [26], a previously published multivariate statistics procedure to define "sites" as a function of their faunal composition ("species"). We began the agglomerative process described above with an unrealistic high number of ecoregions. At every step of the agglomerative process, pathogen strains and ecoregions were ordered by a correspondence analysis, and then analyzed using the "indicator species" method. The procedure runs iteratively, trying to improve the association with further clustering of ecoregions. However, the method does not force a cluster if specifications of cut-off probabilities for ecoregions are violated, and does not assume any a priori condition about the geographical range of any cluster. The procedure is only an indicator that stops when an optimum degree of ecoregion clustering collectively explaining the association of *A. marginale* strains with the environment is reached. Such association may be low or high (in the range 0–1) but is the optimal according to the ecoregion features. After the use of these methods we retained a total of four ecoregion clusters (Figure 1) to which *A. marginale* sequences were unambiguously associated, and the rest were discarded because *A. marginale* strains were not isolated there.

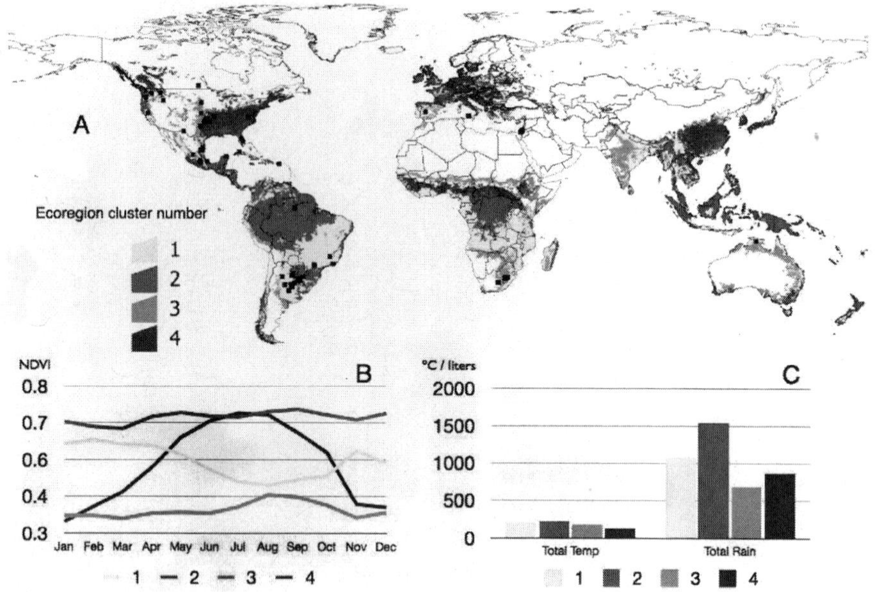

Figure 1. World ecoregions. (A) Clusters of vegetation features of the world computed through an unsupervised classification on Principal Components Analysis decomposition of monthly NDVI values at a resolution of 0.1°. Small islands were removed to improve the presentation. Only the ecoregions providing most information about *A. marginale* distribution according to a Discriminant Analysis were included, grouped in four main clusters and arbitrarily numbered. Overlying records of *A. marginale* are in black. (B) Monthly NDVI values detected for each ecoregion clusters (plotted with the same colors as in A). (C) Monthly accumulated temperature and rainfall recorded for each ecoregion cluster.

Analysis of *A. marginale* MSP1a Repeat Sequences

The conserved amino acid sequences of MSP1a R1 and RL repeats reported in *A. marginale* strains from each ecoregion cluster were used to determine the consensus sequence in the cluster. The distances between groups of sequences ascribed to an ecoregion cluster were then computed, both within and between clusters. The hypothesis herein is that highest affinities between R1 and RL isolates are observed when grouped according to the ecoregion to which they are related. An alternative hypothesis is that isolates cluster according to a geographical background. To check for that alternative hypothesis we examined genetic distances with strains arranged into geographical groups. We used only American strains to overcome the undersampling in Africa. They were grouped as Western USA, Eastern USA, Mesoamerica and the

Caribbean, and South America, and then distances within and between geographical clusters computed again.

Other analyses involved calculation of the percentage of changes between the consensus MSP1a repeat sequences among different ecoregion clusters as a measure of the genetic distance between *A. marginale* strains in these clusters. Some R1 and RL sequences were found associated with more than one ecoregion cluster. Therefore, the percentage of changes in amino acid composition for every sequence and the consensus sequence in each cluster were computed as a measure of similarity of each sequence and consensus sequence for each cluster. MSP1a repeat sequences were aligned for pairwise comparison and determination of non-synonymous (d_N) and synonymous (d_S) substitutions using Mega 4 [27]. The d_N and d_S were determined among all pairwise comparisons of MSP1a repeat sequences within each ecoregion cluster, estimated by the method of Nei and Gojobori [28] with the correction for multiple substitutions [29]. The ratio of the mean d_N/d_S was used as an indicator of the level of selection acting on MSP1a repeat sequences.

As an additional test to verify our hypothesis, we performed a Mantel's test. The explanation of the distribution of the strains in terms of environmental variables may be confounded because the variables are intercorrelated among themselves, and so it may be difficult to ascribe causal mechanisms to the environmental variables. Mantels test is a regression in which the variables are dissimilarity matrices. The operative question is "do strains that have similar sequences also tend to be similar in terms of the environmental variables"? Therefore, we performed a Mantel's test both between a dependent distance matrix (genetic similarities of R1 and RL) and the predictor matrix of geographical distances among strains. A second Mantel's test was done with the same dependent matrix and a predictor dissimilarity matrix based on environmental PCA-derived values. Such correlations will indicate if locations that are closer or locations that are similar environmentally are similar compositionally. Mantel's tests were performed using the Jackard index according to [30].

Analysis of *A. marginale* MSP1a Microsatellite Sequences

The extent of genetic differentiation of *A. marginale* strains at the MSP1a microsatellite was assessed within and among ecoregion clusters using an analysis of molecular variance (AMOVA; [31]) and pairwise population FST significance tests as implemented in ARLEQUIN, version 3.01 [32]. The statistical significance of fixation indices was tested using a non-parametric permutation approach [33] with 20,000 permutations. Ecoregion clusters for which statistically significant subdivision was not detected were pooled to define groups.

The effect of microsatellite size on MSP1a expression was characterized in *A. marginale* strains Wetumka (OK; Genbank accession number AY010247), Okeechobee (FL; AY010244), Idaho (ID; M32868) and HB-A8 (China; DQ811774) sequences. These strains had microsatellite genotypes G (Wetumka and Okeechobee; distance SD-ATG = 23 nucleotides), C (Idaho; SD-ATG = 19 nucleotides) and I (China; SD-ATG = 29 nucleotides). The msp1alpha gene containing promoter sequences active in Escherichia coli [6] was amplified using oligonucleotide primers MSP1aP: 5'GCAT-

TACAACGCAACGCTTGAG3' and MSP1a3: 5'GCTTTACGCCGCCGCCTGC-GCC3' and cloned into pGEM-T vector (Promega, Madison, WI, USA) as reported previously [20]. Three independent clones for each of the MSP1a constructs were transformed in *E. coli* JM109 cells and grown for 15–20 hr at 37°C. Culture volumes of 3 ml were used for RNA and DNA extraction using TriReagent (Sigma, St. Louis, MO, USA) according to manufacturer's instructions. The RNA samples were treated with RNase-free DNase (Invitrogen, Carlsbad, CA, USA) prior to RT-PCR. MSP1a mRNA levels were characterized by real-time RT-PCR using oligonucleotide primers MSP1RT5: 5'ACCAATCGTTGGCAGAAGAG3' and MSP1RT3: 5'ACCT-GCTCCCAAAGTAGCAA3' and normalizing against *E. coli* D-1-deoxyxylulose 5-phosphate synthase gene (*dxs*) [34] and plasmid DNA copy number by msp1alpha PCR using the oligonucleotide primers and conditions described above. Real-time RT-PCR was conducted using the iScript One-Step RT-PCR Kit with SYBR Green and an iQ5 thermal cycler (Bio-Rad, Hercules, CA, USA). Control reactions were performed using the same procedures but without reverse transcriptase to test for DNA contamination in the RNA preparations and without RNA added to detect contamination of the PCR reaction. The normalized mRNA levels were compared between different MSP1a constructs using an ANOVA test ($P = 0.05$).

Anaplasma marginale MSP1a Repeat Sequences Show Ecoregion-specific Signatures

Ecoregion clusters showed different NDVI, temperature and rainfall values (Figure 1). Ecoregion cluster 1 extended over large areas of central Africa and central South America, primarily Argentina and southern Brazil. It involved a region with medium to high NDVI values with a clear seasonal decrease between June and September. This was the ecoregion with the highest recorded temperature and around 1,000 mm of annual rainfall. Ecoregion cluster 2 included vast areas of the Mesoamerican corridor, northern South America and a small territory of eastern South Africa. It consisted of zones with high NDVI along the year without seasonal variability, temperature values similar to those in ecoregion cluster 1 and rainfall around 1,500 mm/year. Ecoregion cluster 3 extended over central South Africa and scattered parts of southern USA and Mexico, with the lowest NDVI values and little change across the year. This ecoregion displayed lower temperature values and the minimum rainfall. Finally, ecoregion cluster 4 extended over large areas of USA and had a clear NDVI signature, very low between November and March and then rising to reach maximum levels around July. This area was the coldest among all the ecoregion clusters and rainfall was around 800 mm/year.

Figures 2 and 3 display the association of the *A. marginale* strains with the four ecoregion clusters. These figures are plotted according to the values of the first two axes derived from PCA on NDVI time series. Figures plot the 80% confidence ellipses of the annual mean NDVI and the seasonal variation of NDVI for each ecoregion cluster, as well as the plot of the isolates in the NDVI envelope. Analysis showed that 77% of MSP1a R1 unique sequences were associated with only one ecoregion cluster (Figure 2). Ten R1 unique sequences (25.6% of the total number of R1 sequences) were reported exclusively in ecoregion cluster 1 and they shared 16 out of 31 amino acids (51.6% of the total number of amino acids; Table 2). Six R1 unique sequences

(12.8%) were reported solely in ecoregion cluster 2 with 64.5% identical amino acids. Twelve R1 unique sequences (30.7%) were found only in ecoregion cluster 3, sharing 64.5% of their amino acids. Only three R1 sequences were exclusively associated with ecoregion cluster 4, with 77.4% identical amino acids. All of the *A. marginale* MSP1a R1 sequences within each ecoregion cluster appeared to be under positive selection as shown by d_N/d_S indexes of 1.83, 1.61, 1.54, and 1.21 for ecoregion clusters 1–4, respectively.

Table 2. MSP1a R1 and RL repeat sequences of unique sequences unambiguously associated to only one ecoregion cluster, including the consensus sequence of the isolates of that cluster.

MSP I a Repeat	Unique sequences	Consensus sequence	Other strains and number of substitutions
R I /Ecoregion I	4, 8, 16, 56, 60, 64, 67, gamma, pi, tau	***SSA***QQ*SSV*S*S** AS*SSQ*G--	A(O), B(O), D(I), T(O), 13(I), 23(I), alfa(O)
R I /Ecoregion 2	28, 48, 53, E, F, epsilon	***SS**GQQQESSV***S*- ASTSSQLG--	A(O), B(O), L(I), T(7), 13(I), 23(I), alfa(O)
R I /Ecoregion 3	I, 3, 5, 6, 27, 33, 34, 39, M, 0, Q, U	**SSSA*GQQQESSV-* QA*TSSQLG--	A(O), D(O)
R I /Ecoregion 4	I, J, K	*D*S*A*GQQQESSVSSQS *QASTSSQLG--	A(O), B(O), L(O), alfa(O)
RUEcoregion 1	8, 9, 12, 15, 59, 61 , 66	***SSSA**QQQES*V*SQS** ASTSSQ*G--	B(O), C(O), M(O), 18(0), 27(0), gamma(O)
RUEcoregion 2	10, 31 , 52, pi, beta	*DSSSA**QQQ*S*V*S*S*- ASTSSQLG--	F(O), H(O), M(O), 27(0), gamma(O)
RUEcoregion 3	3, 7, 35, 37, 38, 44, E, N, P, Q, U, ro	*DSSSAS*QQQESS**S*S *QA**S*Q*G--	B(I), F(O), H, 18(0), gamma(O)
RUEcoregion 4	none		B, C, H

Other strains recorded from more than one ecoregion cluster are included in the last column, stating the number of substitutions in their amino
acid sequences as compared with the consensus sequence in that ecoregion cluster (in parent heses).

Differences were found among the R1 sequences at each ecoregion cluster. Comparison of consensus sequences between clusters 1 and 2 revealed 22.5% of amino acid differences. Ecoregion clusters 1 and 3 differed by 25% of their consensus sequences while ecoregion clusters 2 and 3 had only 19% of different amino acids in their R1 sequences. Five R1 sequences, T, 13, 23, D, and L were found simultaneously in two of the ecoregion clusters (Figure 2, Table 2). Details of their similitude with the consensus sequences of each ecoregion cluster are included in Table 2. Three R1 sequences, A, B, and alpha appeared associated with different ecoregions. Pairwise comparisons between *A. marginale* R1 sequences demonstrated that genetic distances were lower within than between ecoregion clusters (Table 3). However, when strains were compared as clustered geographically, intra-cluster distances were much higher, revealing a clear heterogeneity between geographic strains and rejecting the hypothesis of a pure geographical association. Mantel's test on R1 sequences was 0.82 ($P < 0.001$) when applied to ecoregion clusters using only unique sequences. Mantel's test on R1 sequences dropped to 0.72 ($P < 0.005$) when every sequence in the ecoregion clusters was included. The same test provided a value of 0.31 ($P = 0.145$) for the distances matrix based on geographical association of strains.

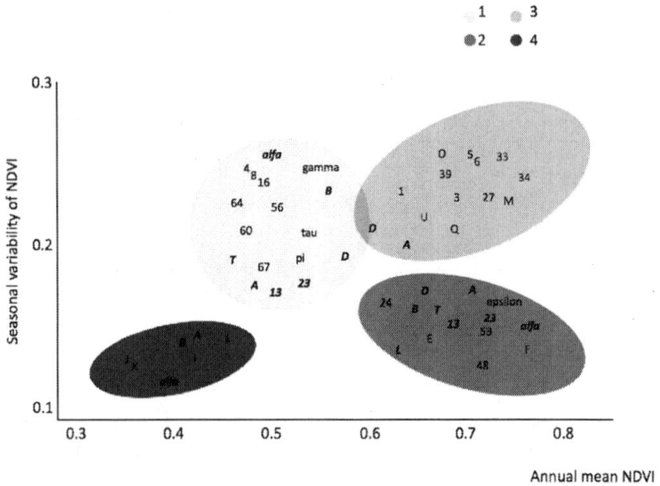

Figure 2. Associations between *A. marginale* MSP1a R1 repeat sequences and ecoregion clusters, plotted along the values of the first two axes derived from Principal Components Analysis on NDVI time series. Figure plots the 80% confidence ellipses of the annual mean NDVI and the seasonal variation of NDVI for each ecoregion cluster, as well as the plot of the isolates in the NDVI envelope. Each letter displays the mean position of the records for that strain. Unique sequences for each ecoregion are displayed in plain type. Sequences recorded in more than one ecoregion cluster are displayed in italic bold.

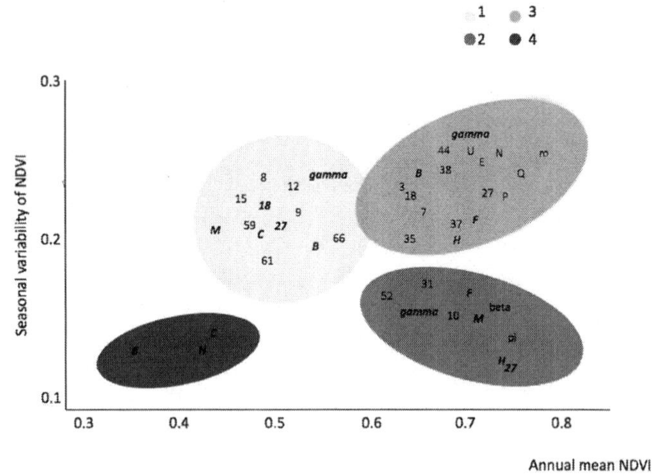

Figure 3. Associations between *A. marginale* MSP1a RL repeat sequences and ecoregion clusters, plotted along the values of the first two axes derived from Principal Components Analysis on NDVI time series. Figure plots the 80% confidence ellipses of the annual mean NDVI and the seasonal variation of NDVI, for each ecoregion cluster, as well as the plot of the isolates in the NDVI envelope. Each letter displays the mean position of the records for that strain. Unique sequences for each ecoregion are displayed in plain type. Sequences recorded in more than one ecoregion cluster are displayed in italic bold.

Table 3. Top table shows the genetic distances among and between the MSP1a R1 repeat sequences reported on each ecoregion cluster.

Ecoregion	Cluster 1	Cluster 2	Cluster 3	Cluster 4
Cluster 1	**0.212**	0.311	0.319	0.337
Cluster 2		**0.199**	0.321	0.371
Cluster 3			**0.175**	0.214
Cluster 4				**0.156**
Geographical cluster	**Western USA**	**Eastern USA**	**Mesoamerican+ Caribbean**	**South America**
Western USA	**0.418**	**0.399**	0.514	0.519
Eastern USA		**0.329**	0.388	0.501
Mesoamerican+ Caribbean			**0.454**	0.455
South America				**0.521**

Discriminant analysis showed that 74.8% of MSP1a RL sequences were found associated with one ecoregion cluster only (Figure 3, Table 2). Seven RL sequences (25% of the total number of RL sequences) were reported exclusively in ecoregion cluster 1, with 61.2% identical amino acids. Five RL sequences (17.8%) were reported solely in ecoregion cluster 2 with 64.5% identical amino acids. Twelve RL sequences (32%) were found only in ecoregion cluster 3, sharing 61.2% of their amino acids. No RL unique sequences were found in ecoregion cluster 4. The differences in RL consensus sequences between ecoregion clusters 1 and 2, 2 and 3, and 1 and 3 were 16%, 26%, and 32%, respectively. In contrast to *A. marginale* MSP1a R1 repeat sequences, the RL sequences within each ecoregion cluster did not appear to be under positive selection as shown by d_N/d_S indexes of 0.89, 0.79, and 0.77 for ecoregion clusters 1 to 3, respectively. Mantel's test on RL sequences was 0.87 ($P < 0.001$) when applied to ecoregion clusters using only unique sequences. This value dropped to 0.82 ($P < 0.001$) when every strain in the ecoregion clusters was included in the analysis. The same test provided a value of 0.27 ($P = 0.208$) for the distances matrix based on geographical association of strains. When the MSP1a Rn repeats, located between R1 and RL were analyzed, a total of 30, 9, 17, and 3 sequences were ascribed to ecoregion clusters 1–4, respectively. The Rn consensus sequences contained 14 (45.2%), 17 (54.9%), 17 (54.9%), and 24 (77.4%) identical amino acids in ecoregions clusters 1 to 4, respectively. The number of R1, Rn and RL repeat sequences varied between ecoregion clusters without a clear pattern.

Anaplasma marginale MSP1a Microsatellite Sequences Have Ecoregion-specific Signatures and Affect Gene Expression in *Escherichia coli*

The analysis of the MSP1a microsatellite sequences resulted in nine different genotypes among *A. marginale* strains (Table 1). The different microsatellite sequences produced SD-ATG distances of between 19 and 29 nucleotides, but the predominant distance was 23 nucleotides in all regions (Table 1).

The analysis of MSP1a microsatellite sequences showed phylogeographic clustering of some genotypes (Table 1). However, only three of the eight microsatellite genotypes included in the analysis were exclusive to a particular ecoregion cluster: genotype A to ecoregion cluster 2, genoytpe B to ecoregion cluster 1 and genotype F to ecoregion cluster 3. A significant genetic differentiation was found between ecoregion clusters for MSP1a microsatellite genotype frequencies, with overall $F_{ST} = 0.16$ ($P < 0.0001$). The ecoregion clusters 1 and 2 were clearly distinct from the other ecoregion clusters, and the largest F_{ST} value (0.28) was found between ecoregion clusters 1 and 4 while significant genetic differences were not found between ecoregion clusters 3 and 4 (Table 4). To control for potential confounding effects of geography, we investigated genetic differentiation between broad geographic regions. This showed that genetic differentiation is mostly dependent on ecoregion-based clustering of the strains, with the highest F_{ST} value being 0.07 for South America and the Mediterranean (see Table 4). The AMOVA showed that 84% of the variance accounted for within-ecoregion cluster differences, while a geographic-based AMOVA showed that within-population variation explained less (73%) of the variance (Table 5).

Table 4. Ecoregion and geographic cluster pairwise FST significance tests (bottom half of each section) and P-values (top half of each section) of *A. marginale* MSP1a microsatellites.

Ecological	**Ecological 1**	**Ecoregion 2**	**Ecoregion 3**	**Ecoregion 4**
Ecological 1	–	0.019 ± 0.001	0.000 ± 0.000	0.001± 0.000
Ecological 2	0.09	–	0.027 ± 0.001	0.032 ± 0.001
Ecological 3	0.22	0.07	–	0.312 ± 0.003
Ecological 4	0.28	0.15	0.016 (NS)	–
Geographical	**North America**	**South America**	**Mediterranean**	**South America**
North America	–	0.324 ± 0.003	0.016 ± 0.001	0.232 ± 0.003
South America	0.004 (NS)	–	0.036 ± 0.001	0.452 ± 0.003
Mediterranean	0.06	0.07	–	0.101 ± 0.002
South Africa	0.01 (NS)	-0.01 (NS)	0.07 (NS)	–

Abbreviation: NS, not sign ificant.

Table 5. Hierarchical analysis of molecular variance (AMOVA) for MSP1a microsatellites.

Source of variation	**d. f.**	**Sum of squares**	**Variance components**	**Percent variation**
Ecological				
Among ecoregions	3	3.99	0.04	16.31
Within ecoregions	109	24.49	0.22	83.69
Total	112	28.48	0.27	...
Geographic				
Among continents	2	1.16	-0.08	-35.70
Among populations within continents	6	8.76	0.15	63.18
Within populations	104	18.57	0.18	72.52
Total	112	28.48	0.25	...

Table shows Wright's fixation index, FsT· Statistical significance was calculated from 20,000 random permutations. Fixation indices and P-values:
(Ecological) FsT = 0.16 (P = 0.00 ± 0.00), (Geographic) FsT = 0.29 (P = 0.00 ± 0.00).

The effect of MSP1a microsatellite size on gene expression was analyzed using the sequence derived from Wetumka, Okeechobee, Idaho and HB-A8 *A. marginale* strains (Table 6). The results showed that MSP1a expression was lower in the construct containing the Idaho strain-derived MSP1a sequence with the lowest SD-ATG distance of 19 nucleotides, while differences were not observed between constructs containing the MSP1a sequences from Wetumka, Okeechobee and HB-A8 *A. marginale* strains with SD-ATG distances of 23 and 29 nucleotides (Table 6).

Table 6. Effect of microsatellite genotype on the expression of *A. marginale* MSP1a in *E. coli.*

A. marginale strain	Microsatellite genotype	SD-ATG distance (nucleotides)	Normalized MSP I a mRNA levels (arbitrary units)
Idaho,ID	C	19	1.57 ± 0.29 (a)
Wetumka, OK	G	23	2.31 ± 0.27 (b)
Okeechobee, FL	G	23	2. 13 ± 0.0 (b)
HB-AS, China	I	29	2.39 ± 0.19 (b)

MSP I a mRNA levels were analyzed by real-time RT-PCR in three independent clones for each of the MSP I a constructs transformed in E coli JM I 09. MSP I a mRNA levels were normalized against E. coli dxs gene and plasmid DNA copy number by msp I a PCR. Normalyzed MSPI a mRNA levels were represented in arbitrary units as average ± SD and compared between different constructs using an ANOVA test (different letters denote significant differences; $P < 0.02$).

DISCUSSION

Vector-borne pathogens have evolved molecular mechanisms of vector-pathogen interactions that involve genetic traits of both the vector and the pathogen [5, 35]. For the tick-borne pathogen *A. marginale*, recent studies have characterized tick and pathogen-derived genes that are involved in tick-pathogen interactions [2, 3, 6-9, 15]. Phylogenetic studies using tick and/or pathogen-derived genetic markers have contributed to our understanding of the evolution of *A. marginale* strains and tick-pathogen relationships [5, 6, 20]. However, the impact of these studies has been limited by the genetic diversity of genes involved in tick-pathogen interactions and thus are likely to reflect pathogen evolution and tick-pathogen relationships [5].

In this study we took a different approach to characterize the evolution of *A. marginale* strains. This is the first study to use remotely sensed vegetation features as a surrogate of an environmental envelope to which genetic variability and structure of a single pathogen is associated. Biogeographic research seeks to identify the processes structuring organism diversity at a variety of geographic and taxonomic scales [36]. Remote sensing is being used increasingly as a tool to discover ecological traits through definite signatures. NDVI is a measure of the vegetation stress, thus a time series of NDVI values over a region reflects the seasonal cycle of vegetation as a surrogate of the seasonal variation in climate. NDVI and other climate features are commonly used to detect ecologically suitable areas for some pathogens and their vectors [37-39]. For example, Randolph and Rogers [40] indicated that climate has directed and constrained the evolution of flaviviruses of the TBE group. Six flaviviruses (SSEV, WTBEV, Russian TBEV, OHFV, and Kyasanur forest virus) fall within a distinct eco-climatic space defined by factors derived from thermal and moisture conditions. Herein, we showed that *A. marginale* MSP1a R1 repeats evolved under positive selection, were associated with specific ecoregion clusters and were not arranged according to geographical features. The different evolutionary pressures operating over different MSP1a repeats were demonstrated previously [6], but the possibility of using the MSP1a R1 repeat as a biogeographical marker has only been suggested [5]. In contrast, MSP1a RL repeat sequences, while still linked to a similar set of ecoregion clusters, did not evolve under positive selection. Consequently, RL repeat sequences were not good genetic markers for the characterization of *A. marginale* biogeography and evolution.

Analysis of MSP1a microsatellite sequences demonstrated that *A. marginale* strains are associated with specific ecoregion clusters, thus corroborating the results obtained with repeat sequences. These results may have a functional significance. It has been shown that the SD-ATG distance between the ribosome binding site (Shine-Dalgarno sequence) and the translation initiation codon affects gene expression in prokaryotes [41]. Little is known about the regulation of gene expression in *A. marginale* [6]. However, as shown here in *E. coli*, the length of the MSP1a microsatellite could affect the expression of MSP1a, which varies during *A. marginale* multiplication in both tick cells and bovine erythrocytes, thus affecting pathogen infection and transmission [42]. Since, MSP1a repeats and microsatellites are unambiguously associated to ecoregion clusters, these results suggested a new factor that may affect the efficiency by which different *A. marginale* strains are transmitted under different environmental conditions.

A. marginale is an obligate intracellular parasite, which alternates between the tick vector and the vertebrate host. Our hypothesis was that the link between pathogen strains and definite portions of the environmental envelope could reflect the effects of climate on the tick vector. Temperature and rainfall, which are indirectly captured by the specific signatures of NDVI, are the main factors affecting the ecology and population dynamics of tick species [43] and these operate at critical levels of selection of tick populations, selecting also specific strains of the pathogen. This framework is further obscured by the "noise" produced by invasive events of the pathogen (by cattle movement or other factors), or by selection of strains transmitted mechanically in areas where ticks are eradicated by acaricide application, contributing to the absence of total consistence between ecoregion clusters and strains.

Adequate reports exist about distribution, seasonal dynamics and abundance of *R. microplus* populations in the study area, allowing a direct comparison with results presented herein. Ecoregion cluster 1 contained the R1 repeats with the lowest percentage of conserved amino acids and the highest positive selection pressure, in areas with high temperature and medium rainfall. *R. microplus* ticks are common in these areas and a strict seasonality in tick population dynamics has been reported, allowing for a high selection of tick populations due to winter mortality [44]. Ecoregion cluster 2 contained sites with constant high temperature and rainfall. In these sites, *R. microplus* ticks are abundant throughout the year without marked seasonality and climate is not a limiting factor in tick mortality [44, 45]. In ecoregion cluster 3, tick populations suffer drastic limitations in effectiveness because of low and inadequate rainfall [46], and this high selection pressure on tick populations might be adverse for pathogen transmission and selection. The R1 repeat sequences in ecoregion clusters 2 and 3 had higher number of conserved amino acids and lower positive selection pressure when compared with R1 sequences in ecoregion cluster 1. Finally, the R1 repeat sequences ascribed to ecoregion cluster 4 had the highest percentage of conserved amino acids and the lowest positive selection pressure, recorded only in sites where *R. microplus* ticks are absent because of the low yearly temperature and thus other tick species act as vectors of *A. marginale* [18]. The analysis of MSP1a microsatellite sequences also supported differences among all ecoregion clusters, except for ecoregion clusters 3 and 4 where *R. microplus* has low prevalence or is absent.

Some R1 repeat sequences such as A, B, D, and alpha as well as microsatellite genotypes C-D, G and H were present across several ecoregion clusters. These sequences appeared in *A. marginale* strains collected in zones where *R. microplus* ticks are common (ecoregion clusters 1 and 2) and in sites, such as central Argentina and southern parts of the USA, where *R. microplus* has been prevalent in the past but has been eradicated [46]. Additionally, these sequences were also found in sites where other tick vectors such as *Dermacentor* spp. are prevalent [18]. These R1 repeats and microsatellite sequences could have evolved from ancestor pathogen strains transmitted by *R. microplus* as the main vector, and then evolved under lower selection pressure, due to pathogen transmission by other tick species or mechanically. The presence of these sequences in sites where *R. microplus* has been historically absent (that is, north-western USA) and now adapted to transmission by *Dermacentor* spp. ticks, could be interpreted as invasive events. The results reported here showed that lowest selection pressure exist in sites where *Dermacentor* spp. ticks are the main biological vectors or where mechanical transmission is predominant because of eradication of *R. microplus*. Therefore, R1 repeats are evolving under high selection pressure only in sites where *R. microplus* is the main vector and is subjected to selection because of climate constraints. This hypothesis did not explain the absence of MSP1a genetic diversity in Australia. Analysis of four *A. marginale* strains in Australia revealed the presence of a single repeat type 8 [11]. We would expect evolution of *A. marginale* MSP1a towards different repeat sequences, sharing the consensus sequence found in ecoregion cluster 1, into which R1 type 8 is ascribed, even in the case of a single invasive event. Reasons accounting for such a lack of diversity are currently unknown, but the combined pressure exerted by tick population structure [47, 48] the *A. centrale* vaccine, acaricide treatments, and cattle movement for pathogen and tick control may have impacted on *A. marginale* genetic diversity in Australia [48].

A. marginale exclusively infects cattle and wild ruminants [2]. Such high host specificity may results in a relatively low impact of vertebrate host factors on the evolution of *A. marginale* strains, thus leaving tick-pathogen interactions as the main contributing factor affecting its biogeography and evolutionary history. However, as previously discussed, cattle movement may have contributed to the genetic diversity of *A. marginale* strains worldwide [11]. Nevertheless, the results reported herein may be relevant in studying the evolution of other vector-borne pathogens. Many vector-borne pathogens, such as some *Babesia, Theileria, Rickettsia, Ehrlichia* and *Plasmodium* species, are also highly host-specific [49] and vector-pathogen interactions may play a crucial role in their evolution and biogeography [35].

CONCLUSION

The results reported herein provided the first evidence that the evolution of *A. marginale* MSP1a repeat and microsatellite sequences was linked to environmental traits, and that strains are not geographically related. Different evolutionary pressures acting on *A. marginale* were found associated with zones where climate and rainfall affect presence, abundance and dynamics of vector populations. We hypothesized that some *A. marginale* strains evolved under conditions of pathogen biological transmission by *R. microplus*, while others may be linked to transmission by other tick species or to mechanical transmission in regions where *R. microplus* is currently absent. The procedures outlined herein could be fundamental toward studying the evolution of other vector-borne pathogens.

KEYWORDS

- **Ecoregion**
- **Mechanical transmission**
- **Microsatellite**
- **Phylogeographic**

AUTHORS' CONTRIBUTIONS

Agustín Estrada-Peña, José de la Fuente and Atilio J. Mangold conceived and designed the study and conducted MSP1a repeat sequence analysis. Agustín Estrada-Peña conducted ecoregion analysis and clustering. Victoria Naranjo, Karina Acevedo-Whitehouse, and José de la Fuente conducted MSP1a expression and microsatellite sequence analyses. Agustín Estrada-Peña, Karina Acevedo-Whitehouse, Katherine M. Kocan, and José de la Fuente contributed to drafting the manuscript. All authors read and approved the final manuscript.

ACKNOWLEDGMENTS

This research was supported by grants from the Ministerio de Ciencia e Innovación, Spain (project BFU2008-01244/BMC), the CSIC intramural project 2008301249 to José de la Fuente and the Oklahoma Agricultural Experiment Station (project 1669) and was facilitated through the Integrated Consortium on Ticks and Tick-borne Diseases (ICTTD-3), financed by the International Cooperation Program of the European Union, coordination action project No. 510561. VN was funded by the European Social Fund and the Junta de Comunidades de Castilla-La Mancha (Program FSE 2007–2013), Spain.

Chapter 7

Cancer Risk Among Residents of Rhineland-Palatinate Winegrowing Communities

Andreas Seidler, Gaël Paul Hammer, Gabriele Husmann, Jochem König, Anne Krtschil, Irene Schmidtmann, and Maria Blettner

INTRODUCTION

To investigate the cancer risk among residents of Rhineland-Palatinate winegrowing communities in an ecological study.

On the basis of the Rhineland-Palatinate cancer-registry, we calculated age-adjusted incidence rate ratios (RRs) for communities with a medium area under wine cultivation (>5–20%) and a large area under wine cultivation (>20%) in comparison with communities with a small area under wine cultivation (>0–5%). In a side analysis, standardized cancer incidence ratios (SIRs) were computed separately for winegrowing communities with small, medium and large area under wine cultivation using estimated German incidence rates as reference.

A statistically significant positive association with the extent of viniculture can be observed for non-melanoma skin cancer in both males and females, and additionally for prostate cancer, bladder cancer, and non-Hodgkin lymphoma (NHL) in males, but not in females. Lung cancer risk is significantly reduced in communities with a large area under cultivation. In the side-analysis, elevated SIR for endocrine-related tumors of the breast, testis, prostate, and endometrium were observed.

This study points to a potentially increased risk of skin cancer, bladder cancer, and endocrine-mediated tumors in Rhineland-Palatinate winegrowing communities. However, due to the explorative ecologic study design and the problem of multiple testing, these findings are not conclusive for a causal relationship.

Some previous studies point to a potential association between pesticide exposure respiratory farming or winegrowing and lymphoma [1-5] or multiple myeloma [6-11], brain cancer [12-14], prostate cancer [15], or bladder cancer [16, 17]. However, the mechanisms of the suspected carcinogenic effects of pesticides are widely unclear.

Among the hypothesis on potential carcinogenic mechanisms from pesticides, the endocrine mediated effects have received much attention. Several pesticides interact with endocrine receptors *in vitro* or have endocrine-mediated effects in laboratory animals *in vivo*: The European Union has listed over 40 pesticides suspected to interfere with the hormone system of humans and wildlife [18]. As endocrine-related mechanisms play an etiologic role in several cancers in humans, the potential association between exposure to pesticides with endocrine activity and cancer incidence has been discussed in the last years. Many epidemiological studies have, for example, examined the relationship between pesticides and breast cancer [19]. However, although endogenous and exogenous estrogens are known to play a causal role in the aetiology of breast cancer, to date epidemiological and experimental evidence is not conclusive for an association between exposure to organochlorine pesticides and breast cancer incidence (for an overview, see [19]). According to Barlow [19], the evidence on other endocrine-related tumor sites (testes, prostate, endometrium) is too sparse to draw any conclusions concerning pesticides.

Rhineland-Palatinate is the federal state with the most extensive winegrowing in Germany: About 3% of the Rhineland-Palatinate area is under wine cultivation. Therefore, a potential pesticide exposure of the residential population might be assumed. Actual deposit measurements in one Rhineland-Palatinate wine district (Moselle region) point to an ongoing insecticide (parathione) and herbicide (atrazine, simazine) exposure of the residential population [20]. Repeatedly, a suspected increase in cancer incidence has been a subject of concern in the mentioned region. The aim of the present ecological study is therefore to investigate the cancer risk among residents of Rhineland-Palatinate winegrowing communities compared to the cancer risk among residents of communities with a small area under wine cultivation.

MATERIALS AND METHODS

Study Population and Study Area

Each Rhineland-Palatinate winegrowing community (n = 503, out of 2,305 communities in Rhineland-Palatinate) was categorized according to the proportion of area under wine cultivation of the whole community area (small: >0–5%; medium: >5–20%; large: >20% area under wine cultivation; see Table 1) based on official data for 1996. The 1.3% of the total area of communities with a small area under cultivation is area under wine, respectively, 12.5% of the total area of communities with a medium area under cultivation, and 31.4% of the total area of communities with a large area under cultivation. Table 1 gives some characteristics of the Rhineland-Palatinate study region.

Table 1. Characteristics of the Rhineland-Palatinate vineyard area.

	Rhineland-Palatinate*	Area under wine (% of community area)		
	Total	>0%, ≤5%	>5%, ≤20%	>20%
Communities	2,305	162	171	170
Total area (ha)	1,984,688	222,736	200,709	129,444
Area under wine (ha)	69,043	2,996	25, 101	40,683
% area under wine	3.5%	1.3%	12.5%	31.4%
Inhabitants (per ha)	4,000,567 (2.02)	564,21 0 (2.53)	526,486 (2.62)	301 , 193 (2.33)
Inhabitants per community (median, min-max)	566 (6-184,752)	1,188 (72-99,750)	1,193 (95-80,535)	984 (84--40,1 1 0)

* All data pertain to Dec 31st. 1996 (Statistisches Landesamt Rheinland-Pfalz 2006)

Cancer Registry Data

This study is based on cancer cases registered in the Rhineland-Palatinate cancer registry which covers a population of approximately 4,000,000 persons. We included all malignant tumors plus benign brain and central nervous system (CNS) tumors and brain and CNS tumors of uncertain behavior. Furthermore, we included malignant bladder tumors plus carcinoma *in situ* and tumors of uncertain behavior of the bladder. Since January, 2000 all Rhineland-Palatinate physicians and dentists are legally obliged to report incident cancer cases to the cancer registry. Therefore, all above mentioned cancers diagnosed between 2000 and 2003 and reported until mid-2005 were included. The following items are registered: diagnosis (ICD-10); topography and morphology (ICD-O-2); staging (TNM); incidence date; most valid basis of diagnosis; occasion of first detection; initial treatment; last occupation and longest held occupation; and date and cause of death (where appropriate). Population figures and data on area under wine cultivation were obtained from the statistical office of Rhineland-Palatinate.

Statistical Methods

Completeness of the Rhineland-Palatinate cancer registry varies with time, region, physician's specialization and type of cancer. This had to be considered in our analysis. Completeness is estimated by the ratio of reported cases to estimated cases for Rhineland-Palatinate calculated from a national pooling of cancer registry data [21, 22]. In communities with a small area under wine cultivation, the completeness (excluding non-melanotic skin cancer) is about 80% in males and 79% in females. Concerning lymphohaematopoetic malignancies, the completeness is considerably lower; in communities with a small area under cultivation, only 62% of NHL in males and 64% in females are reported to the registry.

Primary "Internal" Analysis of Incidence Ratio—Ratios for Communities with a Medium or Large Area Under Cultivation in Comparison with Communities with a Small Area Under Cultivation

To account for regional variations in completeness, in our primary analysis communities with a small area under wine cultivation served as reference. Provided that the completeness does not differ systematically between winegrowing communities with a large area under cultivation and adjoining communities with a small area under cultivation, this allows to calculate valid incidence RRs by Poisson regression.

Population figures are reported in 5 year age categories by the State Statistical Office; due to small numbers, the use of a categorized age variable would have caused numerical problems in the regression analysis. Instead, age was included as a continuous variable in the regression analysis (mid-point of each age category). Many factors, like sociodemographic, lifestyle and environmental factors, might considerably differ between large cities and villages/small cities. Cities with more than 100,000 inhabitants (Mainz, Ludwigshafen/Rhein, Koblenz, Kaiserslautern) were therefore excluded from the analysis. Furthermore, we adjusted for rural (<5,000 inhabitants) versus urban (≥5,000 inhabitants) communities. The proportion of community area under fruit cultivation (another potential source of pesticides exposure) was included in the analyses as dichotomous confounder (<5% vs. ≥5% of community area).

All analyses were preformed in SAS Institute [23], stratified by gender and cancer type. The regression analysis includes cancer rate as dependent variable, and age, wine growing area, rural/urban setting, and fruit cultivation. All analyses were stratified by gender and diagnosis. The results of our initial Poisson regression indicated a possible problem with overdispersion, which is partly due to heterogeneity between communities with respect to unobserved risk factors. We therefore opted to assume a negative binomial distribution for the dependent variable, which allows to estimate a dispersion parameter k for the variance (variance = expected value·(1 + k·expected value)) and includes the Poisson distribution as a special case ($k = 0$). The negative binomial distribution emerges naturally if expected counts (Poisson parameters) vary among communities according to a gamma distribution. The interpretation of RRs stays the same as for Poisson regression. However, results do not substantially differ. For a few rare cancers, the ML fitting algorithm did not converge using the negative binomial distribution. In these cases, estimates from Poisson regression are reported.

Side Analysis of Standardized Incidence Ratios (SIRs) Using German Incidence Rates as Reference

Even in communities with a small area under cultivation, cancer incidence might be elevated, potentially leading to an underestimation of RRs in communities with medium or large area under cultivation. In an additional analysis, we therefore calculated SIRs regardless of the incompleteness of the Rhineland-Palatinate cancer registry. The SIRs were separately computed for winegrowing communities with small, medium, and large area under cultivation using estimated German incidence rates. The expected numbers of cancer (E) for the time period 2000–2003 were compared with the observed numbers (O), calculating SIR as the ratio between the observed and expected numbers. Exact 95%-confidence intervals (CIs) based on the Poisson distribution of O were calculated.

Results of any analysis based on small numbers are difficult to interpret. Therefore, only those results based on at least 10 cases in the respective referent group and 10 cases in both comparison groups combined are reported here.

Tables 2 and 3 present incidence RR for cancer in males and females for winegrowing communities with medium (>5 to ≤20%) and large (>20%) area under cultivation compared to communities with small (>0 to ≤5%) area under cultivation. Significantly increased RR are observed for non-melanoma skin cancer (C44 ICD-10) among men (RR = 1.32 (95% CI 1.20–1.45) for medium and RR = 1.39 (95% CI 1.25–1.54) for a large area under cultivation) as well as among women (RR = 1.40 (95% CI 1.27–1.54) for medium and RR = 1.38 (95% CI 1.23–1.53) for a large area under cultivation).

Table 2. Cancer risks (incidence rate ratios RR) in men with residence in communities with a large or medium area under wine cultivation versus men in communities with low area under wine cultivation.

	Reference* (1 ,665,594 PYt)	Area under wine cultivation > 5, ≤ 20% of community area (1 ,039,435 PYt)			Area under wine cultivation > 20% of community area (612,714 PYt)		
ICD-10 code	Cases	Cases	RR‡§	95% CI	Cases	RR‡§	95% CI
Head & neck (COO-C 14)	369	188	0.91	0.72-1.15	94	0.86	0.65-1.14
Base oftongue (CO I)	35	11	0.53	0.26-1.08	10	0.87	0.40-1.91
Other and unspecified parts of tongue (C02)	35	26	1.22	0.72-2.07	13	1.12	0.56-2.25
Floor of mouth (C04)	42	25	0.98	0.57-1.70	11	0.70	0.34-1.47
Palate (COS)	22	15	1.24	0.62-2.49	5	0.88	0.31-2.54
Other and unspecified parts of mouth (C06)	18	8	0.71	0.29-1.73	3	0.45	0.12-1 .71
Parotid gland (CO7)	14	8	0.92	0.36-2.30	4	0.80	0.24-2.67
Tonsil (C09)	47	22	0.78	0.45-1.34	13	0.79	0.40-1.56
Oropharynx (C I 0)	35	15	0.78	0.39-1.57	5	0.48	0. 17-1.34
Piriform sinus (C 12)	20	10	0.90	0.41-2.02	8	1.67	0.66-4.25
Hypopharynx (C 13)	56	27	0.76	0.46-1.26	17	0.88	0.48-1.62
Oesophagus (CIS)	156	94	0.96	0.73-1.27	42	0.82	0.57-1 .20
Stomach (C 16)	241	166	1.06	0.86-1 .31	92	1.03	0.79-1 .34
Small intestine (C 17)	24	14	0.92	0.45-1.86	2	0.21	0.05-{).95
Colon, sigmoid & rectum (CI8-C21)	1188	806	1.07	0.95-1.21	460	1.10	0.96-1.26
Colon (CIS)	723	473	1.04	0.91-1.20	268	1.06	0.90-1.25
Rectosigmoid junction (C 19)	51	42	1.33	0.89-2.00	30	1.68	1.04-2.71
Rectum (C20)	397	284	1.13	0.94-1.35	157	1.10	0.89-1 .37
Anus and anal canal (C21)	17	7	0.66	0.26-1.64	5	0.85	0.29-2.50
Liver and intrahepatic bile ducts (C22)	141	73	0.94	0.68-1 .30	37	0.88	0.58-1.32
Gallbladder & biliary tract (C23-C24)	76	33	0.67	0.44-1 .02	27	0.95	0.59-1.54
Gallbladder (C23)	17	7	0.58	0.23-1.46	7	0.96	0.36-2.53
Other and unspecified parts of biliary tract (C24)	59	26	0.70	0.43-1.13	20	0.95	0.55-1 .65
Pancreas (C25)	162	99	1.03	0.78-1.36	51	0.96	0.68-1 .37
Nasal cavity and middle ear (C30-C3 I)	18	9	0.73	0.32-1.67	4	0.51	0.16-1.62
Larynx (C32)	135	78	0.94	0.68-1.29	39	0.88	0.59-1.31
Trachea, bronus and lung (C33-C34)	1039	530	0.98	0.84-1.14	232	0.77	0.64-0.92
Bronchus and lung (C34)	1036	530	0.98	0.84-1.15	232	0.77	0.64-0.92
Bone and articular cartilage (C40-C41)	11	7	0.88	0.32-2.42	3	0.49	0.13-1.89
Skin, malignant melanoma (C43)	230	188	1.32	1.08-1 .60	119	1.50	1.18-1.91
Skin, other malignant neoplasms (C44)	1990	1748	1.32	1.20-1.45	959	1.39	1.25-1 .54

Table 2. *(Continued)*

	Reference* (1 ,665,594 PYt)	Area under wine cultivation > 5, ≤ 20% of community area (1 ,039,435 PYt)			Area under wine cultivation > 20% of community area (612,714 PYt)		
ICD-10 code	**Cases**	**Cases**	**RR‡§**	**95% CI**	**Cases**	**RR‡§**	**95% CI**
Mesothelioma (C45)	19	12	1.09	0.52-2.28	5	0.92	0.32-2.68
Other connective and soft tissue (C49)	32	35	1.65	1.00-2.70	9	0.68	0.31-1.48
Breast (C5O)	14	9	1.02	0.43-2.39	3	0.60	0.16-2.25
Penis (C60)	20	15	1.17	0.56-2.44	6	0.71	0.26-1.93
Prostate (C61)	1857	1359	1.26	1.12-1.41	787	1.26	1.11-1.43
Testis (C62)	154	107	1.18	0.88-1 .60	77	1.31	0.94-1.83
Urinary tract (C64-C66+C68)	330	206	1.03	0.85-1 .26	107	0.96	0.75-1.23
Kidney, except renal pelvis (C64)	269	171	1.06	0.86-1.32	89	1.00	0.76-1.31
Ureter (C66)	30	7	0.35	0. 15-0.80	7	0.58	0.24-1 .39
Bladder (C67, 009.0, 041.4)	699	470	1.16	1.01-1.34	266	1.31	1.10-1.55
Eye and adnexa (C69)	15	8	1.02	0.43-2.43	3	0.88	0.23-3.37
Meninges (C70)	22	20	1.45	0.78-2.69	7	0.88	0.35-2.18
Brain, CNS, meninges (C70-C72, 032-33, 042-43)	133	90	1.08	0.81-1.44	53	1.04	0.73-1.49
Brain (C71, 033, 043)	109	70	1.02	0.74-1.41	45	1.06	0.72-1.57
Thyroid gland (C73)	33	23	1.11	0.64-1 .92	16	1.31	0.68-2.53
Hodgkin's disease (C81)	40	24	0.86	0.51-1.46	12	0.64	0.32-1.27
Follicular NHL (C82)	29	21	1.23	0.67-2.25	19	1.98	1.01-3.85
NHL (C82-C85)	197	130	1.12	0.88-1.42	81	1.29	0.96-1.73
Diffuse NHL (C83)	122	66	0.91	0.66-1.24	31	0.80	0.52-1 .23
Peripheral and cutaneous T-celllymphomas (C84)	10	6	0.94	0.33-2.68	8	2.19	0.75--6.37
Other and unspecified types of NHL (C85)	36	37	1.64	1.03-2.59	23	1.84	1.05-3.23
Multiple myeloma (C90)	63	36	0.93	0.60-1.44	21	0.92	0.53-1 .59
Leukaemia (C91-C95)	204	113	0.89	0.69-1.15	61	0.84	0.61-1.16
Lymphoid leukaemia (C91)	116	56	0.74	0.53-1.05	28	0.65	0.41-1.02
Myeloid leukaemia (C92)	77	52	1.09	0.75-1 .59	27	0.99	0.61-1.60
Primary site unspecified	128	83	0.99	0.75-1.30	43	0.88	0.61-1 .28
All malignancies (excluding C44)	7761	5024	1.12	1.05-1.19	2765	1.10	1.03-1.18
All malignancies (including C44)	9751	6772	1.15	1.09-1.22	3724	1.16	1.09-1.23

* Winegrowing communities with >0, <= 5% area under wine cultivation

† PY: Person-Years were approximated by population figures: the sum of population at the end of the year in the years under consideration.

‡adjusted for age, rural or urban environment, and fruit cultivation

§ Poisson distribution of case counts assumed for: C45, CSO, C70

Table 3. Cancer risks (incidence rate ratios RR) in women with residence in communities with a large or medium area under wine cultivation vs. women in communities with low area under wine cultivation.

	Reference* (I ,778,184 PYt)	Area under wine cultivation >5, ≤ 20% of community area (I ,039,435 PYt)			Area under wine cultivation > 20% of community area (634,060 PYt)		
ICD-10 code	Cases	Cases	RR‡§	95% CI	Cases	RR‡§	95% CI
Head & neck (COO-C14)	123	70	1.05	0.7 4-1.50	41	1.14	0.75-1 .74
Other and unspecified parts of tongue (C02)	14	8	1.0 1	0.41-2.48	6	1.56	0.52-4.64
Oropharynx (C10)	19	6	0.53	0.20-1.39	5	0.81	0.28-2.39
Oesophagus (C15)	38	17	0.89	0.46--1.70	9	0.91	0.40-2.08
Stomach (C16)	197	147	1.27	1.01-1.60	61	1.07	0.79-1.47
Small intestine (C17)	24	17	1.17	0.64-2.15	8	1.04	0.45-2.40
Colon, sigmoid & rectum (C 18-C21)	1122	733	1.04	0.92-1.17	346	0.94	0.81-1.09
Colon (CIS)	734	509	1.11	0.97-1.28	214	0.90	0.75-1.08
Rectosigmoid junction (C 19)	64	32	0.78	0.51-1.21	20	0.86	0.50-1.48
Rectum (C20)	301	180	0.95	0.78-1.17	108	1.06	0.83-1 .35
Anus and anal canal (C21)	23	12	0.89	0.44-1.81	4	0.60	0.20-1 .83
Liver and intrahepatic bile ducts (C22)	43	32	1.17	0.72-1.90	19	1.20	0.66--2.19
Gallbladder & biliary tract (C23-C24)	79	58	1.13	0.79-1.61	19	0.64	0.38-1.09
Gallbladder (C23)	39	38	1.44	0.89-2.33	10	0.64	0.30-1 .34
Other and unspecified parts of biliary tract (C24)	40	20	0.78	0.46--1.34	9	0.63	0.29-1 .34
Pancreas (C25)	158	85	0.97	0.72-1.29	40	0.93	0.64-1 .37
Larynx (C32)	19	12	0.99	0.47-2.07	4	0.56	0.18-1.75
Trachea, bronus and lung (C33-C34)	342	168	0.99	0.77-1.27	94	1.19	0.88-1 .59
Bronchus and lung (C34)	340	167	0.99	0.77-1 .27	94	1.19	0.89-1 .60
Skin, malignant melanoma (C43)	274	212	1.17	0.96--1.42	109	1.00	0.78-1.28
Skin, other malignant neoplasms (C44)	1710	1620	1.40	1.27-1.54	807	1.38	1.23-1.53
Retroperitoneum and peritoneum (C48)	10	9	1.72	0.65-4.53	4	1.93	0.53-7.02
Other connective and soft tissue (C49)	30	17	0.98	0.54-1.79	9	1.03	0.46--2.32
Breast (C50)	2525	1527	1.08	0. 98-1.20	779	1.01	0.90-1.12
Vulva (C51)	36	36	0.98	0.64-1.50	22	1.23	0.72-2.10
Vagina (C52)	18	20	1.80	0.94-3.45	5	0.82	0.29-2.33
Cervix uteri (C53)	162	97	1.03	0.79-1.34	47	0.94	0.66--1.34
Corpus uteri, (C54-C55)	382	244	1.15	0.94-1.41	146	1.20	0.95-1.52
Corpus uteri (C54)	370	232	1.13	0.92-1.39	144	1.22	0.97-1.54

Table 3. *(Continued)*

	Reference* (I ,778,184 PYt)	Area under wine cultivation >5, ≤ 20% of community area (I ,039,435 PYt)			Area under wine cultivation > 20% of community area (634,060 PYt)		
ICD-10 code	**Cases**	**Cases**	**RR‡§**	**95% CI**	**Cases**	**RR‡§**	**95% CI**
Uterus, part unspecified (C55)	12	12	1.58	0.70-3 .59	2	0.46	0. 10-2.17
Ovary and other unspecified female genital organs (C56--C57)	297	196	1.09	0.89-1.34	96	0.97	0.75-1.26
Ovary (C56)	284	183	1.07	0.86--1.32	93	0.99	0.76--1.28
Other and unspecified female genital organs (C57)	13	13	1.66	0.76--3.63	3	0.73	0. 19-2.72
Urinary tract (C64-C66+C68)	208	136	1.10	0.86--1 .40	73	1.04	0.77-1.40
Kidney, except renal pelvis (C64)	166	116	1.14	0.88-1.48	63	1.09	0.79-1.50
Renal pelvis (C65)	24	10	0.71	0.33-1.49	4	0.55	0. 18-1.71
Ureter (C66)	16	10	1.02	0.45-2.29	4	0.75	0.23-2.44
Bladder (C67, 009.0, 041.4)	251	158	1.09	0.88-1.34	85	1.19	0.90-1.56
Brain, CNS, meninges (C70-C72, 032-33, 042-43)	167	105	1.11	0.84-1.46	62	1.25	0.89-1.75
Meninges (C70)	57	46	1.33	0.87-2.03	24	1.29	0.75-2.21
Brain (C71 , 033, 043)	107	57	0.98	0.68-1.40	37	1.21	0.79-1.86
Thyroid gland (C73)	102	75	1.20	0.86--1.67	32	0.86	0.56--1.34
Hodgkin's disease (C81)	39	20	0.83	0.48-1.44	9	0.64	0.29-1.39
NHL (C82-C85)	220	114	0.93	0.72-1 .21	52	0.78	0.56--1.09
Foll icular NHL (C82)	50	18	0.58	0.34-1.01	5	0.29	0. 11-0.76
Diffuse NHL (C83)	106	69	1.05	0.76--1.45	25	0.73	0.46--1.17
Other and unspecified types of NHL (C85)	56	24	0.71	0.43-1 .14	19	1.07	0.61-1.89
Multiple myeloma (C90)	68	30	0.72	0.46--1.13	21	0.88	0.51-1.49
Leukaemia (C91-C95)	135	65	0.80	0.59-1.09	43	0.98	0.68-1.43
Lymphoid leukaemia (C91)	67	33	0.78	0.51-1.18	22	0.90	0.54-1 .50
Myeloid leukaemia (C92)	60	32	0.90	0.59-1.37	16	0.92	0.51-1 .66
Primary site unspecified	116	78	1.18	0.90-1 .56	36	1.13	0.76-1.68
All malignancies (excluding C44)	7258	4508	1.09	1.03-1 .17	2293	1.04	0.97-1.11
All malignancies (including C44)	8968	6128	1.14	1.08-1.21	3100	1.10	1.04-1.17

* Winegrowing communities with >0, <= 5% area under wine cultivation

† PY: Person-Years were approximated by population figures: the sum of population at the end of the year in the years under consideration.

‡ adjusted for age, rural or urban environment, and fruit cultivation

§ Poisson distribution of case counts assumed for: C21 , C52, C55, C57, C65, C81

Among men, the RRs for a large versus a small area under cultivation are significantly elevated for the following malignancies: malignant melanoma (C43 ICD-10; RR = 1.50; 95% CI 1.18–1.91), prostate cancer (C61 ICD-10: RR = 1.26; 95% CI 1.11–1.43), bladder cancer (C67 ICD-10; RR = 1.31; 95% CI 1.10–1.55), follicular NHL (C82 ICD-10; RR = 1.98; 95% CI 1.01–3.85), and other and unspecified types of NHL (C85 ICD-10; RR = 1.84; 95% CI 1.05–3.23). In contrast, we find significantly decreased RRs for follicular NHL among women (RR = 0.29 (95% CI 0.11–0.76) for a large vs. a small area under cultivation).

Furthermore RRs are significantly decreased among men for lung cancer (C34 ICD-10; RR = 0.77; 95% CI 0.64–0.92 for a large vs. a small area under cultivation).

Both men and women showed a slightly elevated RR for all malignancies for communities with medium (men: RR = 1.15; 95% CI 1.09–1.22; women: RR = 1.14; 95% CI 1.08–1.21) as well as with a large area under cultivation (men: RR = 1.16; 95% CI 1.09–1.23; women: RR = 1.10; 95% CI 1.04–1.17).

When non-melanotic skin cancer was excluded, among men, risk ratios for all malignancies remained significantly elevated in communities with medium and a large area under cultivation; among women, solely RRs in communities with a medium area under cultivation retained significance.

Tables 4 and 5 present SIR for cancer in males and females for winegrowing communities with small (>0 to ≤5%), medium (>5 to ≤20%), and large (>20%) area under cultivation using estimated incidence of cancer in the national population of Germany as reference. As the incompleteness of the Rhineland-Palatinate cancer registry would tend to result in potentially considerable underestimation, decreased SIR are not mentioned in the following (and should not be interpreted). The SIRs of malignant melanoma remains statistically increased in men (SIR for a medium area under cultivation = 1.17 (95% CI 1.01–1.35), SIR for a large area under cultivation = 1.31 (95% CI 1.08–1.56)). Furthermore, the SIR for prostate cancer remains statistically significant: the SIR is 1.18 (95% CI 1.12–1.24) for a medium area under cultivation and 1.25 (95% CI 1.17–1.34) for a large area under cultivation. The increased incidence of testicular cancer in communities with a large area under wine cultivation is of borderline statistical significance (SIR = 1.26; 95% CI 0.99–1.57). Among women, we find an elevated SIR for endometrial cancer in communities with a large area under cultivation (SIR = 1.43; 95% CI 1.20–1.68). Breast cancer incidence is increased in communities with a medium area under cultivation (SIR = 1.07; 95% CI 1.02–1.12), but not in communities with a large area under cultivation (SIR = 0.99; 95% CI 0.92–1.06).

Table 4. Cancer risks (standardized incidence ratios SIR) in men with residence in communities with planted winegrowing areas with the estimated incidence of cancer in the national population of Germany as reference.

	Area under wine cultivation > 0, ≤ 5% of community area (1,665,594 PYt)				Area under wine cultivation > 5, ≤ 20% of community area (1,039,435 PYt)				Area under wine cultivation > 20% of community area (612,714 PYt)			
IC0-10 code	Ob-served	Ex-pected	SIR	95% CI	Observed	Expected	SIR	95% CI	Ob-served	Ex-pected	SIR	95% CI
Head & neck (COO-CI4)	369	238.00	1.13	1.01-1.25	188	205.94	0.91	0.79-1.05	94	117.54	0.80	0.65-0.98
Stomach (CI6)	241	421.10	0.57	0.50-0.65	166	262.89	0.63	0.54-0.74	92	144.55	0.64	0.51-0.78
Colon, sigmoid & rectum (C18-C21)	1188	1445.33	0.82	0.78-0.87	806	906.16	0.89	0.83-0.95	460	497.93	0.92	0.84-1.01
Trachea, bronchus and lung (C33-C34)	1039	1376.19	0.75	0.71-0.80	530	865.12	0.61	0.56-0.67	232	478.60	0.48	0.42-0.55
Skin, malignant melanoma (C43)	230	257.40	0.89	0.78-1.02	188	160.19	1.17	1.01-1.35	119	91.08	1.31	1.08-1.56
Prostate (C61)	1857	1833.70	1.01	0.97-1.06	1359	1152.26	1.18	1.12-1.24	787	628.58	1.25	1.17-1.34
Testis (C62)	154	164.80	0.93	0.79-1.09	107	101.97	1.05	0.86-1.27	77	61.20	1.26	0.99-1.57
Urinary tract (C64-C66+C68)	330	392.87	0.84	0.75-0.94	206	247.03	0.83	0.72-0.96	107	137.47	0.78	0.64-0.94
Bladder (C67, 009.0, 041.4)	699	718.87	0.97	0.90-1.05	470	451.25	1.04	0.95-1.14	266	246.29	1.08	0.95-1.22
NHL (C82-C85)	197	255.32	0.77	0.67-0.89	130	160.18	0.81	0.68-0.96	81	90. 11	0.90	0.71-1.12
Leukaemia (C91-C95)	204	253.60	0.80	0.70-0.92	113	158.75	0.71	0.59-0.86	61	89.49	0.68	0.52-0.88
All ma-lignancies (excluding C44)	7761	8751.69	0.89	0.87-0.91	5024	5493.56	0.91	0.89-0.94	2765	3041.59	0.91	0.88-0.94

† PY: Person-Years were approximated by population figures: the sum of population at the end of the year in the years under consideration.

Table 5. Cancer risks (standardized incidence ratios SIR) in women with residence in communities with planted winegrowing areas with the estimated incidence of cancer in the national population of Germany as reference.

	Area under wine cultivation > 0, ≤ 5% of community area (1,665,594 PYt)				Area under wine cultivation > 5, ≤ 20% of community area (1,039,435 PYt)				Area under wine cultivation > 20% of community area (612,714 PYt)			
1C0-10 code	Ob-served	Ex-pected	SIR	95% CI	Obse rved	Expe cted	SIR	95% CI	Ob-served	Ex-pected	SIR	95% CI
Head & neck (COO-C14)	123	93.17	1.32	1.10-1.58	70	57.01	1.23	0.96-1.55	41	31.13	1.32	0.95-1.79
Stomach (C16)	197	302.45	0.65	0.56-0.75	147	183.37	0.80	0.68-0.94	61	96.98	0.63	0.48-0.81
Colon, sigmoid & rectum (C18-C21)	1122	1529.06	0.73	0.69-0.78	733	929.83	0.79	0.73-0.85	346	491.03	0.70	0.63-0.78
Trachea, bronchus and lung (C33-C34)	342	429.49	0.80	0.71-0.89	168	263.51	0.64	0.54-0.74	94	142.96	0.66	0.53-0.80
Skin, malignant melanoma (C43)	274	314.29	0.87	0.77-0.98	212	191.68	1.11	0.96-1.27	109	107.56	1.01	0.83-1.22
Breast (CSO)	2525	2332.19	1.08	1.04-1.13	1527	1429.33	1.07	1.02-1.12	779	786.82	0.99	0.92-1.06
Cervix uteri (C53)	162	239.01	0.68	0.58-0.79	97	145.48	0.67	0.54-0.81	47	83.17	0.57	0.42-0.75
Corpusuteri, (C54-C55)	382	309.64	1.23	1.11-1.36	244	190.05	1.28	1.13-1.46	146	102.37	1.43	1.20-1.68
Ovary and other unspecified female genital organs (C56-C57)	297	449.08	0.66	0.59-0.74	196	273.91	0.72	0.62-0.82	96	149.12	0.64	0.52-0.79
Urinary tract (C64-C66+C68)	208	261.56	0.80	0.69-0.91	136	160.51	0.85	0.71-1.00	73	86.15	0.85	0.66-1.07
Bladder (C67, 009.0, 041.4)	251	325.54	0.77	0.68-0.87	158	200.30	0.79	0.67-0.92	85	107.82	0.79	0.63-0.97
NHL (C82-C85)	220	286.04	0.77	0.67-0.88	114	175.38	0.65	0.54-0.78	52	95.33	0.55	0.41-0.72
Leukaemia (C91-C95)	135	228.14	0.59	0.50-0.70	65	139.57	0.47	0.36-0.59	43	75.32	0.57	0.41-0.77
All malignancies (excluding C44)	7258	8285.78	0.88	0.86-0.90	4508	5070.41	0.89	0.86-0.92	2293	2740.00	0.84	0.80-0.87

† PY: Person-Years were approximated by population figures: the sum of population at the end of the year in the years under consideration.

DISCUSSION

In this ecological study, a statistically significant positive association with the extent of viniculture is observed for non-melanoma skin cancer in males and females, prostate cancer, bladder cancer, and NHL in males, but not in females. Lung cancer risk is significantly reduced in communities with a large area under cultivation. Our main hypothesis that pesticides might play a role for the observed associations will be discussed for specific cancer types in the following.

Specific Tumors

Non-melanotic Skin Cancer

Several studies have shown that the lifetime cumulative sun exposure is responsible for the development of non-melanotic skin cancer (for an overview, see [24, 25]). In ecologic studies, squamous cell carcinoma is related more strongly to latitude or measured ultraviolet (UV) radiation than is basal cell carcinoma. As more outdoor workers might be occupied in regions with extensive winegrowing, our finding of an increased non-melanotic skin cancer risk in winegrowing communities appears plausible. In fact, in communities with a large area under cultivation, 14.8% of male skin cancer patients (C44 ICD-10) with known occupation (as recorded in the cancer registry) had worked as an outdoor worker (farmer, winegrower, gardener, forestry worker, or construction worker). In communities with medium and a small area under cultivation, this proportion is 12.2% and 7.5%, respectively. Comparably, the proportion of outdoor workers among female cancer skin cancer patients (C44 ICD-10) is 7.6%, 5.1%, and 2.6% in communities with a large, medium, and small area under cultivation, respectively. Previous arsenic exposure has to be considered as an alternative explanation: arsenical pesticides were applied by Moselle wine growers [26] between 1920 and 1942. The clinical signs of arsenic exposure are arsenical keratoses, which may progress to squamous cell carcinoma or basal cell carcinoma [27]. Moreover, arsenic seems to act as a co-carcinogen with UV radiation [27]. As the latency period of non-melanotic skin cancer is suspected to be very long, an excess in non-melanotic skin cancers might therefore be partly explained by arsenic exposure, however, this explanation appears rather speculative. Moreover, risk estimators for non-melanotic skin cancer do not markedly increase when our analysis is restricted to persons aged 70 or more. The association between sun exposure and melanoma of the skin seems to be more complex: Intermittent sun exposure and sunburn history rather than lifetime cumulative sun exposure plays a role in the aetiology of melanoma of the skin [28, 29]. This complex relationship might explain why our study does not reveal a clearly increased melanoma incidence in communities with a large area under wine cultivation. Moreover, adjusting for potential confounders as, for example, leisure time UV exposure, was not possible in this study.

Brain Cancer

While several epidemiological studies point to an increased brain cancer risk among pesticide exposed persons [13, 14], few studies specifically focus on the residential population in winegrowing regions. In their ecological study in the province of Trento, Italy, Ferrari, and Lovaste [30] find the highest incidence rates of intracranial tumors in regions of intensive fruit and wine cultivation. However, the authors do not indicate the significance level of their findings. Another ecological study among French agricultural workers reveals a significant association between pesticide exposure in vineyards and brain cancer mortality [31]. The results of our ecological study do not support an increased brain cancer risk of residents in winegrowing regions (RR in the primary analysis for large vs. a small area under cultivation = 1.06 (95% CI 0.72–1.57) among men; RR = 1.21 (95% CI 0.79–1.86) among women).

Rectum Cancer

Some previous studies point to a potentially elevated rectum cancer risk [32, 33], other studies find reduced colorectal cancer risks among farmers [34] or farm residents [35]. Altogether, there is very little evidence to date for a possible relationship between pesticide exposure and rectum cancer. Our finding of an increased cancer incidence of the rectosigmoid junction (but not of rectum cancer in all) among males living in winegrowing communities might be alternatively explained by life-style (e.g., dietary) or medical (participation at screening) factors, by inhomogeneous reporting behavior, or by chance.

Non-Hodgkin Lymphoma

The increased NHL incidence among male, but not among female inhabitants of communities with a medium or large area under wine cultivation suggests a potential occupational rather than residential aetiology. However, in communities with a medium or a large area under cultivation, only two male NHL patients (=2% of male NHL patients with known occupation, missing values 55%) and one female NHL patient (=1.3% of female NHL patients with known occupation, missing values 44%) had worked as wine-growers, making an occupational aetiology improbable.

Our finding of an increased NHL incidence among potentially pesticide-exposed residents of winegrowing communities is in accordance with the literature. However, most previous studies are related to agricultural workers in general, not to winegrowing workers. In a large Italian multicenter case-control study [36], orchard, vineyard, and related tree and shrub workers appeared to be at increased risk for hematolymphopoietic malignancies. The carcinogenic effects of pesticides may be associated with their genotoxicity and immunotoxicity [37-39], increased cell proliferation [40], and association with chromosomal aberrations [41]. Because of the lack of a positive association between potential residential pesticide exposure and NHL in females (actually with a significantly decreased RR for follicular NHL in winegrowing communities with a large area under cultivation), our study does not definitely support the hypothesis of an elevated NHL risk among the residential population in Rhineland-Palatinate winegrowing communities.

Bladder Cancer

To date, there is inconclusive evidence for a relationship between pesticide exposure and bladder cancer. In a retrospective cohort study among 32,600 employees of a lawn care company, Zahm [42] finds a significantly increased bladder cancer mortality. However, bladder cancer numbers are very small; furthermore, two of the three observed deaths had no direct occupational contact with pesticides. Rusiecki et al. [16]

evaluate the cancer incidence in atrazine-exposed pesticide applicators among 53,943 participants in the Agricultural Health Study. In their study, assessing atrazine exposure by lifetime days of exposure, the RR for bladder cancer is non-significantly elevated to 3.06 (95% CI 0.86–10.81). Assessing atrazine exposure by intensity-weighted lifetime days, the RR for bladder cancer decreases to 0.85 (95% CI 0.24–2.94). Viel and Challier [17] analyze the mortality from bladder cancer among French farmers. While the mortality among farmers is non-significantly lowered (standardized mortality ratio = 0.96; 95% CI 0.85–1.08), there is a significant association with exposure to pesticides in vineyards (risk ratio = 1.14; 95% CI 1.07–1.22). According to the authors, these results could explain the French south-north gradient in bladder cancer, as vineyards are mainly located in Southern France.

Prostate Cancer

Our finding of an increased prostate cancer risk in potentially pesticide-exposed residents of winegrowing communities is in accordance with the literature. In a recently conducted meta-analysis, van Maele-Fabry et al. [15] include 18 epidemiological studies published between 1984 and 2004. The combined RR for all studies is 1.28 (95% CI 1.05–1.58). According to the authors, no specific pesticide or chemical class is responsible for the increased risk; nevertheless, the strongest evidence consists for phenoxy herbicides possibly in relation with dioxin and furan contamination. Van Maele-Fabry [15] point to the lack of fundamental understanding of the basic biology of human prostate cancer: hormones (both androgens and estrogens) would likely play a role in the etiology or promotion of prostate cancer. Therefore, the authors regard it as plausible that chemicals able to modulate steroid sex hormones as agonists, antagonists or as mixed agonist-antagonist might contribute to the development of prostate cancer through hormone-mediated effects. Several pesticides might interfere with sexual hormones through direct action on receptors but also through indirect non-receptorial mechanisms.

Limitations

We applied an ecologic study design which does not allow a differentiation between residential, occupational, and life-style risk factors for cancer. The chief limitation of ecologic studies is the inability to link exposure with disease in particular individuals. A second major limitation of ecologic studies is the lack of ability to control for the effects of potential confounding factors. Thus, observed risk differences between communities with different area under cultivation may be due not to varying levels of pesticide usage, but rather to the independent effect of other confounding variables on cancer risk. Moreover, our "exposure" categories (small, medium, or large area under cultivation) represent very crude indicators of the individual exposure; the actual individual exposure depends on occupation, place of residence at the time of pesticide spraying, wind direction, and so forth. Furthermore, several tests were performed, introducing a multiple comparison problem (altogether, 270 risk ratios were calculated). In general, our study design should therefore be regarded as exploratory rather than hypothesis testing. Due to small numbers, particularly for cancer cases in communities with a large area under cultivation, the power of the study to detect slight increases in incidence is limited. Many other potential risk factors of occupation and lifestyle from living in agricultural area would need to be discussed to explain the findings, but these would have to be collected in a study using individual information. For instance, data on socioeconomic levels or smoking prevalence were not available on a small scale. The use of 1996 data on agricultural characteristics might be criticized, since a lag time of 4–7 years for cancers occurring 2000–2003 is not plausible. It was not possible to obtain older data, but since the political boundaries did not change and agricultural land use stayed constant, their use seems warranted in the current study.

The completeness of reported cancer cases is still relatively low in Rhineland-Palatinate (about 80% for all cancers). Therefore, the calculation of SIRs for residents of winegrowing communities in comparison with the population of Rhineland-Palatinate might at least partly reflect a higher completeness rather than truly elevated risks. As a probably more reliable approach of calculating cancer risks, we therefore decided to compare the observed cancer cases in communities with a medium or a large area under cultivation with—as a kind of internal reference—the number of cases reported in communities with a small area under cultivation. While we regard the "internal" comparison of winegrowing communities (communities with a medium or large area versus small are under cultivation) as a more reliable approach than the comparison with the Rhineland-Palatinate population, we nevertheless cannot totally exclude a higher (or lower) completeness in communities with a medium or large area under cultivation than in communities with a small area under cultivation.

Increased Incidence of Endocrine-related Tumors with the Estimated Incidence of Cancer in the National Population of Germany as Reference

In our primary analysis, we compared cancer rates in communities with a large respiratory medium area under cultivation with cancer rates in communities with a small area under cultivation. However, in fact even in communities with a small area under cultivation, cancer incidence might be elevated, potentially leading to an underestimation of the results of our primary analysis (concerning RRs in communities with medium or large area under cultivation). In a side analysis, we therefore calculated SIRs regardless of the incompleteness of the Rhineland-Palatinate cancer. Because of the incompleteness of the Rhineland-Palatinate cancer registry, the results of the calculation of SIR tend to underestimate the true cancer risks for incompletely recorded cancer subentities; therefore decreased SIR should not be interpreted. If SIRs were calculated with the estimated incidence of cancer in the national population of Germany as reference, among men we found an elevated SIR for prostate cancer and testicular cancer in communities with a large area under wine cultivation. Among women, we found an elevated SIR for endometrial cancer and (in communities with a medium area under cultivation, but not in communities with a large area under cultivation) for breast cancer incidence. Altogether, the results of our additional SIR analysis are compatible with a potential carcinogenic role of pesticides in the etiology of endocrine-related tumors of the breast, testis, prostate, and endometrium.

CONCLUSION

This ecologic study is the first attempt to examine the relationship between cancer incidence and the area under wine cultivation in Rhineland-Palatinate winegrowing communities. The study results point to a potentially elevated skin cancer risk, bladder cancer risk, and endocrine-related (prostate, testicular, breast, and endometrium) cancer risk of the population in communities with a large area under wine cultivation. Mainly due to the ecologic study design, the problem of multiple testing, and due to the insufficient completeness of the Rhineland-Palatinate cancer registry concerning the considered region, these findings are not conclusive for a causal relationship. There is a need for analytic epidemiologic studies differentiating between environmental and occupational exposures to further clarify the cancer risk associated with pesticide usage in wine cultivation.

KEYWORDS

- **Confidence intervals**
- **Non-Hodgkin lymphoma**
- **Non-melanotic skin cancer**
- **Poisson regression**
- **Rate ratios**
- **Standardized cancer incidence ratios**

AUTHORS' CONTRIBUTIONS

Andreas Seidler conceived the study design, coordinated the study, and drafted the manuscript, Gaël Paul Hammer performed the statistical analysis and participated in the study design and coordination, Gabriele Husmann, Anne Krtschil, and Irene Schmidtmann participated in the design of the study and helped to draft the manuscript, Jochem König participated in the statistical analysis and helped to draft the manuscript, Maria Blettner participated in the coordination of the study and helped to design the study and draft the manuscript. All authors read and approved the final manuscript.

COMPETING INTERESTS

The authors declare that they have no competing interests.

Chapter 8

Risk of Congenital Anomalies Around a Municipal Solid Waste Incinerator

Marco Vinceti, Carlotta Malagoli, Sara Fabbi, Sergio Teggi, Rossella Rodolfi, Livia Garavelli, Gianni Astolfi, and Francesca Rivieri

INTRODUCTION

Waste incineration releases into the environment toxic substances having a teratogenic potential, but little epidemiologic evidence is available on this topic. We aimed at examining the relation between exposure to the emissions from a municipal solid waste incinerator (MSWI) and risk of birth defects in a northern Italy community, using geographical information system (GIS) data to estimate exposure and a population-based case-control study design. By modeling the incinerator emissions, we defined in the GIS three areas of increasing exposure according to predicted dioxins concentrations. We mapped the 228 births and induced abortions with diagnosis of congenital anomalies observed during the 1998–2006 period, together with a corresponding series of control births matched for year and hospital of birth/abortion as well as maternal age, using maternal address in the first 3 months of pregnancy to geocode cases and controls.

Among women residing in the areas with medium and high exposure, prevalence of anomalies in the offspring was substantially comparable to that observed in the control population, nor dose-response relations for any of the major categories of birth defects emerged. Furthermore, odds ratio (OR) for congenital anomalies did not decrease during a prolonged shut-down period of the plant.

Overall, these findings do not lend support to the hypothesis that the environmental contamination occurring around an incineration plant such as that examined in this study may induce major teratogenic effects.

The possibility that atmospheric emissions of contaminants by MSWIs leads to adverse effects on the health of exposed populations, and of carcinogenic and teratogenic effects in particular, has been the object of a limited number of studies which yielded conflicting and inconsistent results [1-5]. In fact, incinerators emit a number of pollutants [6], including some suspected or established teratogens such as polychlorinated dibenzo-p-dioxins and -furans (acting as endocrine disruptors [7]) and heavy metals like chromium, cadmium, lead, mercury, nickel, and arsenic [8-10]. Moreover, residence near these plants was recently associated with induction of genotoxic effects in humans [11]. These issues are of particular interest since waste incineration is widely used in several developed countries and since birth defects monitoring in the exposed populations has been suggested or adopted as a short-term tool to assess health risks associated to waste management options including incineration [1, 2].

In the present study, we studied the extent to which the risk for congenital anomalies varied with maternal exposure to emissions from a modern municipal waste incinerator, by conducting a population-based case-control investigation near a plant with intermittent operation during the study period, and by using a GIS-based approach for exposure assessment and for geographical localization of cases and controls.

MATRIALS AND METHODS

Study Area

A MSWI with a capacity of 70,000 tons/year is located in the city of Reggio Emilia, Emilia–Romagna region (extension 232 km^2, population approximately 150,000). The incinerator consists of two combustion lines that started operating in 1968, and has been equipped since 1992 with a dry scrubbing of flue based on sodium bicarbonate for acidic pollutants gas and since 1994 by an activated carbon device for dioxins, furans, and mercury adsorption. This plant stopped its activity in April 20, 2002 due to abnormalities in the combustion process and excess emissions of carbon monoxide and other contaminants, and started to operate once again in June 16, 2005.

We estimated through a dispersion model the average concentrations of dioxin and furans in the lower part of the atmosphere in the city territory, aiming at identifying municipal areas with different amounts of exposure to incinerator emissions at man's height by using the estimated fall-out of polychlorinated dibenzo-p-dioxins and dibenzo-p-furans (henceforth referred to as "dioxins") as indicators. The dispersion model was computed by using the meteorological database "CALMET", pre-processor developed by the Emilia–Romagna Region Meteorological Service, for the years 1999, 2000, 2001 (data for 1998 were partially corrupted) and from July 1, 2005 to June 30, 2006. We estimated concentration levels through the model WinDimula 3.0 for Windows [12], an air dispersion model initially developed in the 1980s by Enea (Ente per le Nuove Tecnologie, Energia e Ambiente, Rome) and Maind (Maind s.r.l., Milan) and recently updated [12], based on the Gaussian analytic solution of the turbulent diffusion equation. Its main peculiarity is a special algorithm designed to deal with calm wind conditions that are usual in many Italian regions [13]: the model simulates both short-term and climatological concentrations caused by sources with different geometry, and output concentrations are given on regular grids or on user selected receptors. In the present study, the area over which the computations were made was 20 km × 12 km, with a resolution of 100 m, with the emission source located at (7.022; 6.031) km as respect the South-West corner of the domain. We designed maps of different (A-low, B-intermediate, and C-high) ground level exposure to incinerator emissions of dioxins and furans in a geographic information system (GIS) environment, using ArcGIS software (version 9.2, ESRI, Redlands, CA 2006) to implement a project georeferenced in the Italian cartographic system (Gauss Boaga) and in the Modena municipality regional technical map layer (Figure 1) and as cutoff points 5 and 10×10^{-9} µg/m^3 of dioxins, based on maximum incinerator allowed emission (Figure 1). We also computed a dispersion model predicting heavy metals concentrations, using as cutoff values 0.50 and 1.00 µg/m^3: the intermediate exposure area was roughly superimposed to C area for dioxins, and these exposure boundaries were not further used in the analysis.

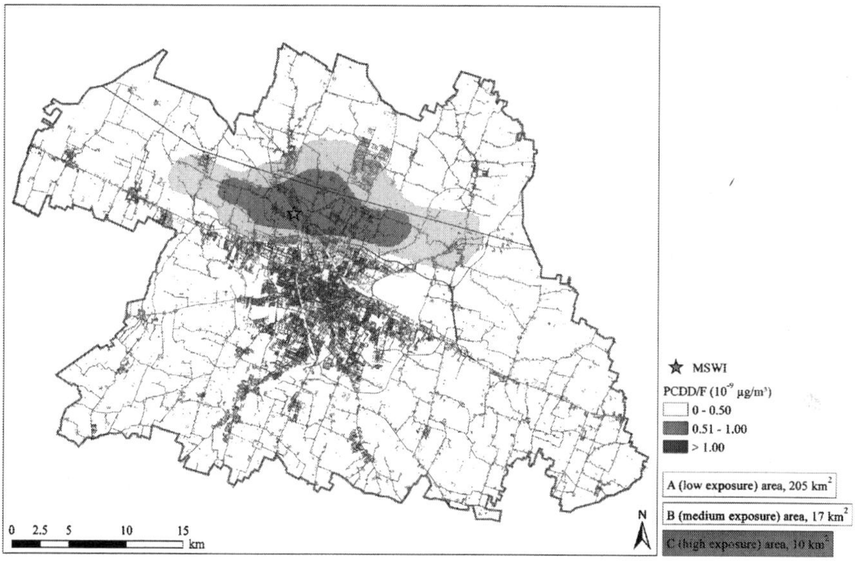

Figure 1. Map of exposure to polychlorinated dibenzo-p-dioxins and dibenzofurans (PCDD/F) in the city of Reggio Emilia, northern Italy, around the municipal solid waste incinerator (MSWI).

Study Population

We attempted to identify all cases of congenital anomalies in the offspring or in aborted foetuses of women residing in the Reggio Emilia municipality since January 1, 1998 until December 31, 2006. To do this, we used data from the population-based registry of congenital malformations of the Emilia–Romagna Region, named IMER and part of the Eurocat EU program [14], recording since 1979 all cases of abnormalities in live- and stillbirths and since 1996 the induced abortions associated with diagnoses of congenital anomaly observed in the regional hospitals. We also used as additional source of data the Hospital Discharge Directory of Emilia–Romagna residents, available at the Emilia–Romagna Region Health Authority since 1996, and in particular the discharges by Italian hospitals to regional residents reporting an ICD9-CM diagnostic codes from 740.0 to 759.9. These diagnoses were further reviewed by a clinical geneticist (F.R.), and all cases of minor malformations, as defined in accordance to the Eurocat guidelines, were removed from the analysis.

We retrieved control births with their corresponding mothers through random selection within the regional Hospital Discharge Directory. Specifically, to each case we associated a control birth randomly selected among the live births without diagnosis of malformations during the same year to women residing in the Reggio Emilia municipality, referred to the same hospital and born in the same year of the matched "case" mother.

We collected information for both case and referent mothers concerning their residence during the 9 months before parturition (or 3 months for women undergoing induced abortion procedure) and their educational attainment at the Reggio Emilia Municipality General Registry Office. In case of change of residence during the gestational period, we considered for exposure assessment in the present study the woman's residence in the first 3 months after the estimated date of conception: in the three case showing a change of residence in this early period of the pregnancy, we attributed exposure status on the basis of the longest period of residence within this time span. We then geocoded the retrieved maternal addresses, using the database made available by the Reggio Emilia Province or, in the few cases in which the addresses could not be found in that database, measuring it on site with a geographical positioning system device (Garmin GPSmap 60CSx, Garmin Int. Corp., Olathe, KS). A mother was considered exposed when her address was comprised within the intermediate and high (B and C) exposure areas, after inputting it in the GIS.

All directly and indirectly nominative data were obtained by the Regional Hospital Discharge Registry and by the Reggio Emilia General Registry Office and subsequently analyzed in accordance with the legal and ethical guidelines for personal data protection in epidemiological and scientific research of the Italian law [15] and with the ethical guidelines of the IMER Registry [16].

Data Analysis

We calculated the prevalence ratio of having a birth or an aborted foetus with a congenital anomaly associated to maternal factors through the OR with its 95% confidence interval (CI) generated by a conditional logistic regression model, entering as predictive variables the area of residence during gestation, and educational attainment level. Since no point estimate could be generated by the conditional analysis for the anomalies characterized by the lowest prevalence, we used, in these cases, unconditional logistic regression adjusting for both maternal age and education. We carried out this analysis for overall congenital anomalies and for single birth defect categories for the entire study period and, when numbers of cases made it possible, also for the normal operation and shut-down periods of operation of the plant. The former period included all births with congenital anomalies occurring in the periods December 1, 1998–October 31, 2002 and April 1, 2006–December 31, 2006, while that ascribed to the shut-down period occurred from February 1, 2003 to December 31, 2005 (Figure 2). The remaining time spans, in the first part of the shut-down period and at the beginning of the reactivation (in 2005) of the plant, were removed from period-specific analysis due to uncertainties in estimating the actual exposure status of women who delivered in these periods. For induced abortions linked to an in utero diagnosis of birth defects, the overall time span considered as "exposed" included the periods January 1, 1998–May 31, 2002 and October 1, 2005–December 31, 2006, while abortions considered as "unexposed" occurred from August 1, 2002 until July 31, 2005, and the events occurred in the remaining periods were removed from analysis. In all these analyses, "period of exposure" of each control birth was made equal to that of the corresponding matched case. For all statistical analyses we used the package Stata-10 [17].

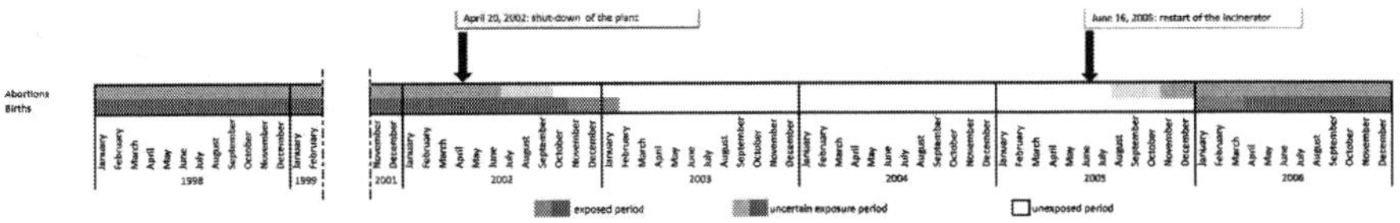

Figure 2. Operation periods of the municipal waste incinerator of Reggio Emilia, Italy, and corresponding exposure status of cohorts of births and of aborted foetuses undergoing a diagnosis of congenital anomaly during the 1998–2006 period.

Overall, we identified 228 cases of congenital anomalies during the study period: 183 live- and stillbirths presenting one or more defects or congenital syndrome and 45 induced abortions of foetuses with single or multiple anomalies. Maternal age of case mothers (at birth or at time of abortion) ranged from 16 to 44 years; regarding educational attainment level, case and referent mothers who had attended elementary school were 19 (8.33%) and 9 (3.95%) respectively, middle school 83 (36.40%) and 78 (34.21%), high school in 90 (39.47%) and 104 (45.61%), and university 36 (15.79%) and 37 (16.23%).

Using a logistic regression model and adjusting for education and maternal age, OR for congenital anomalies was 1.49 (95% CI 0.70–3.19) in the medium exposure group and 0.66 (95% CI 0.25–1.79) in the high exposure group, compared to the remaining municipal population (Table 1), and results were substantially similar when we used a conditional logistic model (Table 2). As summarized in Table 1, grouping together the two highest exposure levels OR in the overall exposed population resulted to be 1.11 (95% CI 0.60–2.04), and when the analysis was carried out for single anomaly categories we found little evidence of excess risk for any disease group and exposure status, with the exception of an increased OR for chromosomal abnormalities in the middle exposure area (OR 2.53, 95% CI 0.88–7.24).

Table 1. Prevalence odds ratio for congenital anomalies associated with maternal exposure to the emissions of the incinerator plant of Reggio Emilia, northern Italy, 1998–2006.

Area	Cases% (n)	Controls% (n)	Odds ratio[1]	
All anomalies				
A (low exposure)	89.0 (203)	90.4 (206)	1.00 (referent)	
B (medium exposure)	7.9 (18)	5.3 (12)	1.49 (0.70-3.19)	
C (high exposure)	3.1 (7)	4.3 (10)	0.66 (0.25-1.79)	
				P trend 0.881
Overall exposed area (B+C)	11.0 (25)	9.6 (22)	1.11(0.60-2,04)	
Cardiovascular system				
A (low exposure)	9.6 (87)	90.4 (206)	1.00 (referent)	
B (medium exposure)	6.3 (6)	5.3 (12)	0.91 (0.36-2.31)	
C (high exposure)	3.1 (3)	4.3 (10)	0.77 (0.22-2.77)	
				P trend 0.666

Table 1. *(Continued)*

Area	Cases% (n)	Controls% (n)	Odds ratio[1]	
Overall exposed area (B+C)	9.4 (9)	9.6 (22)	0.86 (0.40-1.86)	
Nervous system				
A (low exposure)	95.7 (22)	90.4 (206)	1.00 (referent)	
B (medium exposure)	4.3 (1)	5.3 (12)	0.64 (0.83-4.93)	
C (high exposure)	0.0 (0)	4.3 (10)	_2	
				P trend 0.344
Overall exposed area (B+C)	4.3 (1)	9.6 (22)	0.41 (0.05-3.17)	
Chromosomal				
A (low exposure)	85.4 (35)	90.4 (206)	1.00 (referent)	
B (medium exposure)	12.2 (5)	5.3 (12)	2.53 (0.88-7.24)	
C (high exposure)	2.4 (1)	4.3 (10)	0.77 (0.10-6.14)	
				P trend 0.486
Overall exposed area (B+C)	14.6 (6)	9.6 (22)	1.82 (0.70-4.72)	
Genito-urinary				
A (low exposure)	90.4 (19)	90.4 (206)	1.00 (referent)	
B (medium exposure)	4.8 (1)	5.3 (12)	0.65 (0.08-5.05)	
C (high exposure)	4.8 (1)	4.3 (10)	1.08 (0.14-8.68)	
				P trend 0.904
Overall exposed area (B+C)	9.6 (2)	9.6 (22)	0.82 (0.18-3.67)	
Musculoskeletal				
A (low exposure)	87.2 (34)	90.4 (206)	1.00 (referent)	
B (medium exposure)	10.3 (4)	5.3 (12)	1.52 (0.49-4.67)	
C (high exposure)	2.5 (1)	4.3 (10)	0.56 (0.07-4.47)	
				P trend 0. 928
Overall exposed area (B+C)	12.8 (5)	9.6 (22)	1.13 (0.41-3.10)	
Clefts[2,3]				
A (low exposure)	100.0 (4)	90.4 (206)	1.00 (referent)	
Overall exposed area (B+C)	0.0 (0)	9.6 (22)	–	
Eye[2]				
A (low exposure)	85.7 (6)	90.4 (206)	1.00 (referent)	
Overall exposed area (B+C)	14.3 (1)	9.6 (22)	1.78 (0.20-5.65)	
Other and unspecified congenital anomalies[2]				
A (low exposure)	87.5 (14)	90.4 (206)	1.00 (referent)	
Overall exposed area (B+C)	12.5 (2)	9.6 (22)	111 (0.24-5.1 0)	

1 Estimates computed through unconditional logistic regression (95% confidence interval in parentheses)
2 No point estimate could be computed for the B and C areas
3 No point estimate could be computed for the overall exposed area

Table 2. Prevalence odds ratio for congenital anomalies according to maternal exposure to emissions of the municipal solid waste incinerator of Reggio Emilia, northern Italy, 1998–2006.

	Entire study period			Operation period			Shut-down period		
	Cases% (n)	**Controls% (n)**	**OR (95% CI)**	**Cases% (n)**	**Con-trols% (n)**	**OR (95% CI)**	**Cas-es% (n)**	**Con-trols% (n)**	**OR (95% CI)**
All abnormalities									
A area (low expo-sure)	89.0 (203)	90.4 (206)	1.00 (referent)	91.6 (119)	89.2 (116)	1.00 (referent)	83.5 (71)	90.6 (77)	1.00 (ref-erent)
B area (medium exposure)	7.9 (18)	5.3 (12)	1.55 (0.67-3.56)	6.1 (8)	5.4 (7)	1.10 (0.39-3.06)	11.8 (1 0)	5.9 (5)	3.17 (0.65-15.46)
C area (high exposure)	3.1 (7)	4.3 (10)	0.67 (0.25-1 .77)	2.3 (3)	5.4 (7)	0.41 (0. 11-1.61)	4.7 (4)	3.5 (3)	1.30 (0.29-5.82)
			P trend 0.883			P trend 0.321			P trend 0.308
Overall ex-posed area (B+C)	11.0 (25)	9.6 (22)	1.10 (0.59-2.04)	8.4 (II)	10.8 (14)	0.76 (0.34-1.69)	16.5 (14)	9.4 (8)	2.07 (0.71-6.00)
Cardiovascular anomalies									
A area (low expo-sure)	90.6 (87)	88.6 (85)	1.00 (ref-erent)	94.9 (56)	83.0 (49)	1.00 (ref-erent)	81.4 (26)	96.9 (31)	1.00 (ref-erent)
B area (medium exposure)	6.3 (6)	6.3 (6)	0.94 (0.27-3.31)	5.1 (3)	8.4 (5)	0.59 (0. 14-2.49)	9.3 (3)	3.1 (1)	-[2]
C area (high exposure)	3.1 (3)	5.2 (5)	0.58 (0.14-2.45)	0.0 (0)	8.4 (5)	-[2]	9.3 (3)	0.0 (0)	-[2]
			P trend 0.494			P trend 0.64 7			P trend 0. 996
Overall ex-posed area (B+C)	9.4(9)	11.5 (11)	0.76 (0.30-1.96)	5.1 (3)	16.8 (10)	0.29 (0.08-1.08)	18.8 (6)	3.1 (1)	-[2]
Nervous system defects									
A area (low expo-sure)	95.7 (22)	87.0 (20)	1.00 (ref-erent)	100.0 (13)	92.3 (12)	1.00 (ref-erent)	88.9 (8)	77.8 (7)	1.00 (ref-erent)
B area (medium exposure)	4.3 (I)	0.0 (0)	-[2]	0.0 (0)	0.0 (0)	-[2]	1 1.1 (1)	0.0 (0)	-[2]
C area (high exposure)	0.0 (0)	13.0 (3)	-[2]	0.0 (0)	7.7 (1)	-[2]	0.0 (0)	22.2 (2)	-[2]
			P trend 0.200			P trend 0. 995			P trend 0.354
Overall ex-posed area (B+C)	4.3 (1)	13.0 (3)	0.31 (0.03-3.1 I)	0.0 (0)	7.7 (1)	-[2]	1 1.1 (1)	22.2 (2)	0.50 (0.05-5.51)

1 Odds ratio (OR) with 95% confidence interval (CI) computed with conditional logistic regression

2 No point estimate could be computed

In the period-specific analysis, limited to defects with the highest prevalence, we found little evidence of changes in risk during the shut-down of the plant, since the OR among "exposed" women was substantially similar to that computed for the operation period (Table 2).

DISCUSSION

A limited number of epidemiologic studies investigated the risk of congenital anomalies among populations directly exposed to the emissions of waste incinerators [5, 18-23]. Some studies did not find an increased prevalence of overall anomalies or of specific groups of birth defects, while others detected excess risks for nervous system anomalies [21], cardiovascular defects [21], facial cleft [19, 22], urinary defects [22], and overall infant deaths due to congenital malformations [23]. However, methodological limitations considerably hamper the evaluation of results of most investigations: very few studies analyzed maternal residence during the first 3 months of gestation, or adjusted for maternal age and for socioeconomic status. Moreover, distance of maternal residence (at time of delivery or abortion) from the plant was generally considered in exposure assessment, without taking into consideration the characteristics of the plant (chimney heights and widths, type and amount of combusted waste), the amount of waste combusted and the meteorological factors, with the noticeable exception of the study by Cordier et al. [22] who used an exposure index estimated from a Gaussian plume model.

Overall, results of the present study did not suggest the occurrence of an excess teratogenic risk in the vicinity of the incinerator plant, since prevalence increased only in the medium-exposure area and such increase was statistically very unstable, nor any evidence of reduction in risk during shut-down of the plant emerged. Moreover, a more specific analysis for single categories of anomalies did not generally lead us to identify excess risks for any disease group, though these results must be evaluated with caution since most of the computed estimates were statistically unstable. In particular, urogenital anomalies such hypospadias, an abnormality suspected to be associated with parental exposure to environmental endocrine disrupting chemicals [24, 25], did not increase in the exposed population. However, we found an excess prevalence of chromosomal anomalies in middle exposure area, which is difficult to interpret since there risk was not increased in the high exposure area, and no association was found in the two epidemiologic studies which specifically examined this category of birth defects [20, 22]. We therefore consider it useful to further monitor this finding in the study area or in other comparable contexts.

Some degree of exposure misclassification certainly occurred in the present study. First, since chlorinated compounds are contaminants characterized by persistency in the human body and in the environment, assessment of exposure based on residence during gestation and not on long-term residential history or the occupational environment might have biased to some extent the actual exposure burden experienced by the study subjects during the latest years. However, we checked in a random sample of case and referent women (n = 46, 10% of the study population) their residences 3 years before date of delivery or abortion, in order to ascertain the extent of long-term changes in exposure status. We found that 36 (78.3%) were then residing in the same exposure area we assigned in the present investigation according to residence at the beginning of pregnancy and 10 (21.7%) immigrated after that date into the municipality (8 towards the low-exposure area and 2 in the high-exposure area), thus suggesting a limited mobility of the study population across the different exposure areas.

Exposure misclassification might have also occurred due to additional sources of dioxins and of heavy metals in the study area, apart from the incineration plant, through different pathways of intake (inhalation, ingestion, and dermal contact). However, concerning the first category of contaminants, in the study municipality no major industrial sources of dioxins were located such as electric arc furnaces, cement kilns, and copper, and aluminum smelters, while the contribution of vehicle fuel combustion and of domestic wood burning to emissions is expected to be substantially lower that waste incineration and more evenly distributed [26, 27], thus suggesting that waste incineration was by far the major source of environmental contamination with dioxins in the study area, in line with other observations [28, 29]. Concerning exposure to heavy metals, findings from studies examining the specific burden of exposure attributable to waste incineration through measurement of biomarkers of exposure in nearby residents or occupationally-exposed workers are conflicting [30-32], and therefore the risk of exposure misclassification in our study area cannot be entirely ruled out.

Estimating if (and to which extent) exposure to the contaminants emitted by the incinerator actually changed during the study period as a consequence of the interruption in the activity of the incinerator is not easy. Few studies examined the kinetics of dioxins in humans, and half-life of 2,3,7,8-tetrachlorodibenzo-p-dioxin (TCDD) was around 3 years in a Seveso population with a mean age of 24 years [33] and 7–9 years in other selected groups of adults [34]. However, it is unclear if these indications may be extended to dioxins and related compounds other than TCDD, "low-level" exposures might be much less efficient in the TCDD transfer to the foetus [35], and these kinetics appear also to be markedly influenced by the lipid status of the individuals [34], further complicating this issue. Overall, we estimate that the short-term interruption of waste incineration occurred in our study setting had limited effects on exposure status of the local population, also considering the prolonged half-life of dioxins in soils and in locally grown produces, and therefore we emphasize risk estimates based on the overall follow-up period for dioxin exposure. A separate analysis for each operation period might be more meaningful for heavy metals and for their potential teratogenic effects, since it seems likely that interruption of plant activity markedly decreased exposure to elements such as arsenic, lead and mercury having short half-lives in human blood, in the order of 3–40 days [10, 36, 37].

Results of the present study must be extended with caution to other contexts and particularly to older incinerators, to the relevant differences in amounts and types of contaminants which may be released into the environment by these plants, owing to the type of wastes combusted and the air pollution control technologies, as well as to potential differences in susceptibility of the exposed individuals. Moreover, we did not examine other reproductive issues apart from congenital anomaly risk such as altered male-to-female birth ratio, low birth weight or twinning, and therefore our investigation cannot be considered a comprehensive assessment of reproductive health in the exposed population.

CONCLUSION

In this study setting, maternal exposure to the emissions of a MSWI, as estimated through a dispersion model, was not associated with excess risk of congenital anomalies in the offspring. These results might not apply to incinerators emitting higher amounts of pollutants such as dioxins and heavy metals.

KEYWORDS

- **Foetuses**
- **Geographical information system**
- **Logistic regression model**
- **Odds ratio**
- **Teratogenic effects**
- **Teratogens**

AUTHORS' CONTRIBUTIONS

Marco Vinceti and Carlotta Malagoli designed the study protocol, collected information about cases and controls, carried out data analysis, and drafted the manuscript. Sara Fabbi and Sergio Teggi modeled the incinerator emissions and run the GIS database. Rossella Rodolfi selected potential case mothers through the Emilia–Romagna Region Hospital Discharge Registry and did the matching and sampling of control mothers. Livia Garavelli, Francesca Rivieri, and Gianni Astolfi identified the case mothers eligible for the study from the Emilia–Romagna Region birth defect registry "IMER" and reviewed the cases retrieved through the Hospital Discharge Registry. All authors discussed study results, and read and approved the final manuscript.

ACKNOWLEDGMENTS

We acknowledge the cooperation of Elisa Calzolari of the IMER Registry of the Emilia–Romagna region, of Elena Infante, Simona Maggi, and Simona Poli of the Municipality of Reggio Emilia, and of Vincenzo Aronica and Paolo Felisetti of the Santa Maria Nuova Hospital of Reggio Emilia. The study was supported by the Office of the Environment of the Reggio Emilia Municipality, Italy, and by the Pietro Manodori Foundation of Reggio Emilia.

COMPETING INTERESTS

One of the authors (Marco Vinceti) has worked as a consultant of the municipal Gas, Waste, and Water Agency of Reggio Emilia (currently named "ENIA," formerly "AGAC," mainly owned by the Reggio Emilia Province municipalities) for the for health risk assessment of the city incinerator, of a landfill and of drinking water. However, this Agency has not taken any part in designing, carrying out or funding the present study.

Chapter 9

Spatial Analysis of Plague in California

Ashley C. Holt, Daniel J. Salkeld, Curtis L. Fritz, James R. Tucker, and Peng Gong

INTRODUCTION

Plague, caused by the bacterium *Yersinia pestis*, is a public and wildlife health concern in California and the western US. This study explores the spatial characteristics of positive plague samples in California and tests Maxent, a machine-learning method that can be used to develop niche-based models from presence-only data, for mapping the potential distribution of plague foci. Maxent models were constructed using geocoded seroprevalence data from surveillance of California ground squirrels (*Spermophilus beecheyi*) as case points and Worldclim bioclimatic data as predictor variables, and compared and validated using area under the receiver operating curve (AUC) statistics. Additionally, model results were compared to locations of positive and negative coyote (*Canis latrans*) samples, in order to determine the correlation between Maxent model predictions and areas of plague risk as determined via wild carnivore surveillance.

Models of plague activity in California ground squirrels, based on recent climate conditions, accurately identified case locations (AUC of 0.913–0.948) and were significantly correlated with coyote samples. The final models were used to identify potential plague risk areas based on an ensemble of six future climate scenarios. These models suggest that by 2050, climate conditions may reduce plague risk in the southern parts of California and increase risk along the northern coast and Sierras.

Because different modeling approaches can yield substantially different results, care should be taken when interpreting future model predictions. Nonetheless, niche modeling can be a useful tool for exploring and mapping the potential response of plague activity to climate change. The final models in this study were used to identify potential plague risk areas based on an ensemble of six future climate scenarios, which can help public managers decide where to allocate surveillance resources. In addition, Maxent model results were significantly correlated with coyote samples, indicating that carnivore surveillance programs will continue to be important for tracking the response of plague to future climate conditions.

Plague, caused by the bacterium *Yersinia pestis*, is a disease that has played an important role in human history, most notably through the demographic impacts of three major historical pandemics [1]. Plague was introduced to the US during the third pandemic (ca. 1900), and spread from the Pacific coast to its current distribution in the western states. Plague is maintained among wild rodents in distinct geographic foci in the western US [2]. Although the mechanisms by which plague is maintained between epizootic cycles are not well understood, it is generally accepted that the disease cycles between enzootic infections and occasional epizootic outbreaks among susceptible hosts [2]. Humans are presumably at greatest risk of infection during epizootics, when infectious rodent fleas seek a new host. Plague transmission to humans may also occur through contact with infected pets or other animals, through exposure to infected tissue, or via respiratory exposure to infectious air-borne droplets [3, 4].

The incidence of human plague cases is relatively low in the US: for example, a total of 107 cases occurred in the US from 1990 to 2005 [5], compared to over 240,000 cases of Lyme disease, another vector-borne disease, during roughly the same period (1992–2006) [6]. Because of this low incidence, plague surveillance in the western US is often conducted on a limited budget. However, in contrast to Lyme disease, the case-fatality ratio of plague can be high. If antibiotic treatment is not initiated promptly, plague is fatal in 40–70% of bubonic cases and nearly 100% of pneumonic cases [1]. The combination of low incidence with high mortality presents unique surveillance and public health challenges, because early detection through surveillance may not always be feasible and infrequent clinical cases may be misdiagnosed.

In addition, there is concern that certain factors [2, 7-10] could increase the occurrence of plague epizootics as well as the risk of exposure and infection to humans. In particular, the direct and indirect effects of climate change on land use, population distribution, and ecologic character are projected to contribute to an increase in the emergence and incidence of infectious diseases [11], including plague. Climate change may drive plague activity through several pathways (Figure 1), including influences on flea burden, rodent population dynamics, and plague transmission [12-19]. A spatially explicit understanding of how plague risk may shift with changing climate patterns can help not only to direct prevention and control efforts, but can also alert health care providers toward quicker recognition of exposure potential and initiation of appropriate treatment of patients [20], which is critical for improving the health outcome of the individual infected as well as reducing secondary transmission to other people.

Recent studies describing the relationships between future climatic and environmental factors and plague activity in the US have focused on human cases, as well as animal cases in the Southwestern US and Colorado plateau [12, 17, 19, 21-24]; here, we focus on the potential distribution of plague in California. The point inputs to the models developed in this study were derived from plague serology data collected by the California Department of Public Health (CDPH) and other agencies. Because active surveillance had most often been conducted in areas with a known history of plague-positive rodents or human cases, we used ecological niche modeling (ENM) to identify the potential distribution of plague throughout California (including in previously unsampled areas). Niche modeling has most often been applied to predict the potential for plant and animal species occurrences (for example, [25, 26]), and is increasingly being used to identify and map the distribution of diseases, such as Chagas disease [27], filovirus disease [28], Marburg hemorrhagic fever [29], avian influenza [30], and plague [15, 31]. In this study we evaluated Maxent, a presence-only niche modeling technique, to describe the potential distribution of plague foci in California under recent and future climatic conditions.

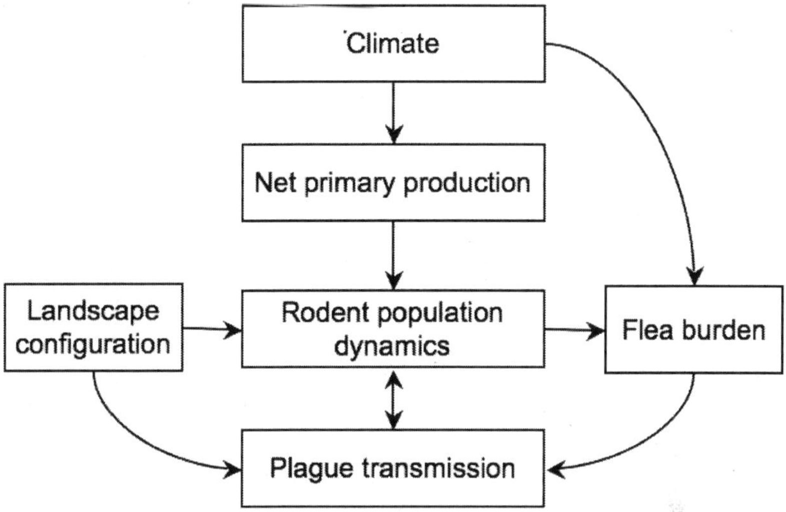

Figure 1. A conceptual model of the mechanisms by which climate influences plague transmission and maintenance. Precipitation and temperature have been linked to plague outbreaks in prairie dogs, and to human cases in the US. A proposed model for this relationship suggests that precipitation and temperature may influence rodent abundance (by influencing rodent survival and food abundance), and that increased rodent populations may affect flea abundance and/or plague transmission rates. In addition to having a positive effect on rodent population dynamics, certain soil moisture, humidity and temperature variables may influence flea ecology and the transmission of the plague pathogen.

MATERIALS AND METHODS

Data

The point inputs to the models developed in this study were derived from plague surveillance data collected by the CDPH and other agencies [32]. Records of approximately 37,000 animals (33 different genera) collected throughout the state of California during 1984–2004 were entered into an access database by public health researchers.

Rodent Point Data

Rodent samples were obtained most often by active surveillance, which was conducted in areas with a known history of plague-positive rodents or human cases [32]. Rodent sera were tested by passive hemagglutination to F1 antigen of *Y. pestis*; specimens with antibody titer $\geq$ 1:32 were considered positive [33-35].

Rodent samples were geocoded based on an address or campsite name, which allowed for location of rodent case point at a<1-km^2 spatial resolution. All rodent records were geo-located using National Geographic TOPO software (National Geographic Society 2001). Locations that could not be reliably located to a campground or address were excluded from this analysis. All geocoded points were projected to Teale Albers, NAD 1983 projection. The geocoded locations of the rodent case points are located along the north-south transect of the Sierra Nevada range, and along the southern coast and inland areas of Southern California. No positive rodents were collected in the Modoc plateau, eastern Mojave, or Colorado Desert bioregions during the surveillance period.

We identified a total of 166 unique locations for positive rodent samples (Figure 2a). The California ground squirrel (*Spermophillus beecheyi*) was the rodent species with the largest total number of specimens (12,546; Table 1) and number of positive specimens (559; Table 1), representing 105 of these unique geocoded locations. Because California ground squirrels are a key indicator species for plague epizootics [36] and human disease risk in California [10], we also ran models for this subset of data only.

Table 1. Number of total samples and positive samples for the rodent species most commonly collected during plague surveillance in California.

Species	Common Name	Total samples	Positive samples	Prevalence (proportion seropositive)
Spermophilus beecheyi	California ground squirrel	12546	559	0.045
Tomias senex	Shadow chipmunk	2701	174	0.064
Spermophilus lateralis	Golden-mantled ground squirrel	2685	19	0.007
Peromyscus maniculatus	Deer mouse	1776	20	0.011
Neotoma fuscipes	Dusky-footed woodrat	1622	10	0.006
Tamias a moe nus	Yellow-pine chipmunk	1014	58	0.057
Tamias speciosus	Lodgepole chipmunk	658	43	0.065
Tamiasciurus douglasii	Douglas' squirrel	475	44	0.093
Spermophilus beldingi	Belding's ground squirrel	408	21	0.051
Tamias quadrimaculatus	Long-eared chipmunk	400	16	0.040
Neotoma lepida	Desert woodrat	307	2	0.007
Tamias merriami	Merriam's chipmunk	277	18	0.065
Neotoma cinerea	Bushy-tailed woodrat	186	6	0.032
Peromyscus boylii	Brush mouse	117	2	0.017
Peromyscus truei	Pinon Mouse	108	0	0.000
T amias minim us	Least chipmunk	96	2	0.021
Peromyscus crinitus	Canon mouse	82	1	0.012
Glaucomys sabrinus	Northern Flying squirrel	71	0	0.000
	Total	25529	995	0.039

Only records of positive rodents were included for niche modeling, as negative samples (Figure 2b) were frequently obtained from areas that had also yielded positive samples, or from which too few specimens had been collected to be considered representative. The 3,788 sampling events had yielded negative samples, but 2,296 of these were at locations where positive samples had also been collected. Of the remaining 1,492 sampling events, only five locations had been sampled more than 20 times, which we estimated as the minimum number of samples that would need to be taken to confirm a location as a true absence.

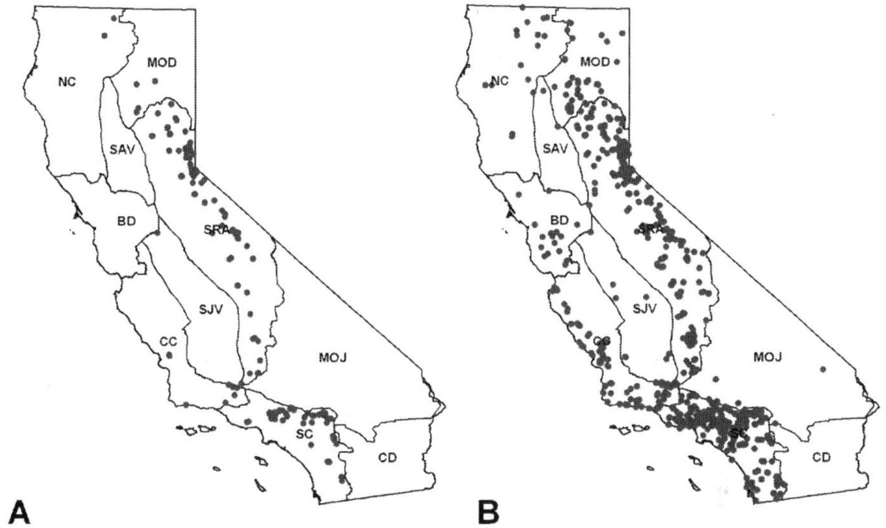

Figure 2. Rodent samples. Study area and geocoded data for (a) 995 positive rodent samples (166 unique locations) and (b) 3,788 negative rodent samples (905 unique locations). Lines designate California bioregions (NC = Klamath/North Coast; BD = Bay Area/Delta; CC = Central Coast; SC = South Coast; MOD = Modoc Plateau; SRA = Sierra; SAV = Sacramento Valley; SJV = San Joaquin Valley; MOJ = Mojave Desert; CD = Colorado Desert).

Coyote Point Data

Sampling for plague in coyotes (*Canis latrans*) was conducted independently from sampling for plague in rodents. Unlike the rodent data, coyote blood specimens were collected opportunistically as part of a depredation control and state-wide plague surveillance partnership between the California Department of Health and the US Department of Agriculture/Wildlife Services. Because coyotes can occupy a home range of up to 80 km^2 [37], the location of capture may not be the location of infection; however, the opportunistic sampling program provides a more complete description of general plague activity throughout the state, albeit at a coarser spatial resolution. Coyote sera were tested by passive hemagglutination to F1 antigen of *Y. pestis*; specimens with antibody titer $\geq$1:32 were considered positive [33-35]. The plague surveillance partnership program and the diagnostic tests that were used are described in detail by [7].

In order to compare environmental niche model results based on rodent/ground squirrel data to data on positive and negative samples of California coyotes, records for 477 positive and 2,250 negative coyotes were identified from the database (Figure 3). Collection sites for coyote samples were geocoded using National Geographic TOPO software (National Geographic Society 2001) based on field records that indicated distance and direction from a town [7, 33, 38]. Collection sites that could not be reliably located were excluded from this analysis. All geocoded points were projected to Teale Albers, NAD 1983 projection. The geocoded locations of the coyote case points were distributed in the northern and southern Sierra Nevada, along the Pacific coast, and across the Modoc plateau.

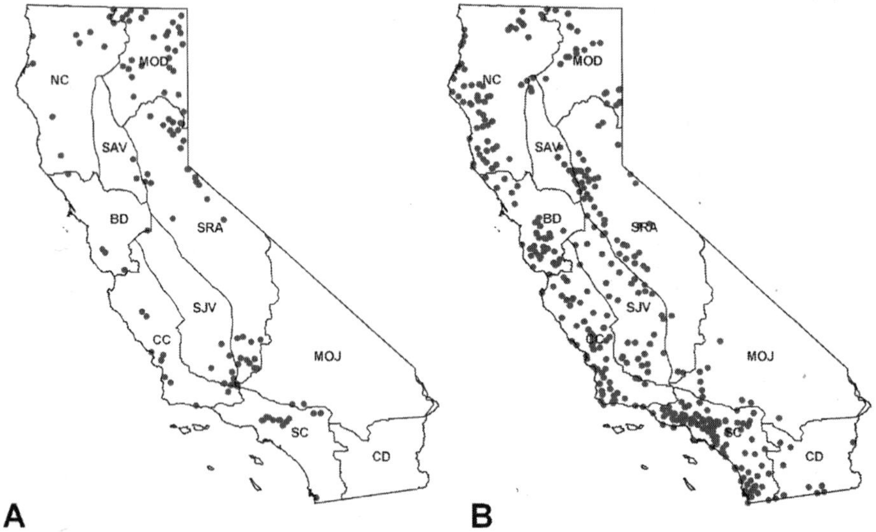

Figure 3. Coyote samples. The 477 plague-positive coyote samples, and 2,250 negative samples were collected.

Environmental Variables

We downloaded the full set of 19 Worldclim bioclimatic variables http://www.worldclim.org (Table 2). These products are derived from monthly weather station measurements of altitude, temperature, and rainfall. They are biologically meaningful variables that capture annual ranges, seasonality, and limiting factors useful for niche modeling (such as monthly and quarterly temperature and precipitation extremes) [39]. The Worldclim data are at ~1-km^2 spatial resolution and have been averaged over a 50-year time period from 1950 to 2000. Elevation was not explicitly used in model construction because it is already used as a covariate in the Worldclim data production. For modeling purposes, all environmental variable layers were masked to fit the extent of the California state outline. These layers were projected to Teale Albers, NAD 1983 projection.

Table 2. Descriptions of BIOCLIM environmental data.

Variable	Description
Biol	Annual Mean Temperature
Bio2	Mean Diurnal Range (Mean of monthly (max temp- min temp))
Bio3	Isothermality (P2/P7) (* I 00)
Bio4	Temperature Seasonality (standard deviation * I 00)
Bio5	Max Temperature of Warmest Month
Bio6	Min Temperature of Coldest Month
Bio7	Temperature Annual Range (PS-P6)
Bio8	Mean Temperature of Wettest Quarter

Table 2. *(Continued)*

Variable	Description
Bio9	Mean Temperature of Driest Quarter
Bio10	Mean Temperature of Warmest Quarter
Bio11	Mean Temperature of Coldest Quarter
Bio12	Annual Precipitation
Bio13	Precipitation of Wettest Month
Bio14	Precipitation of Driest Month
Bio15	Precipitation Seasonality (Coefficient of Variation)
Bio16	Precipitation of Wettest Quarter
Bio17	Precipitation of Driest Quarter
Bio18	Precipitation of Warmest Quarter
Bio19	Precipitation of Coldest Quarter

Because the Worldclim variables are derived from a common set of temperature and precipitation data, they can exhibit multicollinearity [39]. A Spearman rank correlation matrix was created in JMP (SAS Institute) to explore the relationships between the WorldClim bioclimatic variables. We removed the four mean temperature variables (Bio8–Bio11) because they were significantly correlated with minimum and/or maximum temperature variables, and were less likely to be biologically significant in contributing to or limiting plague activity. Of the remaining 15 variables, those that were correlated (Spearman rho > 0.60, p < 0.001) were not used together in the same model. During model runs, a jackknife manipulation was used to assess the relative contribution of each variable, and to remove variables that did not contribute significantly to the model predictions.

Modeling Current and Future Distribution of Plague in California Using Maxent

Models of the current potential (i.e., based on climate conditions) distribution of plague in California were run in Maxent (version 3.1.0). Maxent is a machine learning program that uses presence-only data to predict distributions based on the principle of maximum entropy [40]. Maximum entropy [41] is a method to provide the probability distribution which incorporates the minimum amount of information. Given a set of constraints determined by environmental variables or functions thereof, Maxent outputs the maximum entropy distribution that satisfies these constraints. Among species distribution models, Maxent has been shown to provide better identification of suitable versus unsuitable areas when compared to other presence-only modeling methods [40, 42]. In place of true absences, Maxent uses background points (pseudo-absences) to evaluate commission.

Maxent does not need multiple model runs to be averaged together [40]; thus, for each set of variables, we ran Maxent once. For each Maxent run, 75% of the points were randomly selected for model training and cross-validation, and 25% of the data were set aside for model testing and independent validation. Ten thousand random background points (pseudo-absences) were used to evaluate commission. A regularization setting of two was used for data smoothing and to address spatial autocorrelation. Model results were compared and validated using area under the ROC curve (AUC) statistics. The AUC statistic is similar to the MannWhitney U test and compares the likelihood that a random presence site will have a higher predicted value in the model than a random absence site [42, 43]. One of the appeals of ROC curves is that they do not depend on a user-defined threshold for determining presence versus absence. However, because using a geographical extent that goes beyond the presence environmental domain can lead to inflated AUC scores [44, 45], we limited the study area to the rough geographic extent of the sampling distribution (i.e., the California state boundary). The four most predictive models were used as the final models, and mapped as a cumulative probability output.

To explore the spatial relationship between model predictions and serologic samples of carnivores, we compared the final model results to data on positive and negative specimens from California coyotes. We used prediction values extracted for negative and positive coyote specimens using Hawth's point intersect tool [46]. A one-tailed t-test was performed using JMP (SAS Institute) to test the hypothesis that model predictions at positive coyote points would be significantly higher than model predictions at negative coyote points.

In order to simulate the distribution of plague under possible future climate conditions, we ran Maxent using coupled global climate model data from the IPCC 3rd Assessment (available at http://www.worldclim.org/futdown.htm). These data were originally produced by three different global climate models: CCCma [47], HadCM3 [48, 49], and CSIRO [50], and had been further processed using downscaling procedures in order to match current climate data from Worldclim [39]. We implemented an ArcInfo AML script (freely available at http://www.worldclim.org/mkBCvars.amL) to reformat and substantively convert these future temperature and precipitation data into the same bioclimatic variables that had been used as inputs for current-conditions modeling.

For each model we tested for two different time horizons, 2020 and 2050, and two different emissions scenarios (A2 and B2). The A2 scenario assumes that population growth does not slow down and reaches 15 billion by 2100 [51], with an associated increase in emissions and implications for climate change. The B2 scenario assumes a slower population growth (10.4 billion by 2100) and that precautionary environmental practices are implemented [51], yielding more conservative predictions of anthropogenic emissions. To simulate plague response to climate change, we used the final models that had been developed based on the rodent/ground squirrel data, and ran them with the future climate data.

Four models were selected as the final candidate models predicting plague distribution based on climate variables (Table 3). In all four cases, models based only on California ground squirrel specimens had higher AUC values than their counterpart models that used all rodent samples as case points. Biologically meaningful variables used in these models included two temperature variables (Maximum Temperature of

Warmest Month, and Temperature Annual Range) and four precipitation variables (Precipitation Seasonality, Precipitation of Wettest Quarter, Precipitation of Driest Quarter, and Precipitation of Warmest Quarter). The log response charts for the two most important variables used in models of plague in California ground squirrels (Precipitation in the Wet Quarter and Maximum Temperature of the Warmest Month) reflect a quadratic response to increasing temperatures and precipitation (Table 3).

Table 3. Maxent final models.

Model	AUC (All rodents)	AUC (*S. beecheyi*)	Test omission rate	Variables	% Contribution	Response	p
A	0.876	0.948	0.115	Biol6 Biol8 Biol5 Bio5 Bio7	47.3 17.6 15.2 13.2 6.7	- + + Quadratic +	< 0.0001
B	0.872	0.946	0.115	Biol6 Bio5 Biol7 Biol5 Bio7	48.7 18 14.1 11.9 7.4	- Quadratic Quadratic + +	< 0.0001
C	0.842	0.914	0.192	Biol6 Bio5 Bio7 Biol8	56.5 23.8 10 9.7	Quadratic Quadratic + +	< 0.0001
D	0.835	0.926	0.154	Biol6 Bio5 Biol8	69.5 16.5 13.9	Quadratic Quadratic Quadratic	< 0.0001

Test points were used to evaluate omission and 10,000 background points were used to evaluate commission. The reported test omission rate is for equal sensitivity and specificity. The percent contribution of each variable to the models reflects the increase in regularized gain when added to the contribution of the corresponding variable. P-values are for t-test results for the comparison between positive coyote points and negative coyote points (compared by Maxent prediction).

Models of plague activity in all rodent species (AUC of 0.835–0.88) and in California ground squirrels (AUC of 0.913–0.948) based on recent climate conditions accurately identified case locations. All models predicted the highest plague activity in the Sierra Nevada and along the southern coast under recent climate conditions (Figure 4 and 5). Models using environmental variables based on squirrel data performed well at predicting plague presence in coyotes. All four Maxent models predicted significantly higher values for pixels that overlapped with positive coyote specimens (Table 3).

Under future emissions scenarios, our models indicated that climate conditions will drive (a) an overall decrease in the probability of plague in the state, (b) a subtle shift to higher elevations as well as (c) a subtle shift to higher latitudes. Future climate conditions will support increased plague activity in the northern Sierra and central/north coast counties. However, plague risk associated with climate conditions may decrease in the southern Sierras and southern inland counties (Figure 6).

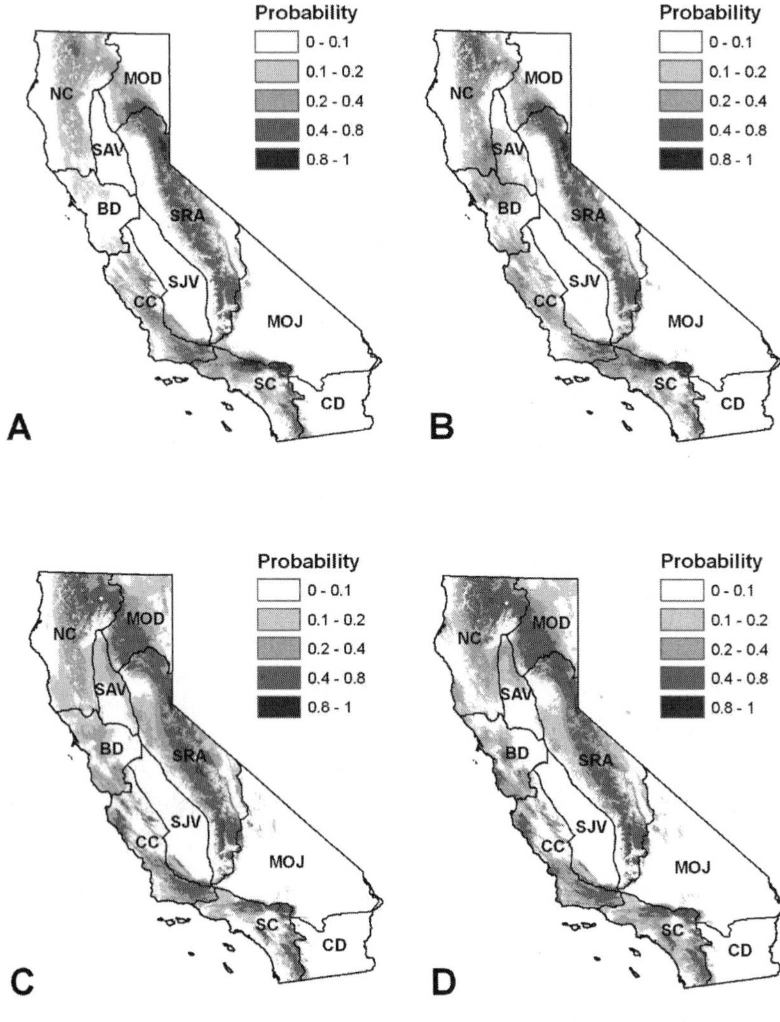

Figure 4. Maxent model results, using all plague positive rodent samples as case points. (a) Model using Precipitation of Warmest Quarter, Precipitation of Wettest Quarter, Precipitation Seasonality, Temperature Annual Range, and the Maximum Temperature of Warmest Month as predictor variables; (b) Model using Precipitation of Driest Quarter, Precipitation of Wettest Quarter, Precipitation Seasonality, Temperature Annual Range, and the Maximum Temperature of Warmest Month as predictor variables: (c) Model using Precipitation of Warmest Quarter, Precipitation of Wettest Quarter, Temperature Annual Range, and the Maximum Temperature of Warmest Month as predictor variables; and (d) Model using Precipitation of Wettest Quarter, Precipitation of Warmest Quarter and the Maximum Temperature of Warmest Month as predictor variables.

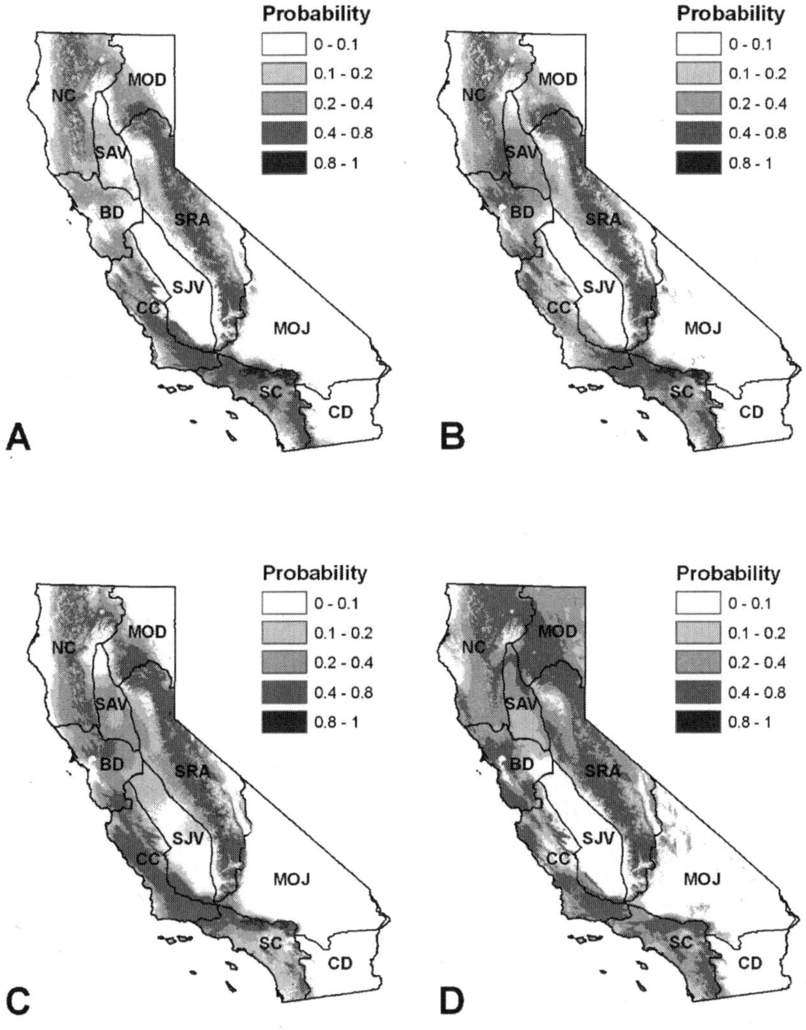

Figure 5. Maxent model results, using positive California ground squirrel samples as case points. (a) Model using Precipitation of Warmest Quarter, Precipitation of Wettest Quarter, Precipitation Seasonality, Temperature Annual Range, and the Maximum Temperature of Warmest Month as predictor variables; (b) Model using Precipitation of Driest Quarter, Precipitation of Wettest Quarter, Precipitation Seasonality, Temperature Annual Range, and the Maximum Temperature of Warmest Month as predictor variables: (c) Model using Precipitation of Warmest Quarter, Precipitation of Wettest Quarter, Temperature Annual Range, and the Maximum Temperature of Warmest Month as predictor variables and (d) Model using Precipitation of Wettest Quarter, Precipitation of Warmest Quarter and the Maximum Temperature of Warmest Month as predictor variables.

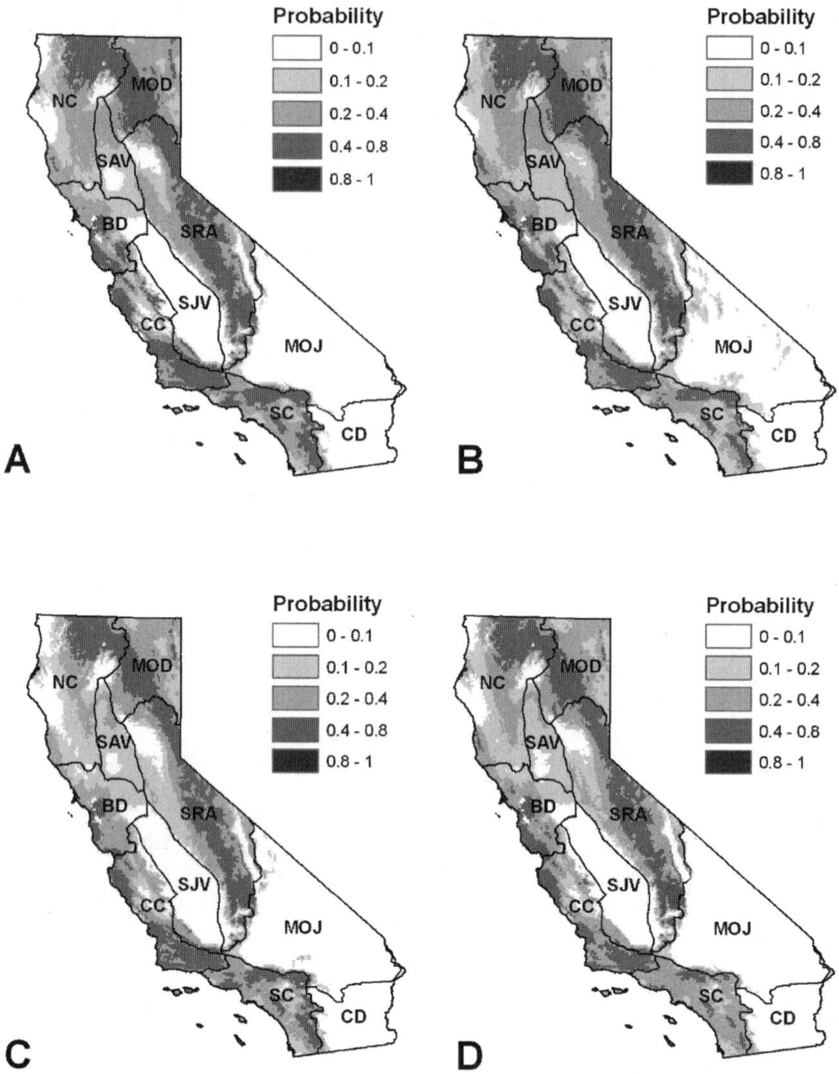

Figure 6. Predicted future plague distributions. Models were developed using data derived from three different global climate models (CCCma, HadCM3, and CSIRO), for two time steps and two emissions scenarios. (a) 2020, A2 scenario, (b) 2020, B2 scenario, (c) 2050, A2 scenario, and (d) 2050, B2 scenario.

DISCUSSION

Climate variables, such as temperature, precipitation, and humidity, can play important roles in vector-borne disease transmission by affecting vector and pathogen development, and by influencing the distribution of disease hosts and habitats [11, 52]. The biologically meaningful variables that were used in the final models we developed included two temperature variables (Maximum Temperature of Warmest Month,

Temperature Annual Range) and four precipitation variables (Precipitation Seasonality, Precipitation of Wettest Quarter, Precipitation of Driest Quarter, and Precipitation of Warmest Quarter). We also found that plague presence exhibits a quadratic response to temperature increases. These results are consistent with other studies [12-14] that have examined the role of temperature and precipitation variables on plague outbreaks in human and animal populations. In addition to having a positive effect on rodent population dynamics, certain soil moisture, humidity and temperature variables may influence flea ecology and the transmission of the plague pathogen [53]. Specifically, while warmer temperatures may in general stimulate plague activity, temperatures above 35°C are associated with a negative effect on flea fecundity, survival, and behavior [13, 18, 54].

Under future emissions scenarios, our models indicate that climate conditions will drive (a) an overall decrease in the probability of plague the state, (b) a subtle shift to higher elevations as well as (c) a subtle shift to higher latitudes. These results are generally consistent with other climate modeling studies that show species movement to higher latitudes and elevations in response to warming [55], and with studies that have examined the historical record of plague response to climate and show a shift to higher latitudes [16, 22]. Several other recent studies have also projected a potential decrease in plague activity in certain areas of the US in response to more frequent hot days 19, 23, 24].

In addition, these results provide insight into the relationship between plague maintenance in carnivore and rodent populations. Carnivores, and particularly coyotes, have been implicated in plague transmission and serve as sentinel species for the disease [7, 56]. Recent studies [57, 58] conducted on the Central Plains Experimental Range and Pawnee National Grasslands (which collectively cover a ~80,000 ha area) link the prevalence of carnivores and rodent hosts in a spatially explicit manner. Our results expand these analyses to a larger scale, by exploring the overlap in predicted plague-positive rodent distributions with positive and negative coyote samples derived through an independent sampling program. Model results demonstrate a link between positive coyote samples and areas of predicted rodent infection, providing additional support for rodent surveillance and follow-up in areas where the carnivore surveillance program identifies plague-positive animals.

California ground squirrels are the rodents that have been the most frequently sampled for plague in California. However, six other species (Douglas' squirrel, Lodgepole chipmunk, Merriam's chipmunk, Shadow chipmunk, Yellow-pine chipmunk, and Belding's ground squirrel) often had higher serum titers than California ground squirrels. This suggests these species may be of interest for further sampling and surveillance, and that additional modeling of these species' distributions could be conducted to explore the spatial heterogeneity of plague foci in California [59]. Maxent models of California ground squirrels fit better than models that used all rodent specimens as training points. Because California ground squirrels occupy a narrower ecologic zone than all rodents collectively, with less variable climatic conditions, these models described a more precise climatic niche for plague.

Current model results matched areas with historical and recent plague activity, including the San Francisco peninsula and San Bruno Mountain, the San Jacinto Mountains, and the Los Padres National Forest area [32]. Models did not yield high prediction values for the Modoc plateau region, which has historically been a focus of plague [10]. Because the low population density and rural nature of this area does not readily lend itself to observation of epizootic events, it is not surprising that no positive rodents were collected in the Modoc plateau during the study period. Thus, no model input points were used from this area, which can present a challenge to niche modeling techniques in terms of extrapolating results to new conditions in geographic and ecologic space [60]. In addition, the low prediction values for the Modoc plateau may be related to the extreme climate profile and characteristics of the plague system in this area, where plague maintenance and transmission is driven by a climate regime and rodent-host complex that differ from the rest of California. Many areas of the Modoc plateau experience plague, but in wood rats (*Neotoma* spp.), as well as in yellow pine chipmunks (*Tamias amoenus*) and their associated fleas.

It is important to keep in mind that by modeling the climatic niche for plague in California, we have modeled a potential distribution for plague that is not the actual or realized distribution. Other important factors, including landscape configuration, biotic variables, and barriers to dispersal likely limit the actual distribution of plague to smaller areas than those predicted using a climatic niche modeling approach [40, 59, 61]. Second, a number of studies have demonstrated that different modeling approaches can yield substantially different predictions [42, 62]. Thus, future work could include modeling plague potential distributions under a suite of different modeling approaches. Additionally, using niche models to predict distributions into expanded temporal and/or spatial domains can result in significant variance inflation [62]. We have attempted to dampen this variability by averaging future model outputs based on three different global climate models. However, the use of global climate models (as opposed to local or regional climate models) may itself be another source of error in niche modeling studies, and thus a potential area of research could explore the effects of different modeling datasets on disease distributions (for example, see [63]). Finally, averaged climate variables dampen seasonal effects and do not capture climatic anomalies, which may be important drivers of plague epizootics [11-13]. Thus multi-temporal modeling is required to elucidate the effects that increased climatic variability will have on vector-borne disease dynamics.

CONCLUSION

Because different modeling approaches can yield substantially different results, care should be taken when interpreting future model predictions. Nonetheless, niche modeling can provide general trends in response to climate conditions. Models of plague activity in California ground squirrels, based on recent climate data, accurately identified plague-positive rodent locations, as well as areas of historical and recent plague activity. Maxent model results were significantly correlated with coyote samples, and suggest that carnivore and rodent plague surveillance programs should be more tightly coupled in California. The final models were used to identify potential plague risk areas based on an ensemble of six future climate scenarios, which can help public managers decide where to allocate scarce surveillance resources.

KEYWORDS

- **Climate data**
- **Geocoded locations**
- **Maxent**
- **Modoc plateau**
- **Plague**
- **Rodent samples**

AUTHORS' CONTRIBUTIONS

Ashley C. Holt, Curtis L. Fritz, James R. Tucker, and Peng Gong designed the study. Ashley C. Holt conducted spatial analysis and Maxent modeling, and contributed to the writing of the manuscript. Daniel J. Salkeld contributed to the analysis of carnivore data and to the interpretation of model results. Curtis L. Fritz and James R. Tucker conducted field sampling and collated the data used in this study, and were instrumental in analyzing model results in the context of recent trends in plague activity in California. All authors read and approved the final manuscript.

ACKNOWLEDGMENTS

The authors would like to thank Lynette Yang for assistance with geocoding. We are grateful to the two anonymous reviewers for their comments and to Mark Novak, Kenneth Gage, and Charles Smith for providing guidance and feedback at various stages of this project. Funding for open-access publication fees was provided by a grant from the Berkeley Research Impact Initiative (BRII).

Disclaimer

The findings and conclusions of this chapter are those of the author(s) and do not necessarily reflect the views of the California Department of Public Health.

COMPETING INTERESTS

The authors declare that they have no competing interests.

Chapter 10

Australia's Dengue Risk: Human Adaptation to Climate Change

Nigel W. Beebe, Robert D. Cooper, Pipi Mottram,
and Anthony W. Sweeney

INTRODUCTION

The reduced rainfall in southeast Australia has placed this region's urban and rural communities on escalating water restrictions, with anthropogenic climate change forecasts suggesting that this drying trend will continue. To mitigate the stress this may place on domestic water supply, governments have encouraged the installation of large domestic water tanks in towns and cities throughout this region. These prospective stable mosquito larval sites create the possibility of the reintroduction of *Ae. aegypti* from Queensland, where it remains endemic, back into New South Wales (NSW) and other populated centers in Australia, along with the associated emerging and re-emerging dengue risk if the virus was to be introduced.

Having collated the known distribution of *Ae. aegypti* in Australia, we built distributional models using a genetic algorithm to project *Ae. Aegypti's* distribution under today's climate and under climate change scenarios for 2030 and 2050 and compared the outputs to published theoretical temperature limits. Incongruence identified between the models and theoretical temperature limits highlighted the difficulty of using point occurrence data to study a species whose distribution is mediated more by human activity than by climate. Synthesis of this data with dengue transmission climate limits in Australia derived from historical dengue epidemics suggested that a proliferation of domestic water storage tanks in Australia could result in another range expansion of *Ae. aegypti* which would present a risk of dengue transmission in most major cities during their warm summer months.

In the debate of the role climate change will play in the future range of dengue in Australia, we conclude that the increased risk of an *Ae. aegypti* range expansion in Australia would be due not directly to climate change but rather to human adaptation to the current and forecasted regional drying through the installation of large domestic water storing containers. The expansion of this efficient dengue vector presents both an emerging and re-emerging disease risk to Australia. Therefore, if the installation and maintenance of domestic water storage tanks is not tightly controlled, *Ae. aegypti* could expand its range again and cohabit with the majority of Australia's population, presenting a high potential dengue transmission risk during our warm summers.

Aedes (Stegomyia) *aegypti* (Linneaus) is an important vector of dengue and other arboviruses. Despite its limited flight dispersal capability [1, 2], its close association with humans and its desiccation-resistant eggs have facilitated many long distance dispersal events within and between continents, allowing it to expand its range globally from its origin in Africa. Its global emergence and resurgence can be attributed to factors including urbanization, transportation, changes in human movement, and behavior, resulting in dengue running second to malaria in terms of human morbidity and mortality [3, 4]. Global historical collections and laboratory experiments on this well studied vector have suggested its distribution is limited by the 10°C winter isotherm [5], while a more recent and complex stochastic population dynamics model analysis suggests the temperature's limiting value to be more towards the 15°C yearly isotherm [6].

While historical surveys in Australia have indicated that *Ae. aegypti* occurred over much of the continent (see Figure 1), its range has receded from Western Australia, the Northern Territory and NSW over the last 50 years. It is now only found in Queensland [7, 8], although recent incursions into the Northern Territory have required costly eradication strategies [8]. The significant reduction in vector distribution has been attributed to a combination of events including the introduction of reticulated water, which reduced the domestic water storage requirements of households that had provided stable larval sites [7, 9], as well as the removal of the railway-based water storage containers hypothesized as being responsible for the long distance dispersal events of *Ae. aegypti* into rural regions in NSW via steam trains [7, 10].

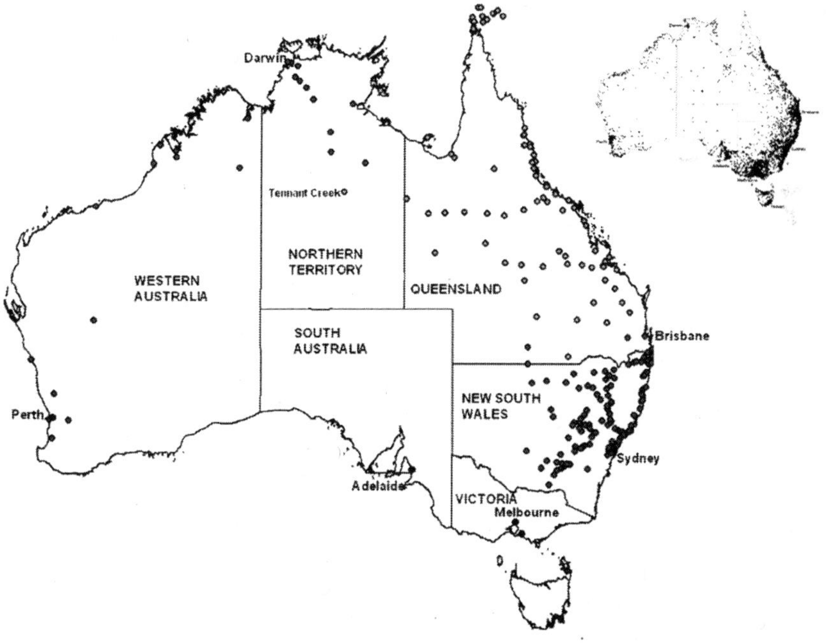

Figure 1. Map of Australia showing the 234 *Ae. aegypti* collection sites. Almost all localities (except site 219 and 220) can be regarded as historical collections while red sites indicate historical sites where *Ae. aegypti* is no longer found and green sites are regarded as contemporary sites, collected since 1980. Top right map displays the current Australia resident population distribution and each dot represents approximately 1,000 people (*Source*: Australian Demographic Statistics (3101.1)).

Today, epidemic dengue is limited to regions of Queensland where *Ae. aegypti* is extant, and the frequency of outbreaks has increased constantly over the past decade [11]. Historically, epidemics of dengue were recorded in northern Queensland in the late 1800s and in southeast Queensland in 1904–1905 [10]. Dengue epidemics in 1926, 1942, and 1943 all extended from Queensland south into NSW, stopping only on the arrival of winter [12]. Derrick and Bicks [12] found that these dengue epidemics ceased when the outside temperature reached a wet bulb isotherm of between 14 and15°C and suggested that a parameter of 14.2°C mean annual wet bulb isotherm (T_W) best represented the limiting parameter for the 1926 epidemic.

The current drying of southeast Australia has placed this region's urban and rural communities on escalating water restrictions, with anthropogenic climate change forecasts suggesting that this drying trend will continue [13]. To mitigate against this regional drying effect and the stress it places on domestic water supply, state government rebate programs have been initiated to encourage the installation of large (>3,000 l) domestic water tanks in towns and cities throughout this region. Data from the Australian Bureau of Statistics [14] records that in 2006, 20.6% of all Australian household dwellings had rainwater tanks.

Given the expansion of domestic rainwater tanks in southern Australia, and assuming these domestic water tanks can provide oviposition sites, we ask this question: can climate be assessed to determine the distributional limits of *Ae. aegypti* and dengue in Australia? We first use a genetic algorithm to develop ecological niche models for the distribution of *Ae. aegypti* in Australia (using data points drawn from both historical and contemporary collection sites) and evaluate the potential distributional limits of *Ae. aegypti* in Australia under today's climate and in future projected climate change scenarios. We map these limits in relation to published experimental and theoretical projections of *Ae. Aegypti's* temperature limits and then compare all projections to dengue transmission climate limits obtained from epidemiological studies of historical dengue epidemics in Australian. We find that human adaptation to climate change—through the installation of large stable water storage tanks—may pose a more substantial risk to the Australian population than do the direct effects of climate change. Additionally, we find that using point occurrence data and environmental parameters of climate and elevation to map the distribution of *Ae. aegypti* in Australia prove deceptive and require interpretation as some *Ae. aegypti* collection sites exist outside our ecological niche models and both theoretical cold temperature limits. This suggests that *Ae. Aegypti's* domestic behavior—with a lifecycle based around human habitation that includes blood-feeding and resting indoors as well as egg laying in artificial containers around houses—plays an influencing role on distribution.

MATERIALS AND METHODS

Distribution of *Aedes aegypti* in Australia

Contemporary collection sites were regarded as those collected since 1980 because most country towns had moved to reticulated water, steam powered trains had been replaced by diesel, and the common railway station water-filled fire buckets were removed [9, 17, 18]. Contemporary sites also include collections made between 1990 and 2005 from southeast Queensland (P. Mottram, unpub. data), and the Northern Territory (P. Whelan, unpub. data).

Base Climate Layers

Raster ASCII grids were generated for Australia at a spatial resolution of 0.025° (approximately 2.5 km) for eight climate variables plus elevation. These included annual mean rainfall and annual mean temperature produced by BIOCLIM using the ANUCLIM software package [19] as well as mean values of maximum temperatures and minimum temperatures for the months of January and July produced by the ESOCLIM component of ANUCLIM. This procedure involved the use of monthly mean climate surface coefficients, generated by the thin plate smoothing spline technique ANUSPLIN [20] from Australian Bureau of Meteorology climate data between 1921 and 1995 [21]. The geographic coordinates of the meteorological stations were used as independent spline variables together with a 0.025° digital elevation model (DEM) for Australia generated with ANUDEM [22] which acted as a third independent variable. As atmospheric moisture is known to be an important factor in terms of the survival and longevity of adult mosquitoes, mean values of dewpoint for January and July were generated with ESOCLIM to provide this.

Climate Change Layers

A further series of ASCII grids were generated from climate change scenarios using OzClim version 2 software [23, 24] at a spatial resolution of 0.25° (approximately 25 km). The scenarios used for this study were for 2030 and 2050 using CSIRO: Mk2 Climate Change Pattern with SRES Marker Scenario A1B and mid climate sensitivity. The output variables corresponded to the predicted change from the base climate for the rainfall and temperature parameters generated with ANUCLIM.

This version of OzClim outputs vapor pressure rather than dewpoint as a measure of atmospheric moisture. For the present study vapor pressure grids for the predicted change from base climate for January and July were generated and the grid cell values were converted to dewpoint by applying the inverse of Tetens' equation [$dp = (241.88 \times \ln(vp/610.78))/(17.558 - \ln(vp/610.78))$]. This mathematical procedure was implemented with the use of ImageJ software (publicly available at http://rsbweb.info.nih.gov/ij) together with the raster operations of TNTmips (MicroImages Inc., Lincoln, Nebraska).

The environmental layers used for climate change modeling were prepared by resampling the OzClim outputs to the geographical extents and grid cell size of the ANUCLIM grids using TNTmips. The resampled outputs were then added to the corresponding ANUCLIM base climate layers to produce the environmental layers predicted for the chosen climate change scenarios.

Ecological Niche Modeling

DesktopGarp version 1_1_6 [25] was used for ecological niche modeling in a manner similar to our earlier studies [26]. Models derived from the historical climate data were generated using the record sites for *Ae. aegypti* as inputs together with the eight base climate layers and elevation (the ANUDEM generated DEM is described above) to model the range of *Ae. aegypti*. Species record sites and the climate change layers for eight environmental parameters were derived from the climate change scenarios for 2030 and 2050 as well as the elevation layer. We utilized the medium sensitivity which corresponds to a global warming of 2.6°C for a doubling of CO_2 from 280 to 560 ppm [27]. The GARP procedure was implemented using half of the species record sites as a training data set for model building and the other half for model testing. Optimization parameters included 100 models for each run with 1,000 iterations per model and 0.01 convergence limits. The best subsets procedure [28] was used to select five models which were added together using TNTmips to produce predicted range maps for each species.

Theoretical Temperature Limits for *Ae. aegypti* Extrapolated Across Australia

Previous studies of the distributional limits of *Ae. aegypti* were used to develop distribution maps for Australia. Christophers [5] hypothesized a climate limit of 10°C winter isotherm based on historical global collection data and laboratory-based experiments. We also evaluated the hypothetical limit from Otero and colleagues [6], who used a complex stochastic population model that incorporates the lifecycle parameters of *Ae. aegypti* to suggest a 15°C annual mean isotherm. Both these values were incorporated into distributional maps of Australia using TNTmips.

Climate Limit of Dengue Transmission in Australia

Dengue transmission maps were developed using data from historical dengue outbreaks in Australia [12]. This work found that these dengue epidemics ceased when the outside temperature reached 14–15°C wet bulb isotherm and that a single parameter of 14.2°C annual mean wet bulb isotherm (T_W) best approximated the limit of the 1926 epidemic—probably as a result of reducing the mosquitoes' feeding activity and the ability of the virus to replicate. This 14.2°C annual mean wet bulb isotherm value was mapped onto Australia for the current climate using TNTmips and three seasonal increments: the annual mean, the warmest quarter (December–February), and the coolest quarter (June–August).

Distributional Projections of *Ae. aegypti*: GARP Modeling

Distribution sites for *Ae. aegypti* in Australia (234 sites) were collated and displayed in a single map using GPS coordinates (Figure 1). Ecological niche models were built with desktop GARP to produce a best subset model that showed agreement with the full complement of *Ae. aegypti* collections in Australia (Figure 2A). In this projection, much of northern, eastern and southeast Australia was projected to present a suitable niche. This model closely tracks an annual rainfall pattern of less than 300 mm. However, the excluded region around central Australia included two *Ae. aegypti* positive collection sites (Meekatharra in central Western Australia and Boulia in Queensland): both collection localities are small regional centers on main inland transport routes.

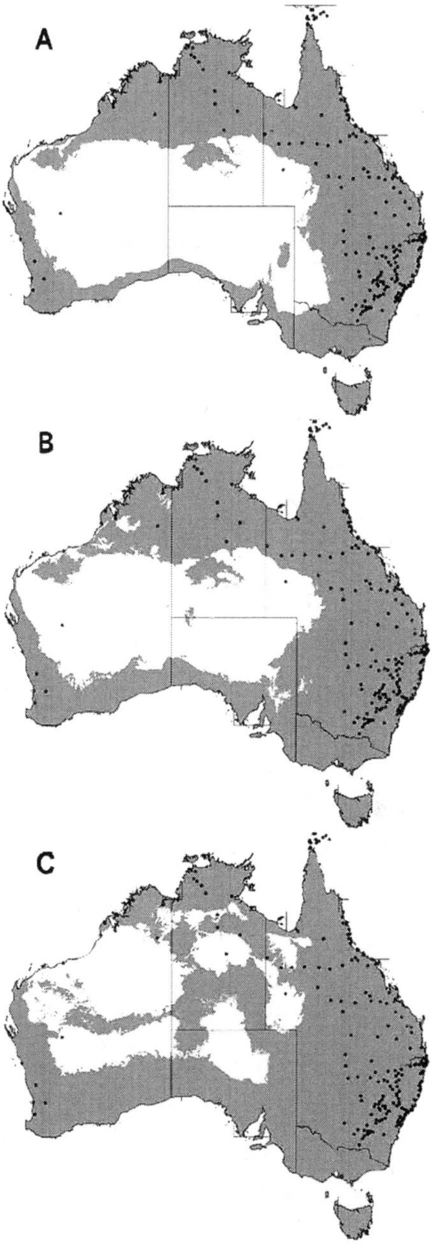

Figure 2. Distributional projections of *Ae. aegypti* in Australia based on 234 collection sites and built using desktop GARP and eight climatic variables. Panel A is the base layer projection (gray region) for the climate of 1995 and is regarded as current climate. Panel B is the projection of the forecasted climate changes for 2030 mid scenario. Panel C is the projection of the forecasted climate changes for 2050 mid scenario.

The projected climate change scenario for 2030 produced distributional models with small expansions of the base model envelope, mostly evident in southern Australia (Figure 2B). Likewise the 2050 model (Figure 2C) extended the 2030 trend, resulting in a reduced niche in north-west Australia's Pilbara region while parts of central Australia opened up as a potential niche.

Theoretical Temperature Limits of *Ae. Aegypti*

The temperature limit parameters of 10°C winter isotherm [5] and 15°C annual isotherm [6] were used to build theoretical isotherm limits for *Ae. aegypti* in Australia (Figure 3). Figure 3A shows a 10°C winter isotherm limit base map for the current climate and OzClim projections were then generated for 2030 and 2050 by adding the projected changes to this base map (3B and 3C respectively). The 15°C annual isotherm limits were similarly generated using a base map and adding the OzClim changes. Both the 10°C (average winter) and 15°C (average annual) limits incorporate the major state capitals cities—Brisbane, Sydney, Adelaide, and Perth. When these isotherm limits were subjected to the climate change scenarios for 2030 and 2050, the projection expanded to include the other mainland state capital, Melbourne (Figures 3B and C).

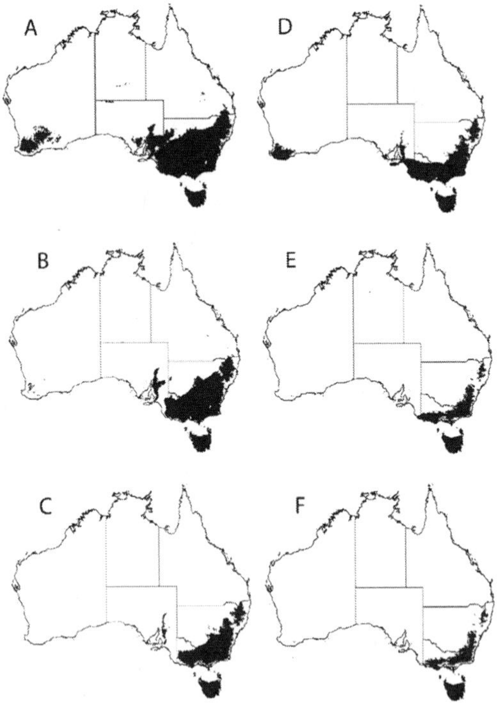

Figure 3. Theoretical distribution limits for *Ae. aegypti* and dengue transmission in Australia. Panels A–C represent the 10°C July isotherm with panel A the base layer projection for the current climate (1995). Panels B and C show the 10°C July isotherm limit of the climate change (mid) scenarios for 2030 and 2050 respectively. Panels D–F show distribution limits of *Ae. aegypti* in Australia based on the climate limit of 15°C annual mean isotherm. Panel D is the current climate (1995), panels E and F show the 15°C annual mean isotherm for climate change mid scenarios 2030 and 2050 respectively.

Several *Ae. aegypti* collection sites occurred well within the two theoretical cold climate limits. Table 1 details six *Ae. aegypti* collection sites as examples where the annual mean temperature and the mean temperature for July (calculated as (mintemp + maxtemp)/2) fall below the theoretical values and range from 12.4–15.4°C and 5.2–7.6°C respectively.

Table 1. Collection sites in NSW that fall below theoretical cool temperature limits.

Site	Locality	Annual mean temp ("C)	Max/min temp July ("C)	Mean temp July (0C)	Elevation (M)
98	Breadalbane	12.4	10.3/0.2	5.25	701
116	Culcairn	14.7	11.9/2.2	7.05	221
187	Wagga Wagga	15.4	12.8/2.4	7.6	177
133	Junee	15.1	12.4/2.2	7.3	295
131	Harden	14.3	12.1 / 1.2	6.65	396
189	Wallendbeen	13.9	11.6/1.0	6.3	468

doi:1 0.1371/journal.pntd.0000429.t001

Theoretical Dengue Transmission Limits

Derrick and Bicks [12] suggested that dengue transmission stopped between the 15°C and 14°C T_W isotherm and suggested that a 14.2°C T_W annual mean isotherm best approximated the temperature limit for transmission in the 1926 dengue epidemic. We applied this isotherm to Australia for the annual mean isotherm (Figure 4A) as well as the warmest quarter isotherm (summer; December–February, Figure 4B) and the coldest quarter isotherm (winter; June–August, Figure 4C). These climate limit maps indicate that if the vector could re-establish itself throughout its former range then much of northern tropical Australia would be receptive to dengue transmission year round and transmission would be possible throughout most of Australia during the summer months.

DISCUSSION

Can the historical distribution of *Ae. aegypti* in Australia provide an insight into the potential distribution potential of this mosquito? Using 234 different spatial data points generated from historical and contemporary collections of *Ae. aegypti* in Australia, we developed ecological niche models to hypothesize the potential range expansion of this mosquito under today's climate and under future climate change scenarios for 2030 and 2050 using OzClim mid sensitivity values that correspond to a global warming of 2.6°C for a doubling of CO_2 from 280 to 560 ppm [27]. In Australia general warming estimates are approximately 1.0°C by 2030 and 1.2–2.2°C by 2050, the latter values being dependent on CO_2 emissions. While rainfall (outside of far north Australia) is estimated to decrease by 2–5%, southern Australia is projected to encounter a 5% reduction in rainfall [13]. Our GARP model for current climate suggested that *Ae.*

aegypti could potentially coexist with over 95% of the Australian population and this distribution did not change significantly, with regard to the Australian population distribution, under either the 2030 and 2050 climate change scenarios.

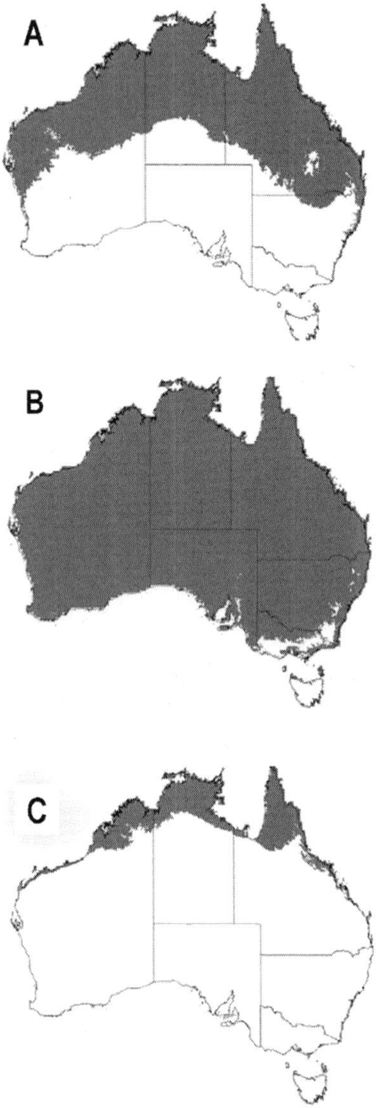

Figure 4. Employing a hypothetical dengue climate limit estimated from epidemics in Australia that stopped on the arrival of winter where the outside temperature fell to a wet bulb isotherm (T_W) of 14–15°C [12], we mapped a 14.2°C T_W isotherm onto Australia using three temporal increments. Panel A represents the 14.2°C annual mean T_W for Australia [12]. Panel B represents the 14.2°C T_W for Australia's warmest quarter (December–February), representing summer transmission. Panel C represents the same isotherm for Australia's coolest quarter (June–August), representing potential year-round transmission.

Only the highly arid central Australian region was excluded from the projection (annual rainfall less than 300 mm). The GARP model did not show southern cold climate thermal limits in Australia, probably due to the presence of several *Ae. aegypti* collection sites from inland NSW that show cool climate parameters. We then mapped two theoretical cool climate limits across Australia—the 10°C winter (July) isotherm [5] and the 15°C annual mean isotherm [6]. Of these two isotherm limits the 15°C annual mean isotherm appeared more representative of the known distribution of *Ae. aegypti* in Australia, although collection sites did exist outside these temperature isotherm limits.

It remains unknown if the cold climate tolerant populations were breeding in the warmer months and surviving the colder winter months as eggs [29], or were surviving as larvae. With regard to these questions, observations have been recorded of viable *Ae. aegypti* larvae in ice encrusted water [5, 7], while experiments have suggested that a water temperature of 1.0°C can be lethal over 24 hr, but larvae can be viable at a constant 7.0°C for over a week [5]. At the other temperature extreme, laboratory experiments show that *Ae. aegypti* larvae perish when the water temperature exceeds 34°C while adults start to die off as the air temperature exceeds 40°C [5]. Domestic water tanks in Australia contain thousands of liters of water that would—in combination with the mosquitoes' domestic (indoor) nature—provide a buffer to temperature extremes and assist mosquito survival in what may appear unsuitable environments. For example, *Ae. aegypti* exists and transmit dengue in India's Thar desert townships in north-western Rajasthan, where the mosquito utilizes household pitchers and underground cement water tanks. [30].

The incongruence between the temperature limits and our ecological niche models highlights the difficulties of using what are essentially sophisticated climate pattern matching procedures to study an organism with a biology and ecology strongly influenced by human activity. Fortunately, we can directly compare our GARP model with a new mechanistic model of the same organism over the same environment [31]. This mechanistic model utilizes biophysical life processes parameters such as the effects of climate on reproduction and larval development. Larval development in both rainwater tanks and smaller containers were assessed and the potential distribution of *Ae. aegypti* was projected across Australia. Projections using rainwater tanks larval development resembled our GARP model for Northern and central Australia, but unlike our projections, a southern cold climate thermal limit was identified which was actually lower than the published parameters displayed in Figure 3 [5, 6]. Apart from showing the clear advantage of a bottom-up approach for modeling this mosquito, this study supports the hypothesis that domestic rainwater tanks contributed for the historical southern distribution of *Ae. aegypti* in Australia.

Humans not only facilitate long distance dispersal events for this mosquito, cohabitation with humans can provide thermal buffers to the outdoor climate as adults rest indoors, and domestic rainwater tanks can provide stable oviposition sites. When the theoretical distributions (GARP models and temperature limits) and actual *Ae. aegypti* distributions are viewed alongside the expansion of domestic water tanks underway in Australia, a trend emerges where *Ae. aegypti* could potentially exist year-round in today's climate throughout the southern Australian mainland. This potential distribution includes the metropolitan areas of Brisbane (pop 1.8 million), Sydney (pop 4.2 million), Adelaide (pop 1.1 million), and Perth (pop 1.5 million). Additionally the climate change temperature limit projections for the mid scenario 2050 see this range expand to include Melbourne (pop 3.6 million). The addition of a theoretical dengue virus transmission limit parameter (we used a 14.2°C wet bulb isotherm) suggests an overlapping dengue risk in many of Australia's metropolitan regions during the summer months (December–February).

The potential for dengue virus introduction to these regions through travelers from endemic regions (including north Queensland) during summer presents a transmission risk that can be inferred by the current incidence of imported and endemic cases of dengue in Australia—many of which enter Australia through national and international transport nodes. For example, for the year to June, 2008 there were 250 dengue notifications for Australia, of which 113 came from Queensland (most via local transmission), 72 from NSW, 15 from NT, 12 from SA, 8 from VIC, and 28 from WA. Notifications from NSW, South Australia, Victoria and Western Australia exceeded the 5-year mean in each jurisdiction suggesting that the frequency of dengue is increasing [32].

Understanding the relationship between climate and dengue transmission is difficult because non-linear relationships exist between the daily survival of *Ae. aegypti*, the extrinsic incubation period (EIP) of the virus, temperature and humidity [33-35]. Forecasted regional warming in Australia may lengthen and intensify the dengue transmission season by shortening the mosquitoes' EIP, although it is important to note that dengue epidemics appear to be more strongly influenced by intrinsic population dynamic (epidemiological) processes than by climate [36]. Even so, any temporal extension effect in the transmission season will follow the expansion of potential larval sites that is now underway in Australia. Thus, while the issue of regional warming is important, the expansion of large rainwater tanks throughout urban regions of Australia is at present a prevailing human adaptation with more immediate possibilities for changing vector distributions in Australia than the direct warming effects projected by anthropogenic climate change scenarios. Whether southern Australia's current drought is due to the region's natural climate variability or part of a changing climate pattern, will continue to be debated by some. Nonetheless, it is important to avoid the cycle where human changes in water storage result in an *Ae. aegypti* range expansion followed by dengue epidemics seeded by viremic travelers [4, 37]. Additionally, domestic water storage can sustain *Ae. aegypti* populations (and dengue transmission) in regions not normally suitable for its survival [38], while active government and community contributions can remove established *Ae. aegypti* populations (and dengue) from areas where it has been endemic [39]—and both of these are human modifications.

In Australia, ineffectively screened domestic rainwater tanks have been identified as key containers with respect to *Ae. aegypti* productivity [40, 41]. The introduction of reticulated water systems in towns and cities throughout Australia is believed responsible for a major range contraction of *Ae. aegypti* over the last 50 years. This trend may now be reversed as humans adapt to climate-change-induced drought conditions—the increased use of domestic water storage in tanks could deliver stable primary larval sites into urban neighborhoods. In Queensland's capital city, Brisbane—which is currently *Ae. aegypti* free—severe water shortages resulted in escalating water restrictions with an eventual prohibition on the use of all outside reticulated water outlets (November, 2007–July, 2008) and 75,000 domestic water tanks being installed by late 2007. This number of tanks represents approximately 21% of households with reticulated water in the Brisbane area (F. Chandler, Brisbane City Council, pers. comm.). Additionally, ad hoc uncontrolled water tanks are now also commonly being used to store rainwater, adding to the potential surfeit of stable breeding sites around Australia that are likely to facilitate the expansion risk of *Ae. aegypti* into urban areas. It is unlikely that any of these water storage tanks—government approved or not—will be maintained sufficiently to prevent mosquito access in the long term.

The flight range for *Ae. aegypti* is understood to be generally small: mark-release recapture experiments show them to have a flight range of only hundreds of meters [42-44]. However, these estimates are limited in time and space, being derived from a snapshot of one or a few gonotrophic cycles which take place in the context of an abundance of ovipositing sites. Longer distance flight range dispersal may be more common, especially when ovipositing sites are rare, but this is difficult to quantify [45, 46]. Human mediated long distance dispersal events are mostly responsible for *Ae. aegypti* movement: their highly domestic nature and desiccation-resistant eggs facilitate successful movement via human transport routes. Surveys in Queensland in the 1990s [17] and 1990–2005 (P. Mottram, unpublished) reveal *Ae. aegypti* collections from over 70 townships and this number is likely an underestimate. As the numbers of individuals and populations of *Ae. aegypti* increase in Queensland towns, the incursion risk beyond these regions via human-induced long distance dispersal events also increases, and with the presence of new stable oviposition sites growing, the expansion of this dengue vector must now be expected.

Operations to remove *Ae. aegypti* incursions are resource-heavy, often requiring both government legislation and widespread community cooperation to reduce adult mosquito populations. A recent example from a 2004 incursion of *Ae. aegypti* into the small Northern Territory town of Tennant Creek (pop 3,200) from Queensland resulted in a 2-year eradication campaign that required 11 personnel and cost approximately $1.5 million and was achieved in 2006 [8].

CONCLUSION

Determining the potential distribution of *Ae. aegypti* in Australia using climatic parameters can be problematic and in this case produced results that neither fully match the known distribution, nor reveal cold climate limits in Australia. Reasons for this may exist in the difficulty of relating the point occurrence data of a species' distribution that is closely tied to humans—unlike native mosquito species in Australia where GARP models appear more representative of known distributions [26, 47]. We must also consider the limited climatic parameters available through the OzClim climate scenario generator that reduced the GARP modeling to a subset of environmental parameters that may have little influence on the organism. Because the GARP models showed no cold temperature limits for *Ae. aegypti* in Australia, we also assessed two published theoretical cold temperature limits across Australia. These temperature limit projections also could not contain all collection sites, which may suggest that in Australia, climate—and in particular temperature—plays a less important role in determining the range of this species due to a combination of its intimate relationship with humans and our propensity to store water. This is where the use of statistical approaches and point occurrence data to evaluate species' distribution may be weak and integrating life processes parameters such as the effects of climate on reproduction and larval development may be more practical and informative.

If it is an assumption that burgeoning domestic water tanks will provide stable larval sites for *Ae. aegypti*, then the synthesis of our GARP modeling, the theoretical climate limits and the historical distribution of this mosquito strongly suggest that a distributional expansion is possible and could expose the majority of Australia's population to this dengue vector. Additionally, viewing this synthesis of *Ae. aegypti* in Australia with dengue transmission climate limits obtained from historical Australian dengue epidemics suggests a real risk of dengue transmission occurring in regions ranging well beyond north Queensland during the summer months.

We conclude that if the installation and maintenance of domestic water storage tanks is not tightly controlled today, *Ae. aegypti* could be spread by humans to cohabit with the majority of Australia's population, presenting a high potential dengue transmission risk during our warm summers.

Current and projected rainfall reduction in southeast Australia has seen the installation of large numbers of government-subsidized and ad hoc domestic water storage containers that could create the possibility of the mosquito *Ae. aegypti* expanding out of Queensland into southern Australian's urban regions. By assessing the past and current distribution of *Ae. aegypti* in Australia, we construct distributional models for this dengue vector for our current climate and projected climates for 2030 and 2050. The resulting mosquito distribution maps are compared to published theoretical temperature limits for *Ae. aegypti* and some differences are identified. Nonetheless, synthesizing our mosquito distribution maps with dengue transmission climate limits derived from historical dengue epidemics in Australia suggests that the current proliferation of domestic water storage tanks could easily result in another range expansion of *Ae. aegypti* along with the associated dengue risk were the virus to be introduced.

KEYWORDS

- ***Aedes aegypti***
- **Epidemic dengue**
- **Ovipositing sites**
- **OzClim outputs**
- **Queensland**

AUTHORS' CONTRIBUTIONS

Conceived and designed the experiments: Nigel W. Beebe. Analyzed the data: Anthony W. Sweeney. Contributed reagents/materials/analysis tools: Nigel W. Beebe, Robert D. Cooper, Pipi Mottram, and Anthony W. Sweeney. Wrote the chapter: Nigel W. Beebe.

ACKNOWLEDGMENTS

The OzClim model used in this study has been jointly developed by the International Global Change Institute (IGCI), the University of Waikato and CSIRO Atmospheric Research.

Chapter 11

Threats from Oil and Gas Projects in the Western Amazon

Matt Finer, Clinton N. Jenkins, Stuart L. Pimm, Brian Keane, and Carl Ross

INTRODUCTION

The western Amazon is the most biologically rich part of the Amazon basin and is home to a great diversity of indigenous ethnic groups, including some of the world's last uncontacted peoples living in voluntary isolation. Unlike the eastern Brazilian Amazon, it is still a largely intact ecosystem. Underlying this landscape are large reserves of oil and gas, many yet untapped. The growing global demand is leading to unprecedented exploration and development in the region.

We synthesized information from government sources to quantify the status of oil development in the western Amazon. National governments delimit specific geographic areas or "blocks" that are zoned for hydrocarbon activities, which they may lease to state and multinational energy companies for exploration and production. About 180 oil and gas blocks now cover ~688,000 km^2 of the western Amazon. These blocks overlap the most species-rich part of the Amazon. We also found that many of the blocks overlap indigenous territories, both titled lands and areas utilized by peoples in voluntary isolation. In Ecuador and Peru, oil and gas blocks now cover more than two-thirds of the Amazon. In Bolivia and western Brazil, major exploration activities are set to increase rapidly.

Without improved policies, the increasing scope and magnitude of planned extraction means that environmental and social impacts are likely to intensify. We review the most pressing oil- and gas-related conservation policy issues confronting the region. These include the need for regional Strategic Environmental Impact Assessments and the adoption of roadless extraction techniques. We also consider the conflicts where the blocks overlap indigenous peoples' territories.

The western Amazon includes parts of Bolivia, Colombia, Ecuador, Peru, and western Brazil (Figure 1). It is one of the most biodiverse areas of the planet for many taxa, including plants, insects, amphibians, birds, and mammals [1-7]. The region maintains large tracts of intact tropical moist forest and has a high probability of stable climatic conditions in the face of global warming [8]. By contrast, the eastern Amazon in Brazil, where much of the global attention has focused, has a high probability of continued massive deforestation [9] and drought risk in the coming decades [10]. The western Amazon is also the home to many indigenous ethnic groups, including some of the world's last uncontacted peoples living in voluntary isolation [11-13].

Figure 1. Study area of the western Amazon.

Underlying this landscape of extraordinary biological and cultural diversity are large reserves of oil and gas, many yet untapped. Record oil prices and growing global demand are now stimulating unprecedented levels of new oil and gas exploration and extraction. It is the nations of the region, and not the indigenous peoples who live on much of the land, who assert their constitutional ownership of subsoil natural resources. National governments delimit specific geographic areas or "blocks" that are zoned for hydrocarbon activities, which they may lease to state and multinational energy companies for exploration and production.

Oil exploration in the western Amazon started as early as the 1920s in Peru [14] and Ecuador [15], with a production boom arriving in the 1970s. The subsequent three decades have seen numerous large projects, such as several oil projects in the central Ecuadorian Amazon, the Urucu gas project in Brazil, and the Camisea gas project in Peru.

Oil and gas development in the western Amazon has already caused major environmental and social impacts 16–19. Direct impacts include deforestation for access roads, drilling platforms, and pipelines, and contamination from oil spills and wastewater discharges. The technologies of the 1970s-era oil operations caused widespread contamination in the northern Ecuadorian [20, 21] and northern Peruvian Amazon [22, 23]. Even the much newer Camisea pipeline, which began operations in the fall of 2004, had 5 major spills in its first 18 months of operation [24]. A 1990s-era oil operation experienced a major spill in Ecuador's Yasuní region as recently as January, 2008 [25]. There are also direct impacts associated with seismic testing activities during the exploration phase of projects [17, 26].

Indirect effects arise from the easy access to previously remote primary forest provided by new oil roads and pipeline routes, causing increased logging, hunting, and deforestation from human settlement [27-29]. For example, much of the extensive deforestation in the northern and central Ecuadorian Amazon followed colonization along the oil access roads [30-32].

Social impacts are also considerable. The national representative organizations of indigenous peoples in Ecuador (CONAIE) and the Peruvian Amazon (AIDESEP) have opposed new oil and gas projects, citing the widespread contamination from previous and current oil projects [33, 34]. In both countries, local residents and indigenous peoples have taken legal actions against US oil companies for allegedly dumping billions of gallons of toxic waste into the forests [35-37]. Intense opposition from indigenous peoples has stopped exploration in two leased blocks in Ecuador (Blocks 23 and 24) for over 7 years [38]. Deforestation and colonization following road building has affected the core territory of several indigenous groups in Ecuador. Oil and gas projects in the territories of indigenous peoples in voluntary isolation have become highly contentious. These peoples, so named due to their decision of avoiding contact with the outside world [11], inhabit remote parts of the western Amazon [11-13] and are extremely vulnerable because they lack resistance or immunity from outsiders' diseases [39]. First contact results in high rates of morbidity and mortality, with mortality estimates ranging between a third and half of the population within the first several years [11].

The extent and intensity of oil and gas exploration and development in the western Amazon may soon increase rapidly. Information on the future of oil and gas activities across the entire region is limited. Here, we quantify and map the extent of current and proposed oil and gas activity across the western Amazon using information from government and news sources. We document how the oil and gas blocks overlap areas of peak biodiversity, protected areas, and indigenous territories. Finally, we discuss policy options that might mitigate the impacts.

There are now ~180 oil and gas blocks covering ~688,000 km^2 of forest in the western Amazon (Figure 2). At least 35 multinational oil and gas companies operate these blocks, which overlap the most species-rich part of the Amazon for amphibians, birds, and mammals (Figure 3). Oil and gas projects affect the forest of all western Amazonian nations, but to varying degrees. For example, in both Ecuador and Peru blocks now cover more than two-thirds of the Amazon, while in Colombia that fraction is less than one-tenth. In Bolivia and western Brazil, historical impacts are minimal, but the area open to oil and gas exploration is increasing rapidly.

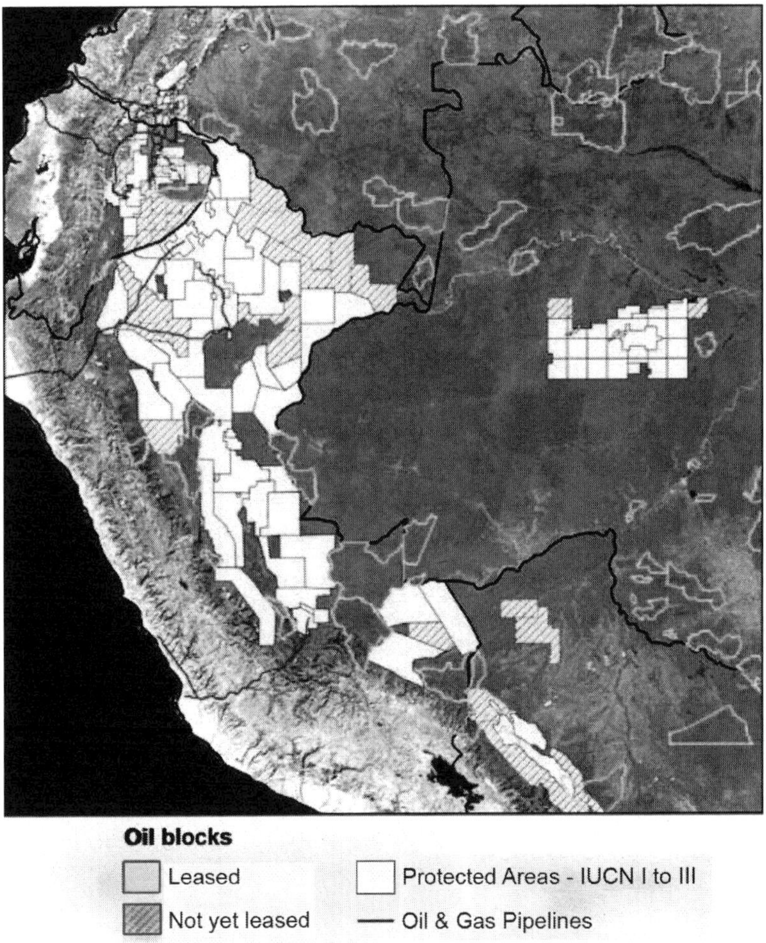

Figure 2. Oil and gas blocks in the western Amazon. Solid yellow indicates blocks already leased out to companies. Hashed yellow indicates proposed blocks or blocks still in the negotiation phase. Protected areas shown are those considered strictly protected by the IUCN (categories I to III).

In 2003, Peru reduced royalties to promote investment, sparking a new exploration boom. There are now 48 active blocks under contract with multinational companies in the Peruvian Amazon (Figure 4). The government has leased all but eight in just the past 4 years. At least 16 more blocks are likely to be signed in 2008. These 64 blocks cover ~72% of the Peruvian Amazon (~490,000 km^2). The only areas fully protected from oil and gas activities are national parks and national and historic sanctuaries, which cover ~12% of the total Peruvian Amazon. However, 20 blocks overlap 11 less strictly protected areas, such as Communal Reserves and Reserved Zones. At least 58 of the 64 blocks overlay lands titled to indigenous peoples. Further, 17 blocks overlap areas that have proposed or created reserves for indigenous groups in voluntary isolation.

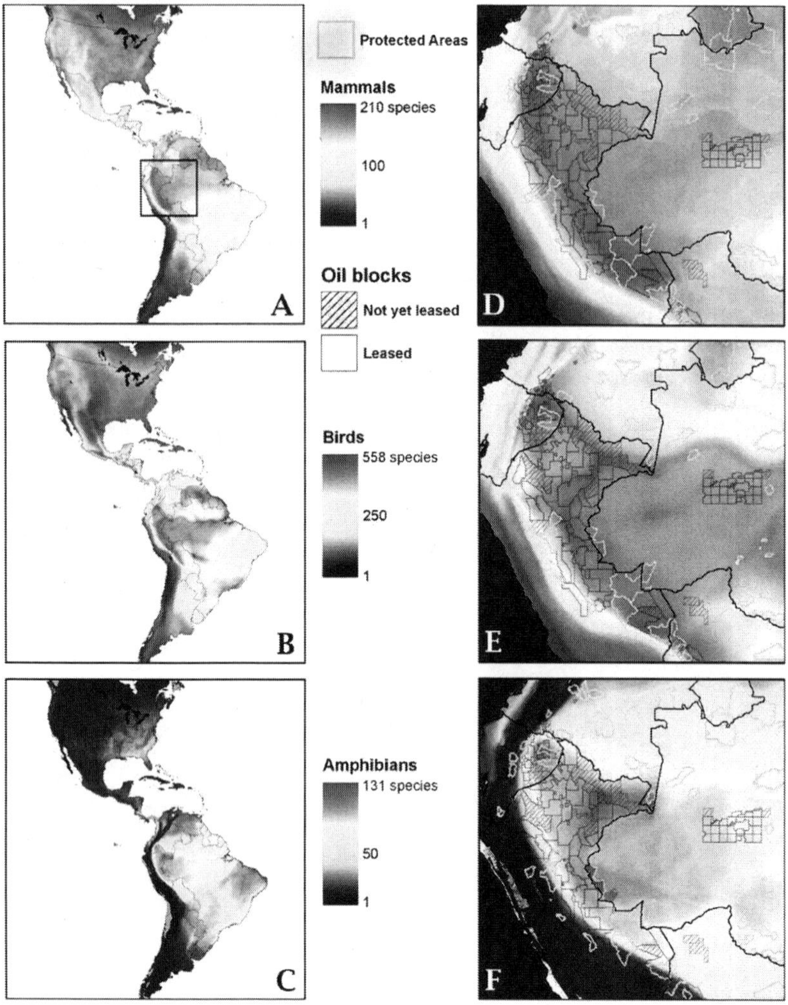

Figure 3. Overlap of oil and gas blocks with biodiversity and protected areas. The number of species of mammals (A), birds (B), and amphibians (C) across the Americas, where the highest diversity occurs in the western Amazon. Detailed view of the western Amazon region, outlined by the box in A, for mammals (D), birds (E), and amphibians (F). In this region hydrocarbon blocks overlap areas of exceptionally high biodiversity. Protected areas shown are those considered strictly protected by the IUCN (categories I to III).

Several large recent oil discoveries in the remote forests on the Peruvian side of the Peru-Ecuador border will likely trigger a new wave of development. Initial estimates indicate over 500 million barrels in Blocks 67 and 39 (labeled in Figure 4), the former of which has recently begun its development phase [40]. Gas development in the Camisea region is likely to continue as well. A new gas discovery in the region announced in January, 2008 brought the proven reserves of the Camisea area to over 15 trillion cubic feet. In addition, a wave of exploration is about to begin as the 40 blocks leased out over the last 4 years begin operations on the ground. In 2007 alone, the government approved the Environmental Impact Studies (EIS, see below) for 10 blocks that are set to begin immediate seismic testing and drilling of exploratory wells.

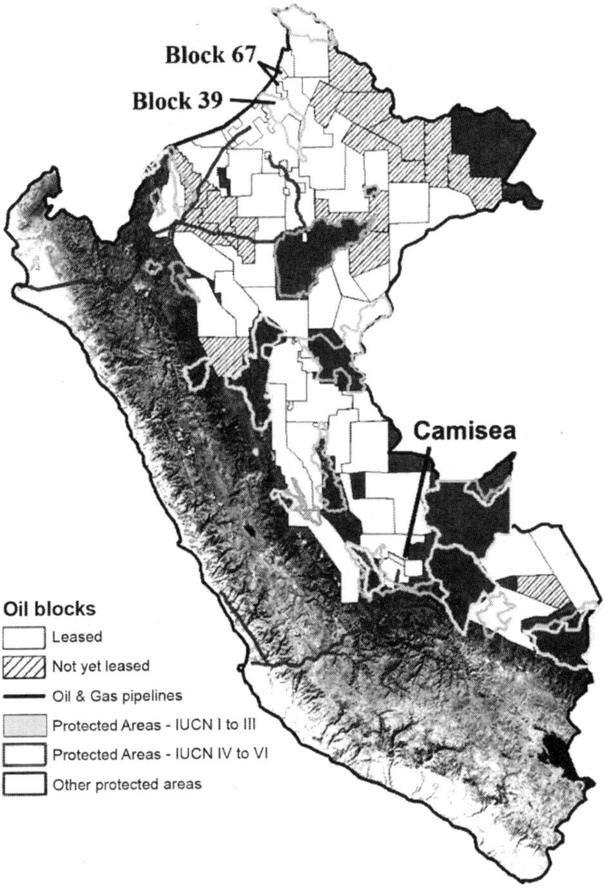

Figure 4. Focus on Peru. Oil and gas blocks in Peru, including all IUCN categorized Amazonian protected areas, protected areas not yet placed in an IUCN category, and key features discussed in the text.

The Ecuadorian government has zoned ~65% of the Amazon for oil activities (~52,300 km^2) (Figure 5). Blocks overlap the ancestral or titled lands of ten indigenous groups. Oil development began in the north in the 1970s. The oil frontier in Ecuador has now shifted south, where a quarter of Ecuador's untapped oil reserves lie in Yasuní National Park, the country's principal Amazonian national park. Unlike Peru, Ecuador permits oil and gas extraction in national parks. In January, 2007, the Ecuadorian government, however, delimited a 7,580 km^2 "Zona Intangible"—an area off-limits to oil, gas, and logging activities—via Presidential Decree in the southern part of Yasuní. It protects a portion of the territory of the Tagaeri and Taromenane, the country's two known indigenous groups in voluntary isolation. To the southwest of Yasuní, intense opposition [38] from indigenous peoples has stopped exploration in two leased blocks (Blocks 23 and 24) for over 7 years. Just to the east of these two blocks, the entire southeastern part of the Ecuadorian Amazon has been zoned into blocks, but not yet offered to multinational oil companies. Newer oil operations from the 1990s and this decade (Blocks 15, 16, and 31) have built new access roads into the primary forests of the Yasuní region. At the time of writing, Ecuador's Constituent Assembly just completed a new Constitution prohibiting extraction in protected areas except by Presidential petition in the name of national interest.

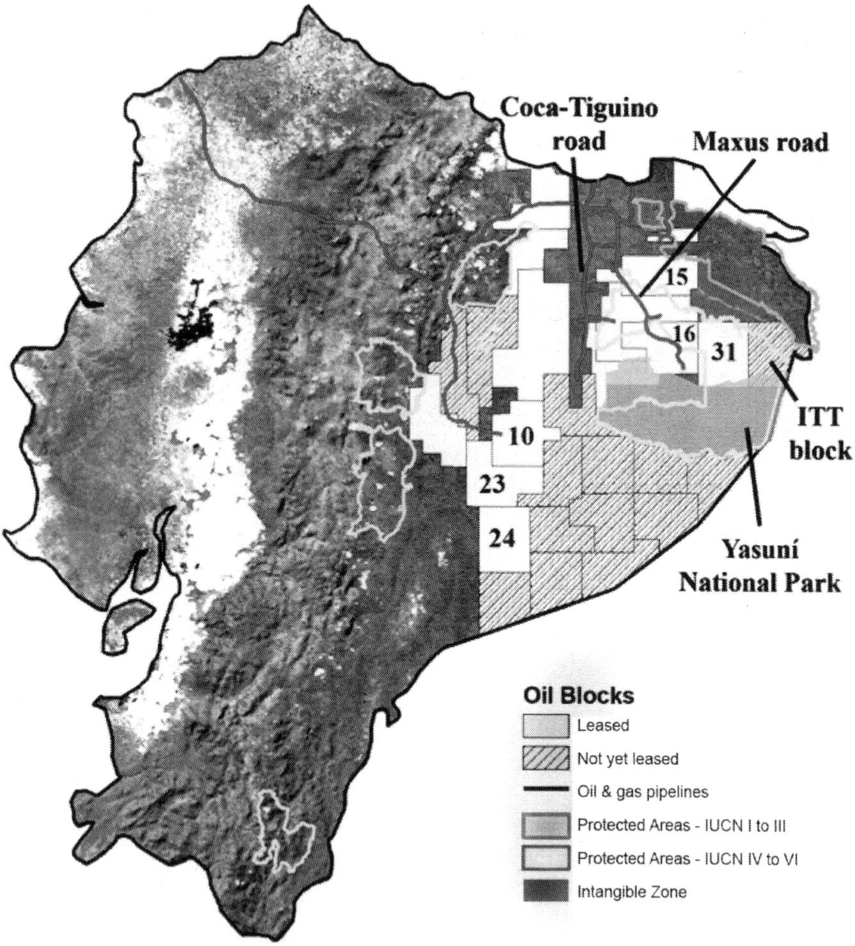

Figure 5. Focus on Ecuador. Oil and gas blocks in Ecuador, including all IUCN categorized Amazonian protected areas and key features discussed in the text. Oil blocks discussed in the text are numbered.

In Bolivia, two leased Amazonian exploration blocks cover ~15,000 km^2, including large parts of Madidi and Isiboro Securé National Parks and Pilon-Lajas Biosphere Reserve. Activity on these blocks has stalled for several years, but recent Bolivian newspaper reports indicate that exploration in this region is imminent [41]. Multinational oil companies operate these blocks, but now the state oil companies of Bolivia and Venezuela are joining forces to explore the region. In August, 2007, Bolivian President Evo Morales and Venezuelan President Hugo Chavez created a new company composed of the state oil companies of the two nations [42]. One of the primary tasks of this new company is to explore for oil in the newly created blocks surrounding Madidi.

In 2005, the Brazilian government leased out 25 contiguous blocks surrounding the Urucu and Jurua gas fields in the state of Amazonas, bringing the total leased area to ~67,000 km^2. These new blocks lie within a largely intact part of the Brazilian Amazon [43]. The Urucu fields already contain producing gas wells, but the Jurua field, discovered in 1978, has yet to be exploited. A nearly 400 km roadless gas pipeline is being constructed to connect the Urucu gas fields to Manaus [44]. Another pipeline has been proposed to carry gas over 500 km to Porto Velho in the state of Rondônia. Brazil's National Petroleum Agency has also recently announced plans to look for oil and gas in the Amazonian state of Acre on the border with Peru and Bolivia [45].

In the Colombian Amazon, 35 exploration and production blocks (~12,300 km^2) are concentrated within and around Putumayo Department on the border with Ecuador. Production in Putumayo peaked years ago, but much of the oil in this region and beyond may be yet untapped or undiscovered [46]. Colombia's Hydrocarbon Agency recently announced a new 2008 bidding round, featuring nine new blocks in Putumayo. Over 90% of the Colombian Amazon is currently free from oil activities.

DISCUSSION

In sum, more than 180 oil and gas blocks now overlap the most species-rich part of the Amazon, including areas having the world's greatest known diversity of trees, insects, and amphibians. The threat to amphibians is of particular concern, not only because so much of their global diversity is concentrated in the western Amazon, but also because they are already the most threatened vertebrate taxa worldwide [5]. Many blocks also cover protected areas—such as national parks in Ecuador and Bolivia and a variety of lower-level protected areas in Peru—that were originally established for biodiversity protection.

Many of the oil and gas blocks are in remote areas and overlap indigenous territories, both titled lands and areas utilized by peoples in voluntary isolation. Moreover, the scope and magnitude of planned activity appears unprecedented. For example, of the 64 blocks now covering the Peruvian Amazon, all but eight have been created since 2004.

Oil and gas development in the western Amazon has already caused major environmental and social impacts. Given the increasing scope and magnitude of planned hydrocarbon activity, these problems are likely to intensify without improved policies.

It is to those policies that we now turn. We consider the impacts of roads, the requirement of free, prior and informed consent (FPIC), the special needs of peoples living in voluntary isolation, the use of strategic environmental impact assessments, and the role of the international community. In each case, the policies adopted will have significant impacts one way or another on the region's biodiversity and the fate of its indigenous peoples. This is not an exhaustive list, but topics our experiences suggest are the most important.

Roads

Roads are one of the strongest correlates of Amazonian deforestation [47, 48]. New access roads cause considerable direct impacts—such as habitat fragmentation—and often trigger even greater indirect impacts, such as colonization [30], illegal logging [49], and unsustainable hunting [27, 28]. Animals often targeted by local and indigenous hunters are involved in key ecological processes such as seed dispersal and seed predation [50]. The overhunting of large primates, for example, has the potential to change the composition and spatial distribution of western Amazon forests due to the loss of these important seed dispersers [51]. Even a rough extrapolation from the oil extraction in previous decades suggests that the planned wave of oil and gas activity may similarly fragment and degrade largely intact forests over huge areas in coming years and decades.

Two Amazonian modeling efforts indicate that deforestation is concentrated in the eastern and southern Brazilian Amazon—areas with high road density—but the western Amazon is largely intact due its remoteness and lack of roads [9, 43]. Oil and gas blocks, however, now fill much of these remote areas. A primary concern is that new oil and gas projects could bring a proliferation of new access routes throughout the western Amazon. Indeed, pending oil and gas projects are currently the primary threat to areas in eastern Ecuador (Blocks 31 and ITT), northern Peru (Blocks 39 and 67), Peru's Camisea region, Brazil's Urucu region, and Bolivia's Madidi region.

Oil access roads are a main catalyst of deforestation and associated impacts. A report from scientists working in Ecuador concluded that impacts along new access roads could not be adequately controlled or managed, particularly in regards to actions of the area's local or indigenous peoples [52]. The report, along with opposition by the Waorani indigenous people, pressured the Ecuadorian government, which banned Petrobras from building a road into Yasuní National Park in July, 2005. The government forced the company to redesign the project without a major access road. As of this writing, Petrobras plans to use helicopters to transport all materials, supplies, equipment, and people to and from the well sites, with oil flowing out via a roadless pipeline. This decision by the Ecuadorian government might set an important precedent for policy no new oil access roads through wilderness areas. A major roadless oil project in Ecuador's Block 10 was the region's first example that such development is possible [53], and Block 15 also features a roadless pipeline with canopy bridges. Elimination of new roads could significantly reduce the impacts of most projects.

Free, Prior and Informed Consent

Governments claim the authority to manage natural resources located on or below indigenous people's territories for the public interest, while indigenous peoples claim that their rights to property and territory allow them the right to FPIC regarding proposed extractive projects on their lands [54, 55].

The key distinction lies between consultation and consent. International law—namely the 1989 International Labour Organization's Indigenous and Tribal Peoples Convention No. 169—clearly mandates that indigenous peoples be consulted about development projects on their territories [56]. Indeed, national regulations in Ecuador and Peru, for example, mandate such consultation [57, 58]. The question is, do indigenous peoples have the right to reject a project planned on their territory after being properly consulted? The latest international instruments indicate "yes." The United Nations Declaration on the Rights of Indigenous Peoples—adopted by the General Assembly in 2007—emphasizes FPIC prior to government approval of any project affecting indigenous lands or territories [59]. Also in 2007, the Inter-American Court on Human Rights issued a landmark ruling, Case of the *Saramaka People v. Suriname*, that the State must ensure the right of local peoples to give or withhold their consent in regard to development projects that may affect their territory [55].

A prerequisite for effective FPIC procedures is that indigenous peoples possess legal title to their traditional lands. The Inter-American Human Rights System has dealt extensively with this issue. In 1998, the Inter-American Commission found that it is a violation of the American Convention on Human Rights (Article 21, Right to Property) for a government to grant an extractive concession without the consent of the indigenous peoples of the area. The Inter-American Court subsequently ruled that this right to property requires the titling of their traditional territory [60]. Although many communities and nationalities have obtained such title, others still have not (or else the process is incomplete). Given that most of the oil blocks in question are in indigenous areas, the resolution of who controls the land and its sub-surface resources will greatly influence the development of the region.

Indigenous Peoples in Voluntary Isolation

The situations in Ecuador and Peru highlight two of the major issues concerning hydrocarbons and indigenous peoples in voluntary isolation: A lack of understanding of the full extent of the territories of peoples in voluntary isolation and debate regarding "intangibilidad"—or untouchablility—of their known territories.

In Ecuador, the government created a Zona Intangible (Untouchable Zone) to protect the territory of its two known isolated groups from oil development in 1999 and delimited the 7,580-km^2 zone via Presidential Decree in January, 2007. However, testimonies from local indigenous Waorani indicate that signs of the Taromenane and Tagaeri are sometimes seen in areas that are covered by oil blocks, north of and outside the Zona Intangible. Moreover, the Taromenane speared to death an illegal logger outside the northern limit of the Zona Intangible in March, 2008 [61], the clearest evidence to date that they range outside the demarcated zone.

In Peru, the Law for the Protection of Isolated Peoples in Voluntary Isolation (Law 28736) was passed in May of 2006, and implementing Regulations were issued by Presidential Decree in October, 2007. The "untouchable" character of protective reserves for peoples in voluntary isolation may be broken for the exploitation of natural resources deemed by the state to be in the public interest, a loophole that allows extraction of oil and gas. Another major issue in Peru concerns hydrocarbon activities in areas formally proposed to be reserves for peoples in voluntary isolation. At least 15 blocks overlap such proposed reserves.

In May, 2006, the Inter-American Commission on Human Rights granted precautionary measures in favor of the two known groups in voluntary isolation in the Ecuadorian Amazon, the Tagaeri and Taromenane, due to threats they face from oil activities and illegal logging. These measures call for the government to prohibit the entry of "third persons"—which would include oil companies—into their territory. In March, 2007, the Inter-American Commission urged the Peruvian government, again through precautionary measures, to protect the indigenous peoples in voluntary isolation in the Madre de Dios region from threats posed by illegal logging. In 2007, indigenous organizations made three more requests to the Inter-American Commission for precautionary measures needed to stem the threats to uncontacted peoples posed by oil and gas projects in Peru.

Strategic Environmental Assessments

Nations of the region require project-specific EIS prior to oil and gas exploration or exploitation projects. The oil companies contract the firms to conduct the studies, a system that clearly lacks independent analysis. Moreover, there are typically no comprehensive analyses of the long-term, cumulative, and synergistic impacts of multiple oil and gas projects across a wider region, generally referred to as a Strategic Environmental Assessment (SEA) [62].

In Peru, hydrocarbon blocks now overlap 20 protected areas. Thirteen of these protected areas preceded creation of the oil blocks and lack compatibility studies required by the Protected Areas Law [63]. An SEA could deal with these types of issues.

For example, in the Napo Moist Forest ecoregion of northern Peru, 28 blocks form a nearly continuous oil zone. There has been almost no regional planning, no analysis of the cumulative and long-term impacts, and no strategic planning for long-term protections of biodiversity and indigenous peoples. No national parks exist in the region, so there are no areas strictly off-limits to oil development. Indeed, the mass of oil blocks overlap two lower-level protected areas, several proposed protected areas, numerous titled indigenous territories, and a proposed Territorial Reserve to protect the indigenous peoples in voluntary isolation living in the core of the region. The development of proper SEAs would potentially reduce the negative impacts across the wider region of the western Amazon.

Role of International Community

In 2006, over half of Ecuador's total oil production went to the US, including nearly 90% of the heavy crude coming out of the controversial OCP pipeline [64, 65]. Much of the oil feeding this pipeline comes from projects in sensitive areas, such as Yasuní National Park. In Peru, American, Canadian, European, and Chinese companies drive the exploration and exploitation of the Amazon.

Ecuador has proposed an innovative opportunity [66] for the world to share in the responsibility of protecting the Amazon. In April, 2007, the President of Ecuador, Rafael Correa, announced that the government's preferred option for the largest untapped oil reserve, located beneath Ecuador's principal Amazonian national park (Yasuní), is to leave it permanently underground in exchange for compensation from the international community. The oil fields, known as Ishpingo–Tiputini–Tambococha (ITT), are within one of the most remote and intact parts of Yasuní National Park, and are part of the ancestral territory of the Waorani.

While the history of oil and gas extraction in the western Amazon is one of massive ecological and social disruption, the future need not repeat the past. Roadless extraction would greatly reduce environmental and social impacts. Proper attention to the rights of indigenous peoples and the outright protection of lands of peoples living in voluntary isolation, who, by definition cannot give informed consent, would bring exploration within widely accepted international norms of social justice. Disinterested, regional scale SEA would prevent piecemeal damage across large areas. Finally, the international community can play a role in widening the options available to the region's nations and its indigenous peoples.

MATERIALS AND METHODS

Most data on oil blocks and pipelines are from government sources and were publicly available online at the time of submission. These include Colombia's Agencia Nacional de Hidrocarburos (http://www.anh.gov.co), Ecuador's Ministerio de Minas y Petróleos (http://www.menergia.gov.ec), Peru's Perupetro (http://www.perupetro.com.pe) and Ministerio de Energía y Minas (http://www.minem.gob.pe/hidrocarburos/index.asp), Bolivia's Ministerio de Hidrocarburos y Energía (http://www.hidrocarburos.gov.bo), and Brazil's Agência Nacional do Petróleo, Gás Natural e Biocombustíveis (http://www.anp.gov.br). When necessary, downloaded maps of boundaries of oil blocks and their attributes were digitized using ArcGIS 9.2.

We also collected information from major newspapers of the region, particularly El Comercio in Ecuador and La Razon in Bolivia.

Boundaries of protected areas are from the World Database of Protected Areas [67]. We digitized the boundaries of Parque Nacional Ichigkat Muja—Cordillera Del Condor, Santiago – Comaina, and Sierra del Divisor from maps available from the Instituto Nacional de Recursos Naturales (http://www.inrena.gob.pe). We divided protected areas into strictly (I–III) and less strictly (IV–VI) protected groups according to the IUCN categories for protected areas [68]. These categories range from I to VI, with lower numbers representing management to maintain natural ecosystems and processes, while higher numbers represent management oriented towards human recreation and sustainable resource extraction.

We converted biodiversity data for birds [69, 70], mammals [71, 72], and amphibians [73] to raster format and analyzed them in ArcGIS. For birds, we used only the breeding range for each species.

Size estimates of blocks were calculated using ArcGIS and verified by comparing to published accounts in government sources.

To calculate the percentage of Ecuadorian and Peruvian Amazon zoned into oil and gas blocks, we used the data in [74] for the size of the Ecuadorian Amazon (81,000 km^2) and in [9] for the size of the Peruvian Amazon (677,048 km^2).

We analyzed indigenous territory maps in Peru [75] and Ecuador [R. Sierra, unpublished data] and recorded the number of overlaps with oil and gas blocks.

KEYWORDS

- **Environmental impact studies**
- **Gas pipeline**
- **Oil and gas blocks**
- **Strategic Environmental Assessment**
- **Western Amazon**

AUTHORS' CONTRIBUTIONS

Conceived and designed the experiments: Matt Finer. Analyzed the data: Matt Finer Clinton N. Jenkins. Contributed reagents/materials/analysis tools: Matt Finer and Clinton N. Jenkins. Wrote the chapter: Matt Finer, Clinton N. Jenkins, Stuart L. Pimm, Brian Keane, and Carl Ross. Formatted figures: Clinton N. Jenkins.

ACKNOWLEDGMENTS

We would like to thank Ellie Happel, Peter Kostishack, Michael Valqui, Tim Killeen, Nigel Pitman, and two anonymous reviewers for helpful comments on earlier drafts of this work. We thank Fernando Ponce and Mariana Vale for the Abstract translations in Spanish and Portuguese, respectively.

Chapter 12

Malaria and Water Resource Development

Delenasaw Yewhalaw, Worku Legesse, Wim Van Bortel, Solomon Gebre-Selassie, Helmut Kloos, Luc Duchateau, and Niko Speybroeck

INTRODUCTION

Ethiopia plans to increase its electricity power supply by 5-fold over the next 5 years to fulfill the needs of its people and support the economic growth based on large hydropower dams. Building large dams for hydropower generation may increase the transmission of malaria since they transform ecosystems and create new vector breeding habitats. The aim of this study was to assess the effects of Gilgel-Gibe hydroelectric dam in Ethiopia on malaria transmission and changing levels of prevalence in children.

A cross-sectional, community-based study was carried out between October and December 2005 in Jimma Zone, south-western Ethiopia, among children under 10 years of age living in three "at-risk" villages (within 3 km from dam) and three "control" villages (5–8 km from dam). The man-made Gilgel-Gibe dam is operating since 2004. Households with children less than 10 years of age were selected and children from the selected households were sampled from all the 6 villages. This included 1,081 children from "at-risk" villages and 774 children from "control" villages. Blood samples collected from children using finger prick were examined microscopically to determine malaria prevalence, density of parasitaemia and identify malarial parasite species.

Overall 1,855 children (905 girls and 950 boys) were surveyed. A total of 194 (10.5%) children were positive for malaria, of which, 117 (60.3%) for *Plasmodium vivax*, 76 (39.2%) for *Plasmodium falciparum* and one (0.5%) for both *P. vivax* and *P. falciparum*. A multivariate design-based analysis indicated that, while controlling for age, sex and time of data collection, children who resided in "at-risk" villages close to the dam were more likely to have *P. vivax* infection than children who resided farther away (odds ratio (OR) = 1.63, 95% CI = 1.15, 2.32) and showed a higher OR to have *P. falciparum* infection than children who resided in "control" villages, but this was not significant (OR = 2.40, 95% CI = 0.84, 6.88). A classification tree revealed insights in the importance of the dam as a risk factor for malaria. Assuming that the relationship between the dam and malaria is causal, 43% of the malaria occurring in children was due to living in close proximity to the dam.

This study indicates that children living in close proximity to a man-made reservoir in Ethiopia are at higher risk of malaria compared to those living farther away. It is recommended that sound prevention and control program be designed and implemented around the reservoir to reduce the prevalence of malaria. In this respect, in localities near large dams, health impact assessment through periodic survey of potential vectors and periodic medical screening is warranted. Moreover, strategies to mitigate predicted negative health outcomes should be integral parts in the preparation, construction and operational phases of future water resource development and management projects.

Malaria is one of the most important causes of morbidity and mortality in tropical and sub-tropical countries. It is responsible for more than 1 million deaths each year [1]. The estimated annual global incidence of clinical malaria is 500 million cases [2]. Recent estimates indicate that more than 2 billion people are exposed to malaria risk in about 100 countries. Close to 90% of all malaria infections occur in sub-Saharan Africa, where malaria causes an estimated 40% of fever episodes [3-5]. More than 90% of the deaths occur in children under 5 years of age in Africa [6]. Most of the infections and deaths in highly endemic areas occur in children and pregnant women, who have little access to health systems [7-9]. Malaria in children is complicated by anaemia, neurological sequels from cerebral compromise, respiratory distress and sub-optimal cognitive and behavioral development [10].

Malaria transmission varies among communities largely due to environmental factors, such as proximity to breeding sites [11]. Many water resources development and management projects result in local outbreaks of malaria and other vector-borne diseases such as schistosomiasis [12], lymphatic filariasis [13] and Japanese encephalitis [14]. These outbreaks can be attributed to an increase in the number of breeding sites for mosquitoes, an extended breeding season and longevity of mosquitoes, relocation of local populations to high-risk reservoir shorelines and the arrival of migrant populations seeking a livelihood around the newly created reservoirs [15-19].

In Ethiopia, approximately 75% of the total area is estimated to be malarious, with 68% of the total population (52 million people) being at risk of infection [16]. According to the national health services statistics, malaria is among the top 10 leading causes of morbidity [16]. Proximity to micro-dams which were constructed for small irrigation development schemes is considered as one of the risk factors for increased malaria incidence [18-20]. The actual malaria cases that occur annually throughout the country are estimated to be 4–5 million [21]. Malaria is responsible for 30–40% of outpatient visits to health facilities, 10–20% of hospital admissions and 10–40% of severe cases in children under 5 years of age [22]. Most transmission takes place following cessation of rains [23]. Previous studies showed that malaria was more prevalent in villages that were close to small irrigation dams than in those farther away [19, 20]. Ethiopia plans to increase its electricity power supply by 5-fold over the next 5 years based on large hydropower dams to fulfill the needs of its people and support the economic growth based on large hydropower dams [24]. Ethiopia's power security is already over 85% dependent on hydropower and could grow to over 95% depending on whether all hydropower dams under construction are commissioned. Eight hydropower dams account for over 85% of Ethiopia's existing 767 MW generating capacity. Five additional hydropower sites with a combined capacity of 3,125 MW are currently under construction. Thus, it is important to look at a variety of impacts from the reservoir as it may create health problems and diseases such as malaria, schistosomiasis and lymphatic filariasis that often increase because reservoirs provide habitat for vectors (e.g., mosquitoes) and intermediate hosts (e.g., snails). Such investigations will also help in planning, designing and monitoring future dams.

Gilgel-Gibe hydroelectric dam, created by impounding the water of the Gilgel-Gibe River in south-western Ethiopia, is currently the largest supply of power (184 MW) in Ethiopia and is operating since 2004. During the construction of the dam, many people were relocated upstream of the reservoir, although some still remain close to the buffer zone (500–800 m from the reservoir edge at full supply level) surrounding the lake. The location of the rural villages near the newly formed reservoir may increase malaria transmission, assuming that this reservoir contributes directly or indirectly to the presence of breeding places for malaria vectors. Studies in various African countries indicate that the flight range of different species of Anopheles ranges from 0.8 km (*An. funestus*) [25] to an average of 1–1.6 km (*An. gambiae* s.s) [26], and the maximum flight range of anopheline vector mosquitoes is about 3 km [27-29].

The current study investigates the possible effects of Gilgel-Gibe hydroelectric dam on malaria transmission and prevalence among children below the age of 10 years, focusing on the distribution of infection in relation to distance of villages from the reservoir shore. Results may further guide the development of appropriate malaria interventions for communities living around the reservoir.

MATERIALS AND METHODS

Study Site and Population

The study area is located 260 km south-west of the capital, Addis Ababa in Oromia Regional State, south-western Ethiopia near Gilgel-Gibe hydroelectric dam. The study area lies between latitudes 7°42'50"N and 07°53'50"N and between longitudes 37°11'22"E and 37°20'36"E, at an altitude of 1,734–1,864 m above sea level. The area has a sub-humid, warm to hot climate, receives between 1,300 and 1,800 mm of annual rainfall and has a mean annual temperature of 19°C. The main socio-economic activities of the local communities are mixed farming involving the cultivation of staple crops (maize, teff, and sorghum), and cattle and small stock raising. The study villages are located in Omo-Nada, Kersa, and Tiro-Afeta districts (weredas) and have similar settlement pattern, have access to health services and are socio-economically similar. Census results taken between August and September 2005 showed a population of 6,985 in the study villages. All the communities residing in the study villages belong to the Oromo ethnic group, which is one of the largest ethnic groups in Ethiopia. The reservoir covers an area of 62 km^2 and is located at an altitude of 1,671 m. There are no other permanent water bodies or impoundments other than the reservoir found around the 6 study villages.

Study Design

A cross-sectional house-to-house survey was conducted between October and December 2005 in 6 villages located around the reservoir created by the newly constructed Gilgel-Gibe hydroelectric dam. Sampling was carried out by stratified cluster survey. Three villages within 3 km of the reservoir (Dogosso, Budo, and Osso) and three villages located 5–8 km from its shore (Shakamsa, Sombo, and Yebo) were randomly selected and designated as "at-risk" and "control" villages, respectively. The selection of "at-risk" and "control" villages was based on the established flight range of anopheline vector mosquitoes as described elsewhere in this chapter [27-29]. The 1,855 children (1,081 and 774 from "at-risk" and "control" villages, respectively) who had lived for at least 6 months in those selected villages were included in the study. Bed net distribution was not started in the study villages until the end of this study but there was malaria control activity through indoor residual spraying, using DDT, and malathion, which stopped 4 months prior to the study in both villages.

Parasitological Investigation

A parasitological study was carried out for 3 months (October–December 2005) to investigate the difference in malaria prevalence between "at-risk" and "control" villages and to characterize malaria in the area. During the survey, socio-demographic data were collected and house-to-house visits were made each month to collect blood samples from every child less than 10 years of age and thick and thin films were prepared directly from finger prick blood samples. Blood sample collection, preparation, staining technique and microscopic identification of Plasmodium species were performed as per standard methods [30]. The thick film served to confirm the presence or absence of the parasite, whereas the thin film was to identify the Plasmodium species. The initial thick films were considered negative if no parasites were seen in at least 100 oil-immersion fields of the thick film [31]. For positive slides, species, and presence or absence of gametocytes was recorded. All blood films were initially read on site or at Omo-Nada District Health Center Laboratory by trained laboratory technicians. Films positive for parasites and a 10% sample of films negative for parasites were subsequently re-examined by an independent senior technician at Jimma University Specialized Hospital Laboratory. The senior microscopist was blinded to the previous microscopy results. The parasite density was counted per 300 leukocytes and was then expressed as number of parasites per microliter by assuming an average leukocyte concentration of 8,000 leukocytes/μl [32]. All Plasmodium positive children were treated according to the national malaria treatment guideline of the Government of Ethiopia [33].

Statistical Methods

Data were entered in and analyzed with the statistical program STATA 10 software package (StataCorp, Texas, USA). Prevalence rates were calculated from monthly positive cases. The prevalence of *Plasmodium falciparum* and *Plasmodium vivax* was calculated across age, village of residence and month of infection. Logistic regressions were conducted to check for any significant differences in the proportions of malaria cases between "at-risk" and "control" communities both in a univariate manner and controlling for age, sex, and month. The clustering at village level was taken into account in the logistic regression models (univariate as well as multivariate) by using a marginal model with the Taylor series linearization method for estimating the variances.

Classification Tree

To investigate the potential complex interactions between the different determinants in explaining the presence of the parasite, classification trees (CART) were used [34]. This technique can be used to investigate how the available determinants can be used in creating homogenous subgroups, with either high or low prevalence's. The CART models are fitted by binary recursive partitioning of a multidimensional covariate space, in which the dataset is successively split into increasingly homogeneous subsets until a specified criterion is stratified. The minimum error tree was selected. The CART provides a predictor ranking (variable importance) based on the contribution predictors make to the construction of the tree. This indicates how important the different independent variables are in determining the division. Importance is determined by playing a role in the tree, either as a main splitter or as a surrogate. Variable importance, for a particular predictor, is the sum across all nodes in the tree of the improvement scores that the predictor has when it acts as a primary or surrogate splitter. It is thus possible that a variable enters the tree as the top surrogate splitter in many nodes, but never as the primary splitter. Such a surrogate splitter will turn out as very important in the variable importance ranking provided by CART. More details on this technique can be found in [35].

Prevalence Fraction

The cross-sectional study allows to compute a prevalence ratio (PR) which is computed as follows: p(D+ IE+)/p(D+ IE-) with p a probability, D+: positive case, E+: living close to the dam and E-: living away from the dam. The "prevalence fraction (exposed)" was calculated using the relation that PrFe = (PR-1)/PR. The PrFe expresses the proportion of disease in exposed individuals that is due to the exposure, assuming that the relationship is causal. Alternatively, the indicator can be viewed as the proportion of infections in the exposed group that would be avoided if the exposure were removed.

Ethical Considerations

Ethical approval for this study was obtained from Jimma University Research and Ethics Committee. Communal consent was first obtained then children were recruited after informed oral consent was sought from their parents or guardians of each child before a child was enrolled in the study.

Of the 1,855 children below the age of 10 years examined in this study, 905 (48.8%) were girls and 950 (51.2%) were boys. The mean age of children was 4.7 years and the number of children surveyed from "at-risk" and "control" communities was 1,081 (58.3%) and 774 (41.7%), respectively. Of the children in "at-risk" communities, 528 (48.8%) were boys and 553 (51.2%) girls while in "control" communities, 377 (48.7%) were boys and 397 (51.3%) were girls. Overall, 194 (10.5%) of the sampled children were positive for malaria, of which, 117 (60.3%) were positive for *P. vivax*, 76 (39.2%) for *P. falciparum* and one (0.5%) for both *P. vivax* and *P. falciparum*.

Highest *P. vivax* (60.7%) and *P. falciparum* (57.9%) positivity rates were observed in October. The *P. vivax* prevalence varied from 5.9% in children<1year of age to

6.4% in those 5–9 years of age. The *P. falciparum* prevalence varied from 4.2% in children<1 year of age to 3.8% in those 5–9 years of age.

Table 1 shows demographic, distance and temporal relationships with malaria infection. The monthly *P. vivax* point prevalence during the 3 months ranged from 0.8% to 10.0% and form 2.3% to 5.9% in "at-risk" and "control" villages, respectively. Monthly *P. falciparum* point prevalence during the 3 months ranged from 2.7% to 6.9% and from 1.2% to 4.0% in "at-risk" and "control" villages, respectively (Table 1). The peak prevalence rate for *P. vivax* was observed in October and gradually decreased during November-December, while the prevalence rate for *P. falciparum* showed a late increase in December (Figure 1).

Table 1. Demographic, distance and temporal relationships with malaria infection, *Plasmodium vivax (Pv)* and *Plasmodium falciparum (Pf)*, in Gilgel-Gibe dam area, south-western Ethiopia, 2005.

	Pv		Pf	
Variable	**Rate**	**Crude OR (95% CI)**	**Rate**	**Crude OR (95% CI)**
Age (years)				
<I 'control'	4/96 (4.2%)	1	3/96 (3 . 1%)	1
'at-risk'	13/190 (6.8%)	1.69 (0.52,5.52)	9/190 (4.7%)	1.54 (0.49,4.79)
1-4 'control'	18/396 (4.5%)	1	13/396 (3.3%)	1
'at-risk'	34/429 (7.9%)	1.81 (1.21,2.71)**	23/429 (5.4%)	1.67 (1.42,6.66)
S-9 'control'	12/282 (4.3%)	1	1 /282 (0.4%)	1
'at-risk'	36/462 (7.8%)	1.9 (0.76,4.77)	27/462 (5.8%)	17.4 (1.22,249.24)**
Village/ groups'control	34/774 (4.4%)	1	17/774 (5.4%)	1
'at-risk'	8311081 (7.7%)	1.81 (1.17,2.79)**	59/ 1 081 (2.2%)	2.57 (1.0 I ,6.57)**
Month				
October 'control'	1 5/253 (5.9%)	1	10/253 (4.0%)	1
'at-risk'	56/559 (1 0.0%)	1.76 (0.88,3.53)*	34/559 (6.1 %)	1.57 (0.32,7.71)
November 'control'	13/260 (5.0%)	1	3/260 (1.2%)	1
'at-risk'	25/262 (9.5%)	2.00 (1.38,2.92)**	7/262 (2.7%)	2.35 (0. 17,32.73)
December 'control'	6/261 (2.3%)	1	4/261 (1.5%)	1
'at-risk'	2/260 (0.8%)	0.33 (0.02,4.96)	18/260 (6. 9%)	4.78 (1.03,22.23) **

* = significant at 0.1 level ** = significant at 0.05 level

The *P. vivax* prevalence was significantly higher in "at-risk" communities compared to the "control" communities in November (OR = 2.00, 95% CI = 1.38, 2.92) and the *P. falciparum* prevalence was significantly higher in "at-risk" communities in December (OR = 4.78, 95% CI = 1.03, 22.23) (Table 1). Differences between the two communities in malaria globally (*P. vivax* and *P. falciparum* together) were obvious and statistically significant in all months ($p < 0.01$).

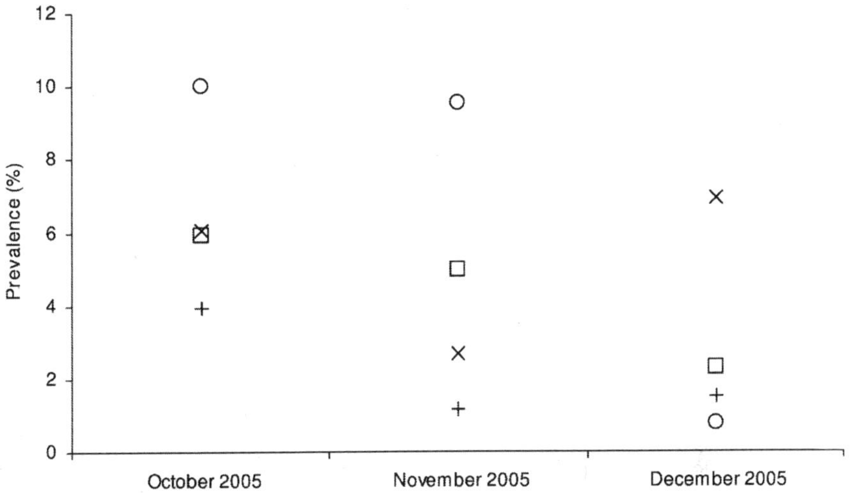

Figure 1. Prevalence rates for *P. vivax* in "at-risk" (squares) and "control" villages (circles) and for *P. falciparum* in "at-risk" (x-signs) and "control" villages (+-signs).

In general, significantly higher malaria prevalence was observed in children living within 3 km from the reservoir than those living farther away (OR = 1.81, 95% CI = 1.17, 2.79 for *P. vivax* and OR = 2.57, 95% CI = 1.01, 6.57 for *P. falciparum*) (Table 1). *P. vivax* prevalence rates differed significantly between "at-risk" and "control" communities among children 1–4 years of age (OR = 1.81, 95% CI = 1.21, 2.71) and *P. falciparum* prevalence rates differed significantly between "at risk" and "control" communities among 5–9 years of age (OR = 17.4, 95% CI = 1.22, 249.24).

Moreover, in a multivariate analysis controlling for age, sex, and time of data collection, it appeared that children who resided in "at-risk" villages close to the dam were more likely to have a *P. vivax* infection than children who resided in "control" villages (OR = 1.63, 95% CI = 1.15, 2.32) (Table 2). The results in Table 2 further indicate a *P. vivax* infection difference between boys and girls, however this was not significant at the 0.05 level (p = 0.054). The multivariate analysis indicated that, while controlling for age, sex, and time of data collection, children who resided in "at-risk" villages close to the dam had a higher OR to have *P. falciparum* infection than children who resided in "control" villages but this was not significant at the 0.05 level (OR = 2.40, 95% CI = 0.84, 6.88). Finally, while controlling for age, sex, and time of data collection, children who resided in "at-risk" villages close to the dam were at a higher risk to have a Plasmodium infection (*P. falciparum* and *P. vivax* combined) than children who resided in "control" villages (OR = 1.97, 95% CI = 1.24, 3.12) (Table 2).

Table 2. Adjusted odds ratios (ORs) using a design-based logistic regression of malaria infection for *Plasmodium vivax* (Pv) and *Plasmodium falciparum (Pf)* by age, gender, month, and village of residence in Gilgel-Gibe dam area, south-western Ethiopia, 2005.

Variable		Adjusted OR Pv	p-value	Adjusted OR Pf	p-value	Adjusted OR Plas-modium postivity	p-value
Village	'at risk'	1.63	0.015 **	2.40	0.085 *	1.97	0.013**
	'control'	1.00	..	1.00.	..	1.00	..
Month	October	1.00	..	1.00	..	1.00	..
	November	0.88	0.428	0.39	0. 199	0.68	0.200
	December	0. 18	0.096 *	0.89	0.828	0.41	0.062*
Age (yrs)	<I	1.00	..	1.00	..	1.00	..
	1-4	1.19	0.209	1.17	0.241	1.20	0.083*
	5-9	1.15	0.710	.94	0.890	1.08	0.837
Sex	Male	1.00	..	1.00	..	1.00	..
	Female	1.79	0.054	0.87	0.541	1.33	0. 146

* = significant at 0.1 level ** = significant at 0.05 level

Figure 2 shows the classification tree for *P. vivax* reproduced by CART. The children are first split into two groups: those sampled in December (prevalence = 1.5%) and those sampled in October–November (prevalence = 8.2%). The group of children sampled in October–November was further split in children living in "at-risk" communities

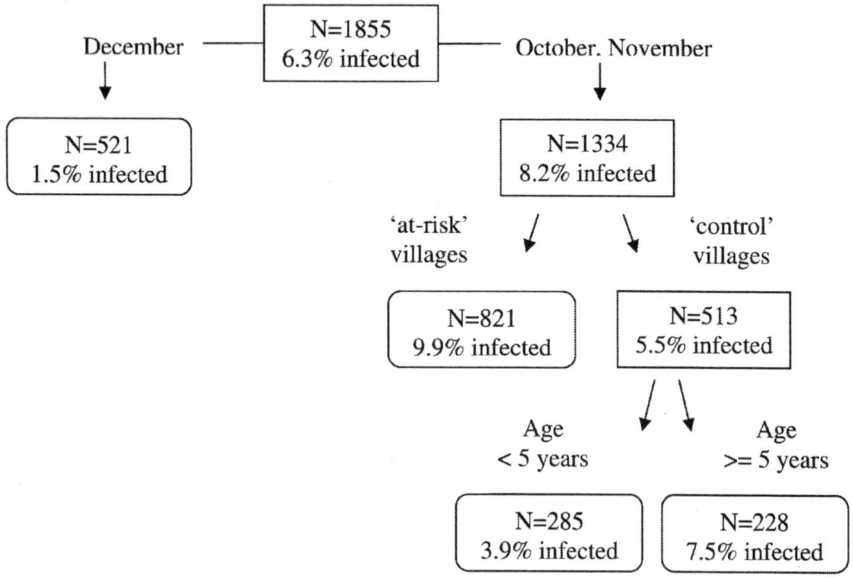

Figure 2. Classification tree of the risk factors for *P. vivax* infection.

(prevalence = 9.9%) and those living in "control" communities (prevalence = 5.5%). The group of children living in "control" communities was further split in children of age below 5 years (prevalence = 3.5%) and above 5 years (prevalence = 7.5%). According to the overall discriminatory power (i.e., the relative importance) in the CART analysis, month emerged as the strongest overall discriminating risk factor for a *P. vivax* infection (Score (Sc) = 100), followed by village type (Sc = 20.21) and age (Sc = 8.73) and sex (Sc = 2.19). The classification tree corresponds well with the *P. vivax* trends in Figure 1. Indeed, the trends show that in December the *P. vivax* prevalences in "at-risk" and "control" communities are both low and that the difference between "at-risk" and "control" communities are especially clear in October-November.

Figure 3 shows the classification tree for *P. falciparum* reproduced by CART. Children are first split into children living in "at-risk" communities (prevalence = 5.5%) and those living in "control" communities (prevalence = 2.2%). The group of children living in "at-risk" communities was further split in children sampled in November (prevalence = 2.7%) and those sampled in October and December (prevalence = 6.3%). According to the overall discriminatory power in the CART analysis, village type emerged as the strongest overall discriminating risk factor for malaria *P. falciparum* infection (Sc = 100), followed by month (Sc = 41.3). The other variables, age, and sex had a zero-Score. The classification tree corresponds well with the *P. falciparum* trends in Figure 1. The trends show that *P. falciparum* prevalences are lower in "control" communities and that in "at-risk" communities the prevalences were lower in November.

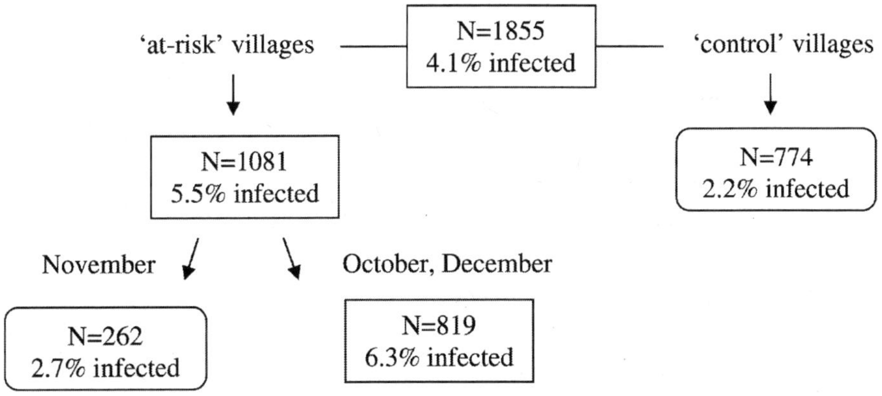

Figure 3. Classification tree of the risk factors for *P. falciparum* infection.

The prevalence fraction (exposed) PrFe, for malaria (*P. vivax* and *P. falciparum* together), measuring the effect of the dam, was calculated using PR p(D+ IE+)/p(D+ IE-) = 7.7/4.4 = 1.75 and is PrFe = (PR-1)/PR. = 0.43. This means that 43% of the malaria occurring in children can be attributed to the dam, assuming that the relationship is causal.

DISCUSSION

In this study, after controlling for age, sex, and time, the prevalence of *P. vivax* (7.7%) in children who reside within 3 km of the reservoir created by the Gilge-Gibe hydroelectric dam was significantly higher than in children living in more distant villages (4.4%) although the villages have a similar eco-topography. *Plasmodium falciparum* prevalence rates in "at-risk" communities (5.4%) were higher than in "control" communities (2.2%), but the difference showed no statistical significance (p-value = 0.085). Overall, the *Plasmodium* prevalence near the reservoir was statistically higher as compared to the *Plasmodium* prevalence in more distant communities (p-value = 0.013). The main reason for the higher prevalence of malaria among children living close to the reservoir may be due to the man-made ecological transformations, which may influence the presence of mosquito-breeding site and might have an impact on the behavior, parity rate and longevity of malaria vectors of the study area. A similar study in Cameroon showed a malaria prevalence of 36% in residents living in close proximity to a man-made lake compared with a prevalence of 25% in a village 14 km away [36]. In India, a 2.4-fold increase in malaria cases and an over 4-fold increase in annual parasite incidence were recorded among children in villages close to a reservoir as compared to more distant villages [37]. A high malaria prevalence, up to 47%, was recorded around the Mantali dam in Senegal, constructed to provide hydropower and irrigation, compared to prevalences of 27.3% and 29.6% in two communities downstream the dam [38]. Risk factors for malaria infections in the Gilgel-Gibe area might be proximity to the dam, as the low socioeconomic status, the health infrastructure and the malaria control methods appear to be similar in "at-risk" and "control" communities.

This study indicates that children between the age of 1 and 4 years tend to have a higher malaria prevalence than children below the age of 1 year (p-value = 0.083, non-significant for *P. vivax* and *P. falciparum* together). This could be because the older children, in contrast to younger children, spend outdoors in the evening when peak biting activities of malaria vector mosquitoes are high or the greater use of anti-malarial drugs in early childhood [39]. A similar study in Gabon showed lower malaria prevalence in children less than 6 months (3.7%) than in children at the age of 47 months (47.5%) [40], which was attributed to low number of children less than 1 year of age and immunity acquired from mothers as difference in the risk of infection among different age groups could be associated to differences in the immunological status. The risk of infection first increases with age and then decreases when the individual reaches a certain degree of immunity due to exposure to the parasite. This was indicated in the study reported in this chapter as well. *Plasmodium* prevalence rates in children between 1 and 4 years, below 1 year and children age of 5–9 years showed no statistically significant differences.

According to several reviews, *P. falciparum* is the dominant species in Ethiopia, followed by *P. vivax*, accounting for 60% and 40% of all malaria cases, respectively [16, 41]. In the present study, the predominant species was *P. vivax* followed by

P. falciparum. *Plasmodium vivax* was found in 117 (60.3%) children, *P. falciparum* in 76 (39.2%) and mixed infection in one (0.5%) child. The other two *Plasmodium* species, *P. malariae* and *P. ovale* were absent. A similar distribution (69% *P. vivax* and 31% *P. falciparum*) was reported in a previous study [42]. But, in central Ethiopia, Woyessa et al. [43] reported the predominance of *P. falciparum* during October while *P. vivax* tends to dominate during November. A parasitological community-based study conducted by Gebreyesus et al. [20] on the impact of small irrigation dams on malaria burden in northern Ethiopia also revealed a predominance of *P. falciparum*. The prevalence of malaria infections varies seasonally, with *P. vivax* dominating in the dry season (March–June) and *P. falciparum* peaking in September–October, after the end of the main rainy season [16]. Hence, the proportion of malaria cases due to the two parasite species can vary across seasons and localities. Ramos et al. [44] reported variability in the distribution of malaria parasites (22.4–54.7% *P. vivax* and 40.9–73.4% of *P. falciparum*) during different seasons.

The classification trees show that using this non-parametric technique allows obtaining a better insight in the data structure and the available interactions between determinants in their influence on (or relation with) malaria. This was also noted by Thang et al. [30]. The classification tree results correspond well with the graphical trend observations, indicating that for *P. vivax*, children can be grouped according to month and children sampled in October–November showed higher prevalences even more when children were living in "at-risk" communities (prevalence = 9.9%). For *P. falciparum*, the children living in "at-risk" communities were grouped together because of higher prevalences. Within "at-risk" communities especially children sampled in October and December showed a higher prevalence of 6.3%.

In conclusion, this study informs that children living in close proximity to the reservoir created by the newly constructed Gilgel-Gibe dam are at a greater risk of Plasmodium infection than children living further away, possibly due to the creation of new vector habitats around the lakeshore. Epidemiological studies focusing on vector dynamics and socioeconomic, demographic and health behavior factors could be conducted to identify underlying causes of the spatial pattern of infection reported in this chapter.

Recommendations

In order to maximize the economic benefits generated by Gilgel-Gibe hydroelectric dam, it is recommended that preventive programs against malaria and other vector-borne diseases be implemented along the periphery of the reservoir. Health Package program, including bed net use and health education, early diagnosis and treatment, residual spraying and environmental management be implemented in an integrated way and strengthened to reduce disease burden from vector-borne diseases in communities living in close proximity to the new reservoir.

KEYWORDS

- **Gilgel-Gibe hydroelectric dam**
- **Malaria**
- **Parasitological Investigation**
- **Prevalence ratio**
- **Socio-demographic data**

AUTHORS' CONTRIBUTIONS

Delenasaw Yewhalaw conceptualized the study design, was involved in the coordination, supervision of data collection, data entry, cleaning, analysis, and drafted the manuscript; Worku Legesse was involved in the design of the survey and reviewed the manuscript; Wim Van Bortel contributed to the discussion and critically reviewed the manuscript; Solomon Gebre-Selassie was involved in the supervision of the laboratory work and in drafting the manuscript; Helmut Kloos contributed to the study design, and reviewed the manuscript; Luc Duchateau was involved in the interpretation of the statistical analysis and reviewed the manuscript; Niko Speybroeck performed the statistical analysis, interpretation and was involved in drafting and revising the manuscript. All authors read and approved the final manuscript.

ACKNOWLEDGMENTS

The authors wish to thank the parents and guardians of children for giving consent to involve their children in the study. We are grateful to the peasant association leaders and health workers in Kerssa, Tiro-Afeta and Omo-Nada districts for their cooperation. Data collectors and laboratory technicians involved in the study are also acknowledged. Financial support for the study was obtained from the Ethiopian Science and Technology Commission, the Flemish Interuniversity Council (VLIR), and the Research and Publication Office of Jimma University.

COMPETING INTERESTS

The authors declare that they have no competing interests.

Chapter 13

Exploring the Molecular Basis of Insecticide Resistance in the Dengue Vector *Aedes aegypti*

Sébastien Marcombe, Rodolphe Poupardin, Frederic Darriet, Stéphane Reynaud, Julien Bonnet, Clare Strode, Cecile Brengues, André Yébakima, Hilary Ranson, Vincent Corbel, and Jean-Philippe David

INTRODUCTION

The yellow fever mosquito *Aedes aegypti* is a major vector of dengue and hemorrhagic fevers, causing up to 100 million dengue infections every year. As there is still no medicine and efficient vaccine available, vector control largely based on insecticide treatments remains the only method to reduce dengue virus transmission. Unfortunately, vector control programs are facing operational challenges with mosquitoes becoming resistant to commonly used insecticides. Resistance of *Ae. aegypti* to chemical insecticides has been reported worldwide and the underlying molecular mechanisms, including the identification of enzymes involved in insecticide detoxification are not completely understood.

The present chapter investigates the molecular basis of insecticide resistance in a population of *Ae. aegypti* collected in Martinique (French West Indies). Bioassays with insecticides on adults and larvae revealed high levels of resistance to organophosphate and pyrethroid insecticides. Molecular screening for common insecticide target-site mutations showed a high frequency (71%) of the sodium channel "knock down resistance" (*kdr*) mutation. Exposing mosquitoes to detoxification enzymes inhibitors prior to bioassays induced a significant increased susceptibility of mosquitoes to insecticides, revealing the presence of metabolic-based resistance mechanisms. This trend was biochemically confirmed by significant elevated activities of cytochrome P450 monooxygenases, glutathione S-transferases, and carboxylesterases at both larval and adult stages. Utilization of the microarray *Aedes Detox* Chip containing probes for all members of detoxification and other insecticide resistance-related enzymes revealed the significant constitutive over-transcription of multiple detoxification genes at both larval and adult stages. The over-transcription of detoxification genes in the resistant strain was confirmed by using real-time quantitative RT-PCR.

These results suggest that the high level of insecticide resistance found in *Ae. aegypti* mosquitoes from Martinique island is the consequence of both target-site and metabolic based resistance mechanisms. Insecticide resistance levels and associated mechanisms are discussed in relation with the environmental context of Martinique Island. These finding have important implications for dengue vector control in Martinique and emphasizes the need to develop new tools and strategies for maintaining an effective control of *Aedes* mosquito populations worldwide.

Every year, 50–100 million dengue infections world-wide causing from 20,000 to 25,000 deaths from dengue and hemorrhagic fever are recorded [1]. As there is still no medicine and efficient vaccine available, vector control by the recourse of environmental management, educational programs, and the use of chemical and biological agents, remains the only method to reduce the risk of dengue virus transmission [1]. Unfortunately, most of dengue vector control programs implemented worldwide are facing operational challenges with the emergence and development of insecticide resistance in *Ae. aegypti* [2] and *Ae. albopictus* [3]. Resistance of *Ae. aegypti* to insecticides has been reported in many regions including Southeast Asia [4, 5], Latin America, [6] and the Caribbean [7].

Inherited resistance to chemical insecticides in mosquitoes is mainly the consequence of two distinct mechanisms: the alteration of target sites inducing insensitivity to the insecticide (target-site resistance) and/or an increased metabolism of the insecticide (metabolic-based resistance) [8]. Metabolic-based resistance involves the biotransformation of the insecticide molecule by enzymes and is now considered as a key resistance mechanism of insects to chemical insecticides [8, 9]. This mechanism may result from two distinct but additive genetic events: (i) a mutation of the enzyme protein sequence leading to a better metabolism of the insecticide, and/or (ii) a mutation in a non-coding regulatory region leading to the over-production of an enzyme capable of metabolizing the insecticide. So far, only the second mechanism has been clearly associated with the resistant phenotype in mosquitoes. Three large enzyme families, the cytochrome P450 monooxygenases (P450s), glutathione S-transferases (GSTs) and carboxy/cholinesterases (CCEs) have been implicated in the metabolism of insecticides [8, 10-12]. The rapid expansion and diversification of these so-called "detoxification enzymes" in insects is likely to be the consequence of their adaptation to a broad range of natural xenobiotics found in their environment such as plant toxins [13]. These enzymes have also been involved in mosquito response to various anthropogenic xenobiotics such as heavy metals, organic pollutants, and chemical insecticides [14-16].

Although identifying metabolic resistance is possible by toxicological and biochemical techniques, the large panel of enzymes potentially involved together with their important genetic and functional diversity makes the understanding of the molecular mechanisms and the role of particular genes a challenging task. As more mosquito genomes have been sequenced and annotated [17, 18], the genetic diversity of genes encoding mosquito detoxification enzymes has been unraveled and new molecular tools such as the *Aedes* and *Anopheles* "detox chip" microarrays allowing the analysis of the expression pattern of all detoxification genes simultaneously have been developed [19, 20]. These specific microarrays were successfully used to identify detoxification genes putatively involved in metabolic resistance in various laboratory and field-collected mosquito populations resistant to insecticides [19-24].

In Latin America and the Caribbean, several *Ae. aegypti* populations show strong resistance to pyrethroid, carbamate, and organophosphate insecticides correlated with elevated activities of at least one detoxification enzyme family [25-28]. In addition, several points of non-synonymous mutations in the gene encoding the trans-membrane voltage-gated sodium channel (*kdr* mutations) have been described and showed to confer resistance to pyrethroids and DDT [27, 29].

Several questions remain concerning the impact of insecticide resistance on the efficacy of vector control operations. In Martinique (French West Indies), high levels of resistance to the organophosphate temephos and the pyrethroid deltamethrin were reported. This resistance was characterized by an important reduction of both mosquito knock-down and mortality levels after thermal-fogging with deltamethrin and P450-inhibitor synergized pyrethroids, indicating that resistance was negatively impacting on control programs and that this resistance was conferred, at least in part, by elevated cytochrome P450 activity [30].

In this study, we explored the mechanisms conferring insecticide resistance in an *Ae. aegypti* population from Martinique island. Larval bioassays and adult topical applications were used to determine the current resistance level of this population to insecticides. The presence of metabolic-based resistance mechanisms was investigated by exposing mosquitoes to enzyme inhibitors prior to bioassays with insecticides and by measuring representative enzyme activities of each detoxification enzyme family. At the molecular level, the frequency of the target-site *kdr* mutation was investigated and a microarray approach followed by quantitative real-time RT-PCR validation was used to identify detoxification genes putatively involved in metabolic resistance. Results from this study will help to implement more effective resistance management strategies in this major disease vector in the future.

Larval bioassays (Table 1) showed that the Vauclin strain is far less affected by temephos than the susceptible Bora-Bora strain (RR_{50} of 44-fold and RR_{95} of 175-fold). In the susceptible strain, temephos toxicity was not significantly increased in the presence of detoxification enzyme inhibitors (PBO, DEF, and DMC). By contrast, the level of resistance to temephos of the Vauclin strain was significantly reduced in the presence of PBO, DEF, and DMC (from 175 to 60, 44 and 109-fold respectively for RR_{95}) indicating the involvement of P450s, CCEs and in a lesser extent GSTs in the resistance of larvae to temephos.

Table 1. Insecticidal activity of temephos with and without enzyme inhibitors on larvae of *Aedes aegypti* Vauclin and Bora-Bora strains.

Strain	Enzyme inhibitor	Slope (± SE)	LC_{50} (j.tg/L) (95% CI)	LC_{95} (j.tg/L) (95% CI)	RR_{50} (95% CI)	RR_{95} (95% CI)	SR_{50} (95% CI)	SR_{95} (95% CI)
Bora-Bora	–	8.49 (0.45)	3.7 (3.6-3.8)	5.7 (5.5-6)	–	–	–	–
	PBO	8.28 (0.67)	4.2 (4-4.4)	6.7 (6.4-7)	–	–	0.87 (0.74-1 .03)	0.87 (0.74-1 .03)
	DEF	8.13 (0.44)	3.3 (3.2-3.4)	5.3 (5. 1-5.6)	–	–	1.10 (0.98-1 .24)	1.10 (0.98-1 .24)
	DMC	11.16 (0.54)	4.3 (4.2-4.4)	6.0 (5.8-6.2)	–	–	0.86 (0.79-0.94)	0.96 (0.81-1.14)
Vauclin	–	2.08 (0.08)	160 (150-180)	1000 (870-1180)	44 (40-48)	175 (150-205)	–	–
	PBO	3.60 (0.24)	140 (130-150)	400 (360-450)	33 (29-38)	60 (51-71)	1.16 (1.05-1.29)	2.52 (2.16-2. 95)

Table 1. *(Continued)*

Strain	Enzyme inhibitor	Slope (± SE)	LC_{50} (j.tg/L) (95% CI)	LC_{95} (j.tg/L) (95% CI)	RR_{50} (95% CI)	RR_{95} (95% CI)	SR_{50} (95% CI)	SR_{95} (95% CI)
	DEF	3.00 (0. 16)	68 (64-72)	240 (210-270)	21 (18-22)	44 (38-52)	2.37 (2.18-2.57)	4.27 (3.64-5)
	DMC	2.05 (0. 11)	103 (92-110)	650 (560-790)	24 (22-27)	109 (92-129)	1.57 (1.39-1.79)	1.57 (1.39-1. 79)

Resistant ratios RR50 and RR95 were obtained by calculating the ratio between the LC50 and LC95 between Vauclin and Bora-Bora strains; Synergism ratios SR50 and SR95 were obtained by calculating the ratio between LC50 and LC95 with and without enzyme inhibitor. (CI): Confidence Interval. Significant RR and SR are shown in bold.

Topical applications of the pyrethroid insecticide deltamethrin on adults of each strain (Table 2) revealed that the Vauclin strain is also highly resistant to deltamethrin (RR_{50} of 56-fold and RR_{95} of 76-fold). In both strains, the toxicity of deltamethrin increased significantly in the presence of detoxification enzyme inhibitors, however only PBO and DMC induced higher synergistic effects in the Vauclin strain than in the susceptible Bora-Bora strain (S_{R50} of 9.94 and 3.76 respectively). In the Vauclin strain, PBO and DMC significantly reduced the resistance level (from 76-fold to 41-fold and 43-fold respectively for RR_{95}), indicating a significant role of P450s and GSTs in the resistance of adults to deltamethrin.

Table 2. Insecticidal activity of deltamethrin with and without enzyme inhibitors on adults of *Aedes aegypti* Vauclin and Bora-Bora strains.

Strain	Enzyme inhibitor	Body weight (mg)	Slope (± SE)	LD_{50} (μg/L) (95% CI)	LD_{95} (μg/L) (95% CI)	RR_{50} (95% CI)	RR_{95} (95% CI)	SR_{50} (95% CI)	SR_{95} (95% CI)
Bora-Bora	–	2.12	3.31 (0.27)	18 (16-19)	55 (47-69)	–	–	–	–
	PBO	2.27	3.65 (0.34)	3.4 (3 . 1-3.7)	9.5 (8. 1-12. 1)	–	–	5.2 (4.52-5.98)	5.79 (4.30-7.81)
	DEF	2.44	2.41 (0.27)	3.4 (3-3.9)	16 (12-25)	–	–	5.12 (4.48-5.86)	3.35 (2.42-4.64)
	DMC	2.39	2.94 (0.22)	7.3 (6.6-8.1)	27 (22-34)	–	–	2.41 (2.11-2.76)	2.09 (1.57-2. 78)
Vauclin	–	2.65	2.61 (0. 19)	990 (880-1100)	4210 (3470-5380)	56 (49-64)	76 (58-99)	–	–
	PBO	2.27	2.78 (0.17)	99 (91-108)	390 (330-470)	29 (26-33)	41 (31-53)	9.94 (8. 79-1 1.23)	10.89 (8.64-13. 72)
	DEF	2.25	2. 14 (0.22)	170 (150-190)	1000 (750-1510)	49 (43-56)	60 (43-86)	5.81 (5.08-6.65)	4.23 (3.16-5.66)
	DMC	2.56	2.57 (0. 16)	260 (240-290)	1150 (950-1460)	36 (32-40)	43 (33-57)	3.76 (3.35-4.23)	3.68 (2.86-4. 72)

Resistant ratios RR_{50} and RR_{95} were obtained by calculating the ratio between the LD_{50} and LD_{95} between Yauclin and Bora-Bora strains; Synergism ratios SR_{50} and SR_{95} were obtained by calculating the ratio between LD_{50} and LD_{95} with and without enzyme inhibitor. (CI): Confidence Interval. Significant RR and SR are shown in bold.

Comparison of constitutive detoxification enzyme activities between the susceptible strain Bora-Bora and the insecticide-resistant Vauclin strain revealed significant differences at both larval and adult stages (Figure 1). The P450 activities were elevated in both larvae and adults of the Vauclin strain (1.57-fold and 1.78-fold respectively with

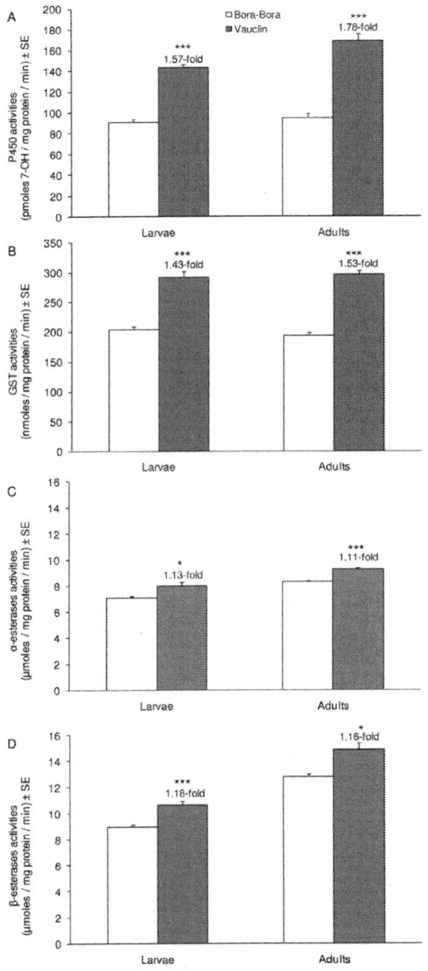

Figure 1. Comparison of detoxification enzymes activities between the insecticide-resistant strain Vauclin and the susceptible strain Bora-Bora. (A) The P450 activities were measured with the ECOD method [63] on 20 µg microsomal proteins after 15 min and expressed as pmol of 7-OH produced/mg microsomal protein/min (± SE). (B) The GST activities were measured with the CDNB method [64] on 200 µg cytosolic proteins during 1 min and expressed as nmol of conjugated CDNB/µg protein/min (± SE). α-esterase (C) and β-esterase (D) activities were measured with the naphthyl acetate method [65] on 30 µg cytosolic proteins after 15 min and expressed as µmol α- or β-naphthol produced/mg protein/min (± SE). For each strain and each life stage, three independent biological replicates were analyzed and measures were repeated 15, 15 and 30 times for P450, GST, and esterase activities respectively. Statistical comparison of enzyme activities between the Vauclin and Bora-Bora strains were performed at each life stage separately with a Mann and Whitney's test (* $p < 0.05$, ** $p < 0.01$, *** $p < 0.001$).

$P < 0.001$ at both life stages). Similarly, GST activities were found elevated in larvae and adults of the Vauclin strain (1.43-fold and 1.53-fold respectively with $P < 0.001$ at both life stages). Finally, α- and β-carboxylesterase activities were also found slightly elevated in the Vauclin strain in larvae (1.13-fold and 1.18-fold with $P < 0.05$ and $P < 0.001$ respectively) and adults (1.11-fold and 1.16-fold with $P < 0.001$ and $P < 0.05$ respectively).

Sequencing of the voltage-gated sodium channel gene conducted on the Vauclin strain showed the presence of the *kdr* mutation at position 1016 (GTA to ATA) leading to the replacement of valine by an isoleucine (V1016Ile) at a high allelic frequency (f(R) = 0.71, n = 24) with RR = 12, RS = 11 and SS = 1. Conversely, no *kdr* resistant allele was detected in the susceptible Bora-Bora strain (n = 30).

We used the microarray "*Aedes Detox Chip*" (Strode et al., 2007) to compare the transcription levels of all *Ae. aegypti* detoxification genes between the insecticide-resistant strain Vauclin and the susceptible strain Bora-Bora in larvae and adults. Overall, 224 and 214 probes out of 318 were detected consistently in at least three hybridizations out of six in larvae and adults respectively. Among them, 31 detoxification genes were significantly differentially transcribed (transcription ratio > 1.5-fold in either direction and corrected P value < 0.01) in larvae or adults (Figure 2). Most of these genes encode P450s (*CYPs*) with 4 of them being differentially transcribed in the Vauclin strain at both life stages (*CYP9J22, CYP6Z6, CYP6M6 and CYP304C1*).

In larvae, 18 genes (15 *CYPs*, 1 *GST*, and 2 *CCEs*) were found significantly differentially transcribed between the insecticide-resistant strain Vauclin and the susceptible strain Bora-Bora (Figure 2A). Among them, 14 genes were over-transcribed in the Vauclin strain while only four genes were under-transcribed. Most over-transcribed genes were represented by *CYP* genes with a majority belonging to the *CYP6* subfamily (*CYP6BB2, CYP6M6, CYP6Y3, CYP6Z6, CYP6M10,* and *CYP6AA5*). Three *CYP9s* were also over-transcribed in larvae of the Vauclin strain (*CYP9J23, CYP9J22,* and *CYP9J9*) with a strong over-transcription of *CYP9J23* (5.3-fold) together with 2 *CYP4s* (*CYP4J15* and *CYP4D23*). Among other over-transcribed genes, 2 carboxy/cholinesterases (*CCEunk7o* and *CCEae2C*) and 1 glutathione *S*-transferase (*AaGSTE7*) were slightly over-transcribed in the Vauclin strain. Lastly, four *CYPs* (*CYP9M9, CYP9J20, CYP304C1* and *CYP6AG8*) were under-transcribed in insecticide-resistant larvae comparatively to susceptible Larvae.

In adults, 18 genes (12 *CYPs*, 1 *GST*, 3 *CCEs*, and 2 *Red/Ox*) were found differentially transcribed in the insecticide-resistant strain Vauclin comparatively to the susceptible strain Bora-Bora (Figure 2B). As in larvae, most of the over-transcribed genes belong to the *CYP6* and *CYP9* subfamilies (*CYP6CB2, CYP6M11, CYP6Z6, CYP6M6* and *CYP9J22, CYP9M9, CYP9J6*) with only two additional CCEs (*CCEae3A* and *CCEae4B*) being moderately over-transcribed in the Vauclin strain. Nine genes were under-transcribed in Vauclin adults, including five *CYPs* (*CYP304C1, CYP9M6, CYP325Q2, CYP325V1* and *CYP6P12*), 1 CCE (CCEunk6o), 1 GST (GSTS1-1) and 2 thioredoxin peroxidases (*TPx4* and *TPx3B*). Interestingly *CYP304C1* and *TPx4* were both found strongly under-transcribed (14.1 and 10.4-fold respectively) in insecticide-resistant adults.

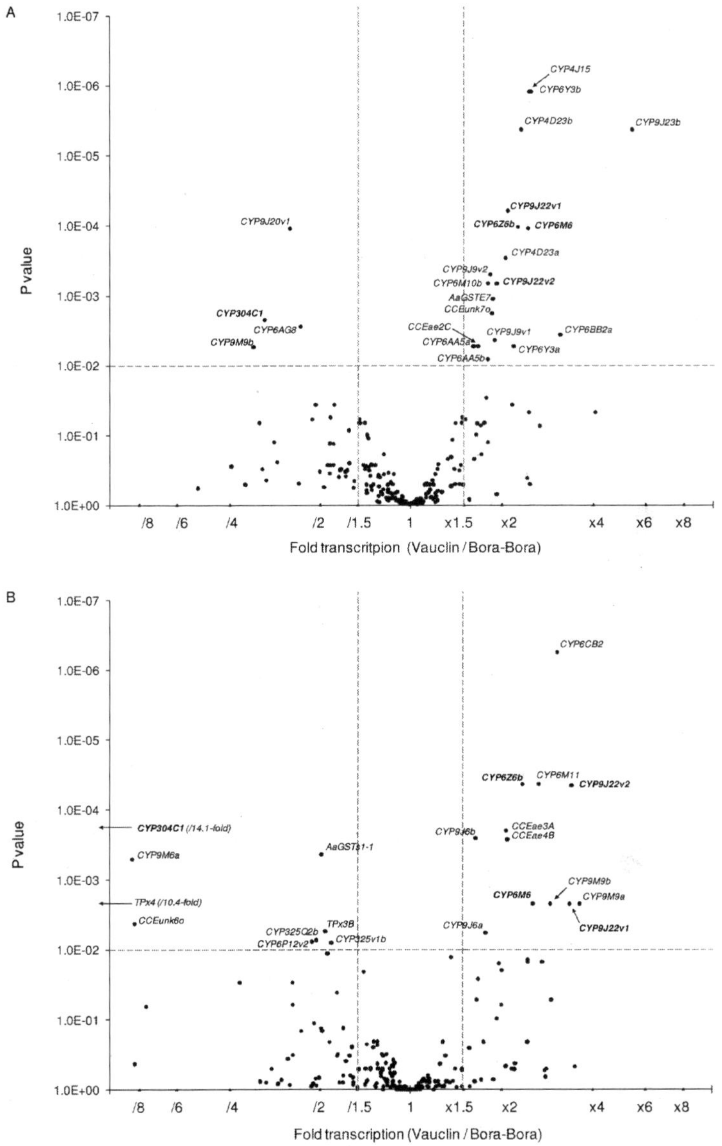

Figure 2. Microarray screening of detoxifications genes differentially transcribed in the insecticide-resistant strain Vauclin. Differential transcription of detoxification gens was investigated separately in 4th-stage larvae (A) and 3-days old adults (B). For each life stage, differences in gene transcription are indicated as a function of both transcription ratio (Vauclin/Bora-Bora) and ratio's significance (t-test P values). For each comparison, only probes showing consistent data in at least three hybridizations out of six were considered. Vertical lines indicate 1.5-fold transcription difference in either direction. Horizontal line indicates significance threshold ($p < 0.01$) adopted for the one sample t-test after Benjamini and Hochberg multiple testing correction procedure. Probes showing both more than 1.5-fold differential transcription and a significant P value are named. Probes that were found under- or over-transcribed in both larvae and adults are shown in bold. Suffixes *a* and *b* represent two different probes of the same gene while suffixes *v1* and *v2* represent two different alleles of the same gene.

Validation of microarray data was performed by real-time quantitative RT-PCR on 10 detoxification genes identified as over-transcribed in larvae or adults of the Vauclin strain (Figure 3). The over-transcription of genes identified from microarray experiments were all confirmed by quantitative RT-PCR in both life stages, although expression ratios obtained from RT-PCR were frequently higher than those obtained from microarray experiments.

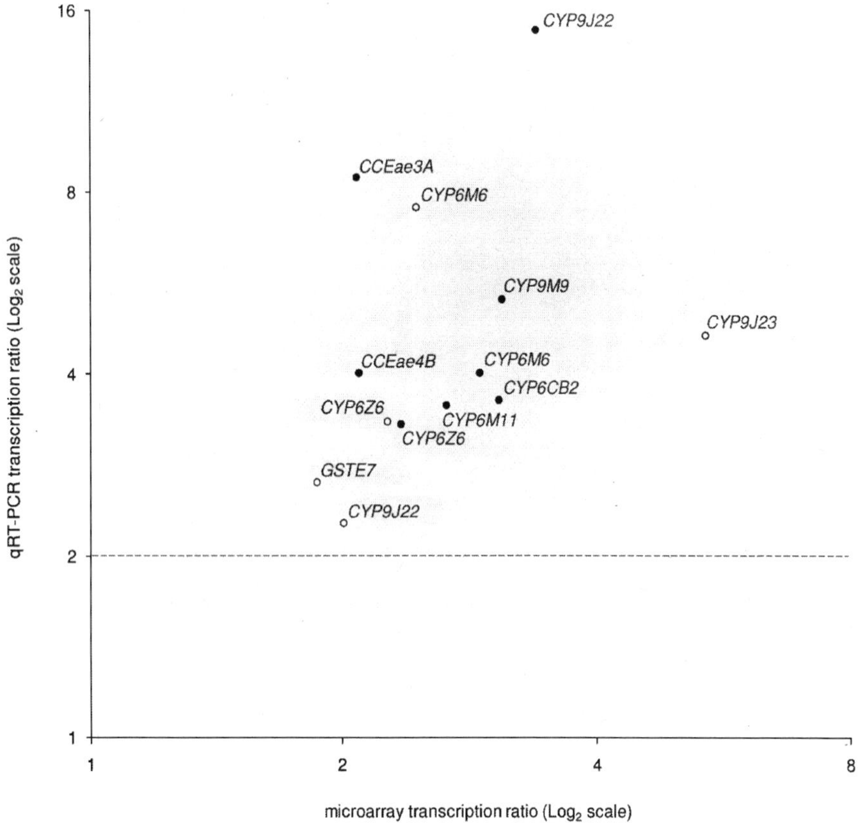

Figure 3. Real-time quantitative RT-PCR validation of microarray data. Validation of differential transcription between the two strains was performed on 11 selected genes in 4th-stage larvae (white dots) and 3-days old adults (black dots). Transcription ratios obtained from real-time quantitative RT-PCR experiments were normalized with the two housekeeping genes *AeRPL8* and *AeRPS7* and shown as mean value over 3 independent biological replicates.

DISCUSSION

The aim of the present study was to investigate insecticide resistance mechanisms of *Ae. aegypti* mosquitoes from Martinique (French West Indies).

Toxicological results confirmed the high level of resistance of the Vauclin strain from Martinique to the organophosphate temephos at the larval stage and to the pyrethroid deltamethrin at the adult stage [30]. The use of specific detoxification enzyme inhibitors suggested that resistance of larvae to temephos is linked to carboxylesterases and to a lesser extent P450s and GSTs. In adults, resistance to deltamethrin appeared principally linked to P450s and GSTs. Comparison of global detoxification enzyme activities between the two strains revealed elevated P450s, GSTs, and in a lesser extent CCEs activities in the Vauclin strain at both life-stages, confirming the importance of metabolic resistance mechanisms in Martinique.

Carboxylesterases based-resistance mechanism is a major mechanism for organophosphate resistance in insects [12]. Several examples of *Ae. aegypti* resistance to organophosphates in the Caribbean linked to elevated carboxylesterases activities have been described [25, 31]. Our toxicological and biochemical data confirms these observations despite a moderate elevated level of CCEs activities in the Vauclin strain. Among detoxification enzymes, P450s have been shown to play a major role in pyrethroid resistance in insects [8, 10, 32]. In Martinique, Marcombe et al. [30] suggested the involvement of P450s in the reduced efficacy of deltamethrin space-spray operations. Elevated GST levels have also been frequently associated with insect resistance to insecticides such as DDT and pyrethroids [33-35]. Our toxicological and biochemical data support the role of P450s and GSTs in insecticide resistance in Martinique.

At the molecular level, several mutations in the voltage-gated sodium channel gene have been associated with pyrethroid resistance in *Ae. aegypti* from Asian, Latin American, and Caribbean countries [27, 29, 36]. Our results revealed a high frequency (71%) of the V1016I *kdr* mutation in *Ae. aegypti* populations from the community of Vauclin. The role of this mutation in pyrethroid resistance was clearly demonstrated by genotype-phenotype association studies [37]. The high frequency of the mutation, together with the incomplete effect of enzyme inhibitors in adults, supports a contribution of this *kdr* mutation in deltamethrin resistance.

Acetylcholinesterase (AChE) is critical for hydrolysis of acetylcholine at cholinergic nerve synapses and is a target for organophosphate and carbamate insecticides [38]. Altered AchE is an important resistance mechanism to organophosphates in many insects. Following the methods of Alout et al. [39] and Bourguet et al. [40], the AChE activities of Vauclin mosquitoes were determined to investigate the presence of the G119S and/or F290V mutations. No insensitive AChE phenotypes were found in any of the mosquitoes tested (Corbel V., unpublished data), suggesting that organophosphate resistance of the Vauclin strain is rather due to detoxification enzymes unless other mutations occurred elsewhere in the Ace gene.

Our microarray screening identified 14 and nine over-transcribed detoxification genes in larvae and adults of the Vauclin strain respectively. Among them, four P450s (*CYP6M6, CYP6Z6, CYP9J23* and *CYP9J22*), the glutathione S-transferase *GSTe7* and the carboxy/cholinesterase *CCEae3A* were all confirmed to be over-transcribed at both life-stages, supporting their involvement in insecticide-resistance. Other genes appeared more highly over-transcribed in adults (*CYP9J22, CYP9M9, CYP6M11, CCEae3A*) or in larvae (*CYP6M6*), suggesting that particular enzymes might be more specifically involved in resistance to one insecticide during a particular life-stage as argued by Paul et al. [41]. Validation of transcription profiles by real-time quantitative RT-PCR was successful for the 10 genes tested although expression ratios obtained with RT-PCR were often higher. The underestimation of transcription ratios obtained from microarray data is likely due to technical issues and has been previously evidenced in other studies [14, 42].

Over-transcription of genes encoding P450s has been frequently associated with metabolic-based insecticide resistance mechanisms in insects [10]. In mosquitoes, the *CYP6Z* subfamily has been previously associated with response to pyrethroid, carbamates, and organochlorine insecticides. In *Ae. aegypti*, *CYP6Z9* has been found 4-fold over-transcribed in a permethrin-resistant strain collected in Northern Thailand [20]. In two recent studies, *CYP6Z8* was also identified as inducible by permethrin and other pollutants [14, 15]. In *An. gambiae*, *CYP6Zs* have been frequently found constitutively over-transcribed in permethrin- and DDT-resistant strains [19, 21, 43]. Recent studies demonstrated that the enzyme encoded by *An. gambiae CYP6Z1* can metabolize the insecticides carbaryl and DDT while *CYP6Z2* with a narrower active site, can only metabolize carbaryl [44, 45]. Recently, another *An. gambiae* P450 (*CYP6P3*), was shown to be able to degrade pyrethroid insecticides [22]. The over-transcription of *CYP6Z6* in the Vauclin strain may indicate the involvement of *Ae. aegypti CYP6Zs* in insecticide resistance in Martinique. However, the decisive demonstration of their capability to metabolize insecticides requires further investigations.

The association of *CYP6Ms* with metabolic resistance to pyrethroids has also been previously described in mosquitoes. In *Ae. aegypti* larvae, *CYP6M6* and *CYP6M11* were found inducible by permethrin and pollutants [14]. Although no *Aedes CYP6Ms* have been found constitutively over-transcribed in other insecticide-resistant strains, *An. gambiae CYP6M2* was found significantly over-transcribed in various strains resistant to pyrethroids [21, 46]. Recent studies indicate that *CYP6M2* is able to metabolize pyrethroid insecticides (Stevenson B. personal communication). Our results suggest that *Ae. aegypti CYP6M6* and *CYP6M11*, with protein sequences similar to *An. gambiae CYP6M2*, might also be involved in resistance of *Ae. aegypti* to pyrethroids in Martinique.

Finally, the glutathione S-transferase *GSTE7* and the carboxy/cholinesterase *CCEae3A* were both found over-transcribed in both life-stages of the Vauclin strain. The role GSTs in resistance to chemical insecticides has been previously evidenced in insects with the enzyme encoded by *An. gambiae GSTE2* metabolizing DDT [35, 47, 48] and the housefly *MdGST6-A* metabolizing two organophosphate insecticides [49]. In *Ae. aegypti, GSTE2* also metabolises DDT and is over-transcribed in a pyrethroid and DDT-resistant strain from Thailand [35]. In 2008, Strode et al. [20] also revealed the over-transcription of *GSTE7* in pyrethroid-resistant mosquitoes. Our results confirm that *GSTE7* might have a role in insecticide resistance in *Ae. aegypti.* Over-production of carboxylesterases has been showed to play an important role in resistance to organophosphate insecticides in mosquitoes [50-53]. Elevated esterase activities conferring resistance to organophosphate insecticides has usually been linked to genomic amplification of specific alleles although gene over-transcription may also be involved [12]. Considering the high resistance of larvae of the Vauclin strain to temephos, over-transcribed CCEs represent good candidates for organophosphate metabolism in *Ae. aegypti*.

It has been suggested that insecticide resistance could be accentuated by the exposure of mosquito populations to pollutants and pesticides used in agriculture [14, 15, 54, 55]. In Martinique, bananas, sugar cane, and pineapple represent important cultured surface areas often localized near mosquito breeding sites. These cultures have been submitted for decades to heavy use of insecticides such as the organochlorates aldrin, dieldrin and chlordecone and herbicides such as the triazine simazine, the pyridines paraquat, and glyphosate [56]. This particular situation is likely to have contributed to the high resistance of *Ae. aegypti* to chemical insecticides and to the selection of particular detoxification genes in Martinique.

MATERIAL AND METHODS

Mosquito Strains

Two strains of *Ae. aegypti* were used in this study. The susceptible reference Bora-Bora strain, originating from Bora-Bora (French Polynesia) is free of any detectable insecticide resistance mechanism. An *Ae. aegypti* colony was established from wild field-caught mosquito larvae collected from individual houses in the community of Vauclin in Martinique (Vauclin strain). Larvae and adults obtained from the F1 progeny were used for bioassays, biochemical, and molecular studies.

Insecticides and Detoxification Enzyme Inhibitors

Two technical grade compounds were used, representing organophosphate and pyrethroid classes of insecticides, temephos (97.3%; Pestanal™, Riedel-de-Haën, Seelze, Germany) and deltamethrin (100%; AgreEvo, Herts, United Kingdom). In addition, three classical detoxification enzyme inhibitors were used for larval and adult bioassays; piperonyl butoxide (PBO; 5-((2-(2-butoxyethoxy)ethoxy) methyl)-6-propyl-1,3-benzodioxole; 90% Fluka, Buchs, Switzerland) an inhibitor of mixed-function oxidases, tribufos (DEF; S,S,S-tributyl phosphorotrithioate; 98.1% Interchim, Montluçon, France) an inhibitor of carboxylesterases and in a lesser extent of GSTs and chlorfenethol (DMC; 1,1-bis (4-chlorophenyl) ethanol; 98% Pestanal™, Riedel-de-Haën, Seelze, Germany) a specific inhibitor of GSTs.

Larval Bioassays

Larval bioassays were performed using a standard protocol described by the World Health Organization [58]. Bioassays were carried out using late third and early fourth-instar larvae of the Bora-Bora and Vauclin strains. For each bioassay, 20 larvae of each strain were transferred to cups containing 99 ml of distilled water. Five cups per concentration (100 larvae) and 5–8 concentrations of temephos diluted in ethanol leading to 0–100% mortality were used. For each concentration, 1 ml of temephos at the desired concentration was added to the cups. Control treatments of 1 ml of ethanol were performed for each test. Temperature was maintained at 27°C ± 2°C all over the duration of bioassays, and larval mortality was recorded 24 hr after exposure. Three replicates with larvae from different rearing batches were made at different times and the results were pooled for analysis. Larvae were then exposed to the insecticide plus each enzyme inhibitor for 24 hr. Dose of enzyme inhibitors were determined according to preliminary bioassays showing that the sub lethal concentrations of inhibitors were 1 mg/l, 1 mg/l and 0.008 mg/l for PBO, DMC, and DEF respectively.

Topical Applications

The intrinsic activity of deltamethrin against adult mosquitoes was measured using forced contact tests to avoid any side effects linked to the insect behavior as recommended by the World Health Organization [59]. A volume of 0.1 μL of insecticide solution in acetone was dropped with a micro-capillary onto the upper part of the pronotum of each adult mosquito that was briefly anaesthetized with CO_2 and maintained on a cold table. Doses were expressed in nanograms of active ingredient per mg of mosquito body weight. A total of 50 individuals (non blood fed females, 2–5-days old) were used per insecticide dose and for controls, with at least five doses leading to 0 to 100% mortality. Each test was replicated twice (n = 100 per dose) using different batches of insects and insecticide solutions. After treatment, mosquitoes were maintained at 27°C ± 2°C and 80% ± 10% relative humidity in plastic cups with honey solution provided. Mortality was recorded after 24 hr. To assess the effect of detoxification enzyme inhibitors, each adult female was exposed to sub lethal doses of PBO (1000 ng/female), DEF (300 ng/female) and DMC (500 ng/female) 1 hr prior to deltamethrin topical application following the same protocol described above.

Mortality Data Analysis

Larval and adult mortality levels were corrected by the formula of Abbott [60] in case of control mortality > 5%, and data were analyzed by the log-probit method of Finney [61] using the Probit software of Raymond et al. [62]. This software uses the iterative method of maximum likelihood to fit a regression between the log of insecticide concentration and the probit of mortality. The goodness of fit is estimated by a weighted χ^2. It also estimates the slope of the regression lines and the lethal concentrations (LC_{50} and LC_{95} for larvae) or dosages (LD_{50} and LD_{95} for adults) with their 95% confidence intervals. Bora-Bora and Vauclin strains were considered as having different susceptibility to a given pesticide when the ratio between their $LC_{50/95}$ or $LD_{50/95}$ (resistance ratio: $RR_{50/95}$) had confidence limits excluding the value of 1. A mosquito strain is considered susceptible when its value of RR_{50} is less than 5, moderately resistant when RR_{50} is between 5 and 10, and highly resistant when RR_{50} is over 10. For detoxification enzyme inhibitors, synergism ratio's (SR_{50} and SR_{95}) were obtained by calculating the ratio between the LC_{50} (or LD_{50}) and LC_{95} (or LD_{95}) of each insecticide with and without each enzyme inhibitor. A SR significantly higher than 1 indicated a significant effect of enzyme inhibitor and synergist effects were considered different between the two strains when their confidence interval (CI) were not overlapping.

Detoxification Enzyme Activities

The P450 monooxygenase activities were comparatively evaluated between susceptible and resistant strains in both larvae and adults by measuring the 7-ethoxycoumarin-O-deethylase (ECOD) activity on microsomal fractions based on the microfluorimetric method of De Sousa et al. [63]. One gram fresh fouth stage larvae or 3 days-old adults (50% males and 50% females) were homogenized in 12 mL of 0.05 M phosphate buffer (pH 7.2) containing 5 mM DTT, 2 mM EDTA and 0.8 mM PMSF. The homogenate was centrifuged at 10000 g for 20 min at 4°C and the resulting supernatant was ultracentrifuged at 100,000 g for 1 hr at 4°C. The microsomal fraction was then resuspended in 0.05 M phosphate buffer and the microsomal protein content was determined by the Bradford method. Twenty μg microsomal proteins were added to 0.05 M phosphate buffer (pH = 7.2) containing 0.4 mM 7-ethoxycoumarin (7-Ec, Fluka) and 0.1 mM NADPH for a total reaction volume of 100 μL and incubated at 30°C. After 15 min, the reaction was stopped and the production of 7-hydroxycoumarin (7-OH) by P450 monooxygenases was evaluated by measuring the fluorescence of each well (380 nm excitation, 460 nm emission) with a Fluoroskan Ascent spectrofluorimeter (Labsystems, Helsinski, Finland) in comparison with a scale of 7-OH (Sigma). The P450 activities were expressed as mean pmoles of 7-OH per mg of microsomal protein per min ± SE. Statistical comparison of P450 activities between the two strains was performed by using a Mann and Whitney test (N = 15).

Glutathione S-transferase activities were comparatively measured on 200 μg of cytosolic proteins from the 100,000 g supernatant (see above) with 1-chloro-2,4-dinitrobenzene (CDNB, Sigma) as substrate [64]. Reaction mixture contained 2.5 ml of 0.1 M phosphate buffer, 1.5 μM reduced glutathione (Sigma), 1.5 μM CDNB and 200 μg proteins. The absorbance of the reaction was measured after 1 min at 340 nm with a UVIKON 930 spectrophotometer. Results were expressed as mean nmoles of conjugated CDNB per mg of protein per min ± SE. Statistical comparison of GST activities between the two strains was performed by using a Mann and Whitney test (N = 15).

Carboxylesterases activities were comparatively measured on 30 μg of cytosolic proteins from the 100,000 g supernatant (see above) according to the method described by Van Asperen et al. [65] with α-naphthylacetate and β-naphthylacetate used as substrates (α-NA and β-NA, Sigma). Thirty μg cytosolic proteins were added to 0.025 mM phosphate buffer (pH 6.5) with 0.5 mM of α-NA or β-NA for a total volume reaction of 180 μL and incubated at 30°C. After 15 min, reaction was stopped by the addition of 20 μL 10 mM Fast Garnett (Sigma) and 0.1 M sodium dodecyl sulfate (SDS, Sigma). The production of α- or β-naphthol was measured at 550 nm with a Σ960 microplate reader (Metertech, Taipei, Taiwan) in comparison with a scale of α-naphthol or β-naphthol and expressed as mean μmoles of α- or β-naphthol per mg of cytosolic protein per min ± SE. Statistical comparison of esterase activities between the two strains was performed by using a Mann and Whitney test (N = 30).

Kdr Genotyping

Genomic DNA was extracted from whole adult mosquitoes of the Bora-Bora and Vauclin strains by grinding tissues with a sterile micro-pestle in DNA extraction buffer (0.1 M Tris HCl pH 8.0, 0.01 M EDTA, 1.4 M NaCl, 2% cetyltrimethyl ammonium bromide). The mixture was incubated at 65°C for 5 min. Total DNA was extracted with chloroform, precipitated in isopropanol, washed in 70% ethanol, and resuspended in sterile water. The *kdr* genomic region was amplified by PCR using Dip3 (5'-ATCATCTTCATCTTTGC-3') and Dip2A (5'-TTGTTGGTGTCGTTGTCGGCCGTCGG-3') primers. PCR steps included an initial denaturation step at 95°C for 3 min, followed by 45 cycles at 95°C for 30 sec, 48°C for 30 sec, and 72°C for 45 sec, and a final extension step at 72°C for 6 min. PCR products were gel-purified with the QIAquick Gel Extraction Kit (Qiagen) before sequencing on an ABI Prism 3130 XL Genetic Analyzer (Applied Biosystems) using the same primers.

Microarray Screening of Differentially Transcribed Detoxification Genes

The *Aedes detox* chip DNA-microarray, initially developed by Strode et al. [20] and recently updated with additional genes, was used to monitor changes in the transcription of detoxification genes between the Vauclin and the Bora-Bora strains in 4th-stage larvae and 3 days-old adults. This microarray contains 318 probes representing 290 detoxification genes including all cytochrome P450 P450s, GSTs, CCEs, and additional enzymes potentially involved in response to oxidative stress from the mosquito *Ae. aegypti*. Each probe, plus six housekeeping genes and 23 artificial control genes (Universal Lucidea Scorecard, G.E. Health Care, Bucks, UK) were spotted four times at different positions on each array.

The RNA extractions, cRNA synthesis and labeling reactions were performed independently for each biological replicate. Total RNA was extracted from batches of 30 4th-stage larvae or 30 3 days-old adults (15 males and 15 females) using the PicoPure™ RNA isolation kit (Molecular Devices, Sunnyvale, CA, USA) according to manufacturer's instructions. Genomic DNA was removed by digesting total RNA samples with DNase I by using the RNase-free DNase Set (Qiagen). Total RNA quantity and quality were assessed by spectrophotometry using a Nanodrop ND1000 (LabTech, France) and by using a Bioanalyzer (Agilent, Santa Clara, CA, USA). Messenger RNAs were amplified using the RiboAmp™ RNA amplification kit (Molecular Devices) according to manufacturer's instructions. Amplified RNAs were checked for quantity and quality by spectrophotometry and Bioanalyzer. For each hybridization, 8 μg of amplified RNAs were reverse transcribed into labeled cDNA and hybridized to the array as previously described by David et al. [19]. For each life-stage, 3 pairwise comparisons of Vauclin strain versus Bora-Bora strain were performed with different biological samples. For each biological replicate, two hybridizations were performed in which the Cy3 and Cy5 labels were swapped between samples for a total of six hybridizations per comparison in each life-stage.

Spot finding, signal quantification and spot superimposition for both dye channels were performed using Genepix 5.1 software (Axon Instruments, Molecular Devices, Sunnyvale, CA, USA). For each data set, any spot satisfying one of the following conditions for any channel was removed from the analysis: (i) intensity values less than 300 or more than 65,000, (ii) signal to noise ratio less than 3, (iii) less than 60% of pixel intensity superior to the median of the local background ± 2 SD. Data files were then loaded into Genespring 7.2 (Agilent Technologies, Santa Clara, CA USA) for normalization and statistic analysis. For each array, the spot replicates of each gene were merged and expressed as median ratios ± SD. Data from dye swap experiments were then reversed and ratios were log transformed. Ratio values below 0.01 were set to 0.01. Data were then normalized using the local intensity-dependent algorithm Lowess [66] with 20% of data used for smoothing. For each comparison, only genes detected in at least 50% of all hybridizations were used for further statistical analysis. Mean transcription ratios were then submitted to a one-sample Student's t-test against the baseline value of 1 (equal gene transcription in both samples). Genes showing a transcription ratio > 1.5-fold in either direction and a t-test P value lower than 0.01 after Benjamini and Hochberg multiple testing correction [67] were considered significantly differentially transcribed between the two strains.

Real-Time Quantitative RT-PCR Validation

Transcription profiles of 10 detoxification genes in 4th-stage larvae and adults were validated by reverse transcription followed by real-time quantitative RT-PCR on the same RNA samples used for microarray experiments. Four μg total RNAs were treated with DNAse I (Invitrogen) and used for cDNA synthesis with superscript III (Invitrogen) and oligo-dT_{20} primer for 60 min at 50°C according to manufacturer's instructions. Resulting cDNAs were diluted 125 times for PCR reactions. Real-time quantitative PCR reactions of 25 μL were performed in triplicate on an iQ5 system (BioRad) using iQ SYBR Green supermix (BioRad), 0.3 μM of each primer and 5 μL of diluted cDNAs according to manufacturer's instructions. For each gene analyzed, a cDNA dilution scale from 5 to 50,000 times was performed in order to assess efficiency of PCR. Data analysis was performed according to the ΔΔCT method taking into account PCR efficiency [68] and using the genes encoding the ribosomal protein L8 [GenBank DQ440262] and the ribosomal protein S7 [GenBank EAT38624.1] for a dual gene normalization. For each life-stage, results were expressed as mean transcription ratios (± SE) between the insecticide-resistant strain Vauclin and the susceptible strain Bora-Bora. Only genes showing more than 2-fold over- or under-transcription in the Vauclin strain were considered significantly differentially expressed.

CONCLUSION

We have identified multiple insecticide resistance mechanisms in *Ae. aegypti* mosquitoes from Martinique (French West Indies) significantly reducing the insecticidal activity of insecticides used for their control. Microarray screening identified multiple detoxification genes over-transcribed at both life-stages in resistant mosquitoes, suggesting their possible involvement in insecticide-resistance. Further experimental validation by using enzyme characterization and RNA interference will allow confirming the role of these genes in the resistance phenotype. As previously shown in mosquitoes [57], the epistasis between the *kdr* mutation and particular P450s genes is likely to contribute to the high level of resistance to pyrethroids in *Ae. aegypti* from Martinique and might seriously threatens the control of dengue vectors in the future. A better understanding of the genetic basis of insecticide resistance is an essential step to implement more effective vector control strategies in the field in order to minimize dengue outbreaks.

AVAILABILITY

Data Deposition

The description of the microarray "*Aedes Detox Chip*" can be accessed at ArrayExpress http://www.ebi.ac.uk/arrayexpress acc. No. A-MEXP-623.

All experimental microarray data can be accessed at http://funcgen.vectorbase.org/ExpressionData/.

KEYWORDS

- ***Aedes aegypti***
- ***Aedes detox* chip**
- **Hemorrhagic fevers**
- **Temephos**

AUTHORS' CONTRIBUTIONS

Sébastien Marcombe participated in toxicological and biochemical studies together with microarray screening and *kdr* genotyping and helped to draft the manuscript. Rodolphe Poupardin participated in biochemical studies, microarray screening and RT-qPCR. Frederic Darriet participated in toxicological studies. Stéphane Reynaud participated in RT-qPCR and helped to draft the manuscript. Julien Bonnet participated in toxicological studies. Clare Strode participated in microarray study. Cecile Brengues participated in *kdr* genotyping and sequencing. André Yébakima coordinated field mosquito collection in Martinique and helped to draft the manuscript. Hilary Ranson helped to draft the manuscript and coordinated the microarrays studies. Vincent Corbel conceived of the study and participated in its design and coordination and helped to draft the manuscript. Jean-Philippe David participated in the design of the study and its coordination, performed microarray data analysis and conceived the manuscript. All authors read and approved the final manuscript.

ACKNOWLEDGMENTS

The present research project was funded by the French Institut de Recherche pour le Développement (IRD), the French Agency for Environmental Health and Safety (grant AFSSET N° 13-12-2007 to Vincent Corbel) and the Laboratory of Alpine Ecology of Grenoble (French National Research Agency grant ANR MOSQUITO-ENV N° 07SEST014 to Jean-Philippe David and Stéphane Reynaud). We thank Manuel Etienne and Said Crico for help on mosquito collection in Martinique. We thank Prof. A. Cossins, Dr. M. Hughes and the Liverpool Microarray User Community for microarray printing. We thank Dr. B. MacCallum and Vectorbase community for valuable help with microarray data deposition. We are grateful to Prof. P. Ravanel for useful comments on the manuscript and J. Patouraux for technical help.

Chapter 14

Solar Drinking Water Disinfection (SODIS) to Reduce Childhood Diarrhea

Daniel MΔusezahl, Andri Christen, Gonzalo Duran Pacheco, Fidel Alvarez Tellez, Mercedes Iriarte, Maria E. Zapata, Myriam Cevallos, Jan Hattendorf, Monica Daigl Cattaneo, Benjamin Arnold, Thomas A. Smith, and John M. Colford, Jr.

INTRODUCTION

Solar drinking water disinfection (SODIS) is a low-cost, point-of-use water purification method that has been disseminated globally. Laboratory studies suggest that SODIS is highly efficacious in inactivating waterborne pathogens. Previous field studies provided limited evidence for its effectiveness in reducing diarrhea.

We conducted a cluster-randomized controlled trial in 22 rural communities in Bolivia to evaluate the effect of SODIS in reducing diarrhea among children under the age of 5 year. A local nongovernmental organization (NGO) conducted a standardized interactive SODIS-promotion campaign in 11 communities targeting households, communities, and primary schools. Mothers completed a daily child health diary for 1 year. Within the intervention arm 225 households (376 children) were trained to expose water-filled polyethyleneteraphtalate (PET) bottles to sunlight. Eleven communities (200 households, 349 children) served as a control. We recorded 166,971 person-days of observation during the trial representing 79.9% and 78.9% of the total possible person-days of child observation in intervention and control arms, respectively. Mean compliance with SODIS was 32.1%. The reported incidence rate (IR) of gastrointestinal illness in children in the intervention arm was 3.6 compared to 4.3 episodes/year at risk in the control arm. The relative rate (RR) of diarrhea adjusted for intracluster correlation was 0.81 (95% confidence interval (CI) 0.59–1.12). The median length of diarrhea was 3 d in both groups.

Despite an extensive SODIS promotion campaign we found only moderate compliance with the intervention and no strong evidence for a substantive reduction in diarrhea among children. These results suggest that there is a need for better evidence of how the well-established laboratory efficacy of this home-based water treatment method translates into field effectiveness under various cultural settings and intervention intensities. Further global promotion of SODIS for general use should be undertaken with care until such evidence is available.

Globally, 1.8 million people die every year from diarrheal diseases the vast majority of whom are children under the age of 5 year living in developing countries [1]. Unsafe water, sanitation, and hygiene are considered to be the most important global risk factors for diarrheal illnesses [2].

Recent systematic reviews concluded that interventions to improve the microbial quality of drinking water in households are effective at reducing diarrhea, which is a principal source of morbidity and mortality among young children in developing countries [3-5]. One widely promoted water disinfection method with encouraging evidence of efficacy in laboratory settings is SODIS [6]. Global efforts are underway to promote SODIS as a simple, environmentally sustainable, low-cost solution for household drinking water treatment, and safe storage (www.who.int/household_water, www.sodisafricanet.org). The SODIS is currently promoted in more than 30 countries worldwide (www.sodis.ch) and in at least seven Latin American countries through the SODIS Foundation including in Bolivia.

Despite this widespread promotion, evidence of the effectiveness of SODIS from field studies is limited. The three reported SODIS trials to date implemented the intervention at the household level, two of them in highly controlled settings that ensured very high compliance [7-9]. The highest reduction in incidence (36%) was recorded in a trial carried out among 200 children in an urban slum in Vellore, India [9].

Because SODIS is a behavioral intervention designed to reduce infectious diarrhea, disease transmission and its interruption likely have community-level dynamics [10]. In addition, because SODIS is typically rolled out in practice through community rather than household level promotion, there is an urgent need for effectiveness data from such settings. We conducted a community-randomized intervention trial to evaluate the effectiveness of SODIS in decreasing diarrhea in children <5 year in rural communities in Bolivia.

MATERIALS AND METHODS

Ethics Statement

The study was approved by the three human subjects review boards of the University of Basel, Switzerland, the University of California, Berkeley, and the University of San Simon, Cochabamba, Bolivia. The Cochabamba and Totora municipal authorities also approved the study and informed consent was obtained from community leaders and male and female household heads prior to implementation of the study. Informed consent was obtained before randomization to the treatment arms (Figure 1). Mildly ill children from households participating in the study were provided with and instructed to use oral rehydration ORS, or they were referred by field staff to the local health system where clinical services were provided free of charge. The project provided transport and treatment costs for those patients. All project staff completed training on research ethics (www.fhi.org/training/sp/Retc/). Project staff comprised all project personnel of all project partners. Field staff comprised all personnel working in our laboratories and at our Totora field station including data enumerators and data- and project-management staff, supervisors, and community-based field workers living in the study communities. The trial protocol and the CONSORT statement checklist are available online as supporting information.

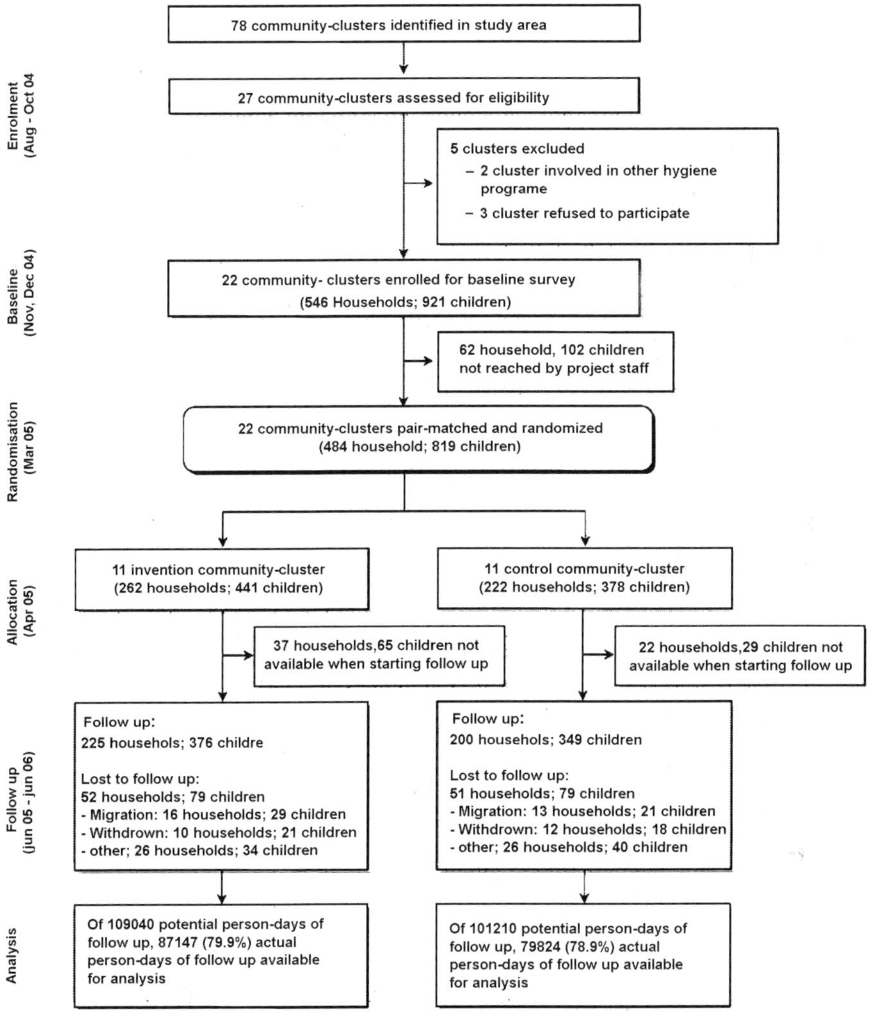

Figure 1. Community-randomized trial flow diagram on point-of-use SODIS in Totora District, Bolivia.

Site and Population

Our trial, the Bolivia Water Evaluation Trial (BoliviaWET), was conducted in an ethnically homogeneous Quechua setting in rural Totora District, Cochabamba Department, Bolivia. Our study was part of a comprehensive SODIS roll-out program in collaboration with Project Concern International, a NGO. Most of the local residents are farmers, typically living in small compounds of three buildings with mud floors, with five or more persons sleeping in the same room. Our own surveys showed that 15% of homes have a latrine or other sanitary facilities and that most residents defecate in the nearby environment.

Drinking water is typically stored in 10-liter plastic buckets or open jerry cans of 5–20 l in the household. Baseline assessments of the drinking water quality in the home indicated a median contamination of thermotolerant coliforms (TTC) of 32 TTC/100 ml (interquartile range (IQR) = 3–344; n = 223). Samples of at least one water source per community were tested for *Giardia lamblia* and *Cryptosporidium parvum*. The two parasites were detected in 18/24 and 11/23 water samples, respectively.

Parasites were detected by using immunomagnetic separation and polymerase chain reaction (PCR) techniques [11]. Piped water, when available, is not chlorinated.

Design

Twenty-seven of 78 communities in the study area fulfilled the selection criteria (geographically accessible all year round; at least 30 children <5 year; reliance on contaminated drinking water sources). Two communities were excluded because of other ongoing health and hygiene campaigns, and three communities withdrew participation before baseline activities because of a change in political leadership. Community health workers undertook a census and identified households with at least one child <5 year. All children <5 year were enrolled in the participating villages.

We pair-matched communities on the incidence of child diarrhea as measured in an 8-week baseline survey [12]. The intervention was then assigned randomly to one community within each of the 11 consecutive pairs. This assignment was done during a public event because key political stakeholders were worried about possible backlash, public outcry, or a drop-off in group participation, which would result from providing some members with a new benefit while others got "nothing." It was agreed that a public drawing event was necessary to increase perceived fairness among the participating district and municipal authorities. Three authorities, the district head (Alcalde), representatives of the Ministries of Health and Education, and the deputy of the farmers union (Central Campesina), each drew one of two balls (with community codes inscribed that were randomly assigned beforehand) representing paired communities from a concealed box. It was agreed that the first draw assigned the community to the intervention arm. The group allocation was immediately recorded in a protocol by an independent witness. Subsequently, the witness disclosed the sequence, informed the community members and the authorities present in the town hall, and all drawers signed the protocol.

We explicitly chose community-level randomization because important components of the intervention (i.e., community efforts to encourage adoption of the SODIS-method) would occur at the community level. Randomization below the community level would not reflect the reality of scale-up program implementation, and we would not have captured the potential community-level reinforcement of the behavior change. Furthermore, community-level randomization is considered ethically optimal, because participants expect to equally benefit from interventions within their community [13-15]. Additionally, we believed cross-contamination (of the intervention) between the intervention and control communities was minimized by vast geographical dispersion of the communities. Control communities knew from the beginning of the study that they would receive the intervention as part of the NGO's development plans after study completion. It was not possible for the NGO to carry out the intervention in all the communities at the same time, thus making randomization feasible and acceptable to the three ethical review boards overseeing the study.

Sample size was calculated according to methods outlined by Hayes and Bennett [16], assuming an IR in the control villages of five episodes/child/year [17], and accounting for clustering, the number of episodes, and the expected effect. We assumed a coefficient of between-cluster variation (k) of similar studies, between 0.1 and 0.25 (as cited by Hayes and Bennett) and a minimum of 10 child-years of observation per cluster [16]. We calculated that nine pairs of clusters were required to detect a difference of at least 33% in the IR between the control and intervention arms with 80% power, $k = 0.20$ and an alpha level of 0.05. Anticipating a drop-out of at least one cluster per arm and a loss of follow-up of individuals, the final sample size was adjusted to 11 pairs with 30 children per community cluster. We powered the study to detect a 33% reduction in diarrhea incidence after reviewing the evidence base for point-of-use water treatment at the time of the study's inception in 2002 [18].

Implementation of the Intervention

The SODIS intervention was designed according to the published guidelines for national SODIS dissemination (http://www.sodis.ch/files/TrainingManual_sm.pdf). Promotion activities were targeted at primary caregivers and all household members (biweekly), whole communities (monthly), and primary schools (three times) by the NGO as part of its regional community development program. Eleven communities (262 households and 441 children) were randomized to the intervention; 11 communities (222 households, 378 children) served as a control group (Figure 1). For a period of 15 month an intensive, standardized, and repeated interactive promotion of the SODIS method was implemented in the intervention communities beginning 3 month before the start of follow-up.

Within the intervention arm, participating households were supplied regularly with clean, recycled PET bottles. The households were taught through demonstrations, role plays, video, and other approaches to expose the water-filled bottles for at least 6 hr to the sun. The NGO staff emphasized the importance and benefits of drinking only treated water (especially for children), explained the germ—disease concept, and promoted hygiene measures such as safe drinking water storage and hand washing as they relate to the understanding of drinking water and the faecal—oral route of transmission of pathogens. During household visits the NGO staff encouraged all household members to apply the method, answered questions, and assisted mothers and primary caregivers to integrate the water treatment into daily life. The same intervention (in terms of contents and messages) was supplied to the communities in the control arm by the NGO-staff at the end of the study.

Outcome

The primary outcome was the IR of diarrhea among children <5 year, defined as number of diarrhea episodes per child per year obtained from daily assessment of individual diarrhea occurrence. We applied the WHO definition for diarrhea of three or more watery bowel movements or at least one mucoid/bloody stool within 24 hr [19, 20]. We defined a new episode of diarrhea as the occurrence of diarrhea after a period of 3 d symptom-free [20-22]. An episode of diarrhea was labeled "dysentery" if signs of blood or mucus in the stool were recorded at any time. We also calculated the longitudinal prevalence (number of days a child suffered diarrhea divided by the number of days of observation) because of its closer relation to severity, growth faltering, and mortality than diarrhea incidence [19, 23]. Severe diarrhea (SD) was defined as the occurrence of diarrhea on more than 10% of the observed days [24].

Data Collection and Field Staff

The primary outcome was measured by community-based field workers who were recruited nearby and who lived one per community during data collection periods. The field workers were extensively trained in interviewing and epidemiological observation techniques, data checking, recording, and in general approaches to community motivation. Community-based field workers were randomly rotated between communities every 3 month. Child morbidity was reported by the closest caregiver using the vernacular term "K'echalera," which had been established previously to correspond to the WHO definition of diarrhea [25]. Mothers or closest caretakers kept a 7-d morbidity diary recording daily any occurrence of diarrhea, fever, cough, and eye irritations in study participants [25]. Community-based field workers visited households weekly to collect the health diaries, and supervisors revisited an average 7% of homes. Discrepancies between supervisors and community-based field workers' records were clarified during a joint home revisit. Child exposure risks were also assessed by community-based staff interviewing mothers once during baseline and twice during the 1-year follow-up.

Compliance with the SODIS method was measured using four different subjective and objective indicators. Three of the indicators were assessed by field staff independent from the implementing NGO: (i) the number of SODIS-bottles exposed to sunlight and, (ii) the number of bottles ready-to-drink in the living space, and (iii) the personal judgment about families' user-status was provided by community-based field workers living among the families in the intervention arm. Judgment criteria for this main compliance indicator study included observing regular SODIS practice and bottles exposed to sun or ready to drink in the kitchen and being offered SODIS-treated water upon request. The fourth SODIS-use indicator was based on self-reporting and caregivers' knowledge of and attitudes toward the intervention that was assessed at the beginning (i.e., 3 month after start of the intervention) and at the end of the 12-month follow-up period.

Statistical Analysis

An intention-to-treat analysis was applied comparing the IR of diarrhea between children <5 year in intervention and control communities. Diarrhea prevalence (PR) and SD were additionally analyzed. Generalized linear mixed models (GLMMs) were fitted to allow for the hierarchical structure of the study design (pair-matched clusters). In contrast to our original trial protocol we selected the GLMM approach rather than generalized estimating equations (GEEs) because recent publications indicated that the latter method requires a larger number of clusters to produce consistent estimates [26].

The crude (unadjusted) model included only the design factors and the intervention effect [12, 27]. Further models included potential confounders (selected a priori: child's age, sex, child hand-washing behavior, and water treatment at baseline). Following an evaluation of the best fit, the GLMM included the log link function for negative binomial data (IR) and logit for binomial data (PR and SD). Denoting the link function of the outcome Y by $g(E(Y))$, the crude and adjusted models were: $g(E(Y_{ijk})) = \mu+B_i+\tau_j+\xi_{ij}$, and $g(E(Y_{ijk})) = \mu+B_i+\tau_j+\xi_{ij}+x'b$ where Y_{ijk} denotes the observed outcome value for the kth individual from a community allocated to the jth intervention, in the ith pair, μ is the general mean, B_i is the random effect of the ith pair $\approx N(0, \sigma^2_p)$, τ_j is the fixed effect of the SODIS intervention, and ξ_{ij} is the random effect of the interaction of the ith pair with the jth intervention applied to the community $\approx N(0, \sigma^2_{pt})$ (signifying the within-pair cluster variance and used as error term for τ_j), x is the vector of potential confounding factors, and b the vector of the corresponding regression coefficients.

The intracluster correlation coefficient (ICC) and the coefficient of between-cluster variation (k) were calculated after data collection to validate the degree of clustering and our assumptions for the sample size. The ICC and k were estimated from the unscaled variance of the IR's GLMM. To estimate the uncertainty of ICC and k, we obtained the 95% credible region (Bayesian equivalent of 95% CI) through an analogous Bayesian hierarchical regression [28]. Noninformative priors were used. The statistical analyses were performed using SAS software v9.1 (PROC GLIMMIX, SAS Institute Inc.) and WinBUGS v1.4 (Imperial College and MRC).

Participant Flow and Recruitment

Among the 1,187 households in the 22 communities there were 546 that met the inclusion criteria (Figure 1). The median number of participating households with children <5 year per community was 22. Because of political unrest and national election campaigns in 2005 a period of 6 month passed between the baseline and the start of follow-up. Subsequently, 62 households (102 children) were no longer traceable before randomization, and 59 households (37 intervention, 22 control) were lost before data collection had started. The loss to follow-up was balanced in intervention and control arms. Data were obtained from 376 children (225 households) in the intervention and 349 children (200 households) in the control arm, thus reaching our originally planned sample size.

Follow-up started in June, 2005 and ended in June, 2006. During the 51 week of the study, information on the occurrence of diarrhea was collected for 166,971 person-days representing 79.9% and 78.9% of the total possible person-days of child observation in intervention and control arms. We excluded from the potential observation time the experience of 94 children who dropped out before the start of follow-up. National festivities, holidays, and political unrest over the entire year amounted to further 9 week during which outcome surveillance needed to be suspended. The main reasons for incomplete data collection were migration (28%) and withdrawal (67%). Supervisors reevaluated the outcome during 984 unannounced random home visits, and discrepancies between community-based field workers' and supervisors' records were found for five (0.5%) of all visits.

Baseline Characteristics

At baseline the households in the different study arms were well balanced on multiple other factors suggesting successful randomization (Table 1). The main types of water sources for household chores and drinking were similar in both arms as was the distance to the source (median distance 50 m and 30 m in the control and intervention arms, respectively). Storing water for longer than 2 d was more common among the intervention (26.8%) than the control arm (13.9%). Nearly 30% of all households reported treating water regularly before drinking. Boiling was the most common water treatment before the trial (20.2% in both arms).

Table 1. Baseline community and household characteristics of a community-randomized trial of SODIS.

Category	Description	*n* Children or Households	Control 11 Clusters	*n* Children or Households	Intervention 11 Clusters
Demography	Community size: n of households [mean (SD)]	–	50 (20)	–	58 (20)
	Household size: n of household members [mean (SD)]	N=222	6.2 (2.1)	N=262	6.3 (2.6)
	n of children < 5 y per household [mean (SD)]	–	1.8 (0.7)	–	1.7 (0.8)
	n of children < 5 y per community [mean (SD)]	–	35.3 (6.6)	–	41.4 (9.9)
	Female household head [n (%)]	–	20 (9.0)	–	14 (5.4)
	Closest child caregiver (female)	–	223 (99.5)	–	266 (99.6)
	Age of closest child caregiver (y) [mean (SD)]	–	31(9)	–	30 (10)
	n of children < 1 y	–	65 (4.7)	–	67 (4.1)
	n of children < 5 y	–	369 (26.6)	–	426 (25.9)
Education	Household chief: reported years of education [mean (SD)]	N=167	4.1 (2.6)	N=178	4.2 (2.4)
	Closest child caregiver: reported years of education [mean SD)]	N=179	2.5 (1.9)	N=198	2.7 (1.8)
Socio-economic variables	Main occupation of the household chief as farmer	N=208	180 (86.5)	N=228	207 (90.8)
	Ownership of truck, car, or motorbike	–	12 (5.8)	–	14 (6.2)
	Ownership of radio	–	129 (86.1)	–	194 (85.1)
	Ownership of bicycle	–	109 (52.4)	–	121 (53.1)
	Ownership of television	–	24 (11.5)	–	15 (6.6)
	n of rooms in the house [mean (SD)]	–	2.9 (1.4)	–	2.8 (1.2)

Table 1. *(Continued)*

Category	Description	*n* Children or Households	Control 11 Clusters	*n* Children or Households	Intervention 11 Clusters
Water management	Spring as source of drinking water	N=208	100 (48.1)	N=228	136 (59.6)
	Tap as source of drinking water	–	108 (51.9)	–	129 (56.6)
	River as source of drinking water	–	46 (22.1)	–	29 (12.7)
	Rain as source of drinking water	–	31 (14.9)	–	71 (31.1)
	Dug well as source of drinking water	–	31 (14.9)	–	37 (16.2)
	Distance to water source (m) [median (Q1, Q3)]	–	50 (7.5, 1 00)	–	30 (6, 150)
	Container for water collection: plastic bucket	–	189 (90.9)	–	205 (89.9)
	Container for water collection: jerry can	–	165 (79.3)	–	156 (68.4)
	Container for water collection: bottles	–	32 (15.4)	–	36 (15.8)
	Container for water collection: jar/ pitcher	–	13 (6.3)	–	20 (8.8)
	Container for water collection: barrel	–	10 (4.8)	–	25 (10.9)
	Child's consumption of untreated water (glasses/ day) [mean (SD)]	M=318	1.2 (1.2)	M=359	1.2 (1.4)
	Treat water before drinking	N=208	59 (28.4)	N=228	67 (29.4)
	Store water for > 2 d	–	29 (13.9)	–	61 (26.8)
	Water storage container: jerry can	–	23(11.1)	–	49 (21.5)
	Water storage container: plastic bucket	–	17 (8.2)	–	37 (16.2)
	Water turbidity in water storage container > 30 NTU	–	13 (11.2)	–	24 (18.8)
Sanitation	Reported n of interviewee's hand washing per day [mean (SD)]	N=177	3.8 (1.7)	N=200	4.1 (1.8)
	Reported n of child hand washing per day [mean (SD)]	M=348	2.5 (1.2)	M=376	2.6 (1.4)
	Child washes hands: before eating	–	228 (65.5)	–	270 (71.8)
	Child washes hands: when hands are dirty	–	62 (17.8)	–	56 (14.9)
	Child washes hands: other occasions	–	58 (16.7)	–	50 (13.3)
	Latrine present	N=208	27 (13.0)	N=228	38 (16.7)
	Use of latrine by the interviewee (day or night)	–	15 (7.2)	–	20 (8.8)
	Feces visible in yard	N=202	121 (59.9)	N=219	124 (56.6)

Data shows numbers and percentages unless otherwise specified. Baseline data from December 2004. Abbreviations: 30NTU, threshold for efficacious pathogen-inactivation of the SODIS method; M, number of children; N, number of households; NTU, nephelometric units; SD, standard deviation.
doi:1 0.1371/"ournal. med.1 000125.t001

Intervention and Attendance

The NGO conducted 210 community events and 4,385 motivational household visits in intervention communities; 3,060 visits occurred in the households with children <5 year followed up and analyzed for the study, and 1,325 household visits took place in homes that were not taking part in the study. Study households attended a median of nine community events (IQR = 5 – 12) and were visited by the SODIS-program team a median 11 times at home (IQR = 7 – 18). To ensure a sufficient number of PET bottles, the NGO provided as many SODIS-bottles as required by participants (mean 955 bottles/community).

Diarrheal Illness in the Control and Intervention Arm

Children in the SODIS-intervention arm reported a total of 808 episodes or a mean of 3.6 per child per year-at-risk (Table 2). In the control arm there were 887 episodes and an annual mean of 4.3 per child per year. In both arms median length of episodes was 3 d. The unadjusted RR estimate (0.81, 95% CI 0.59–1.12) suggested no statistically significant difference in the number of diarrhea episodes between the SODIS and control arms of the study (Table 3). In an analysis of the longitudinal prevalence of diarrhea we found no significant treatment effect (odds ratio (OR) = 0.92, 95% CI 0.66–1.29). Furthermore, no strong evidence was detected for the reduction of odds of SD cases (OR = 0.91, 95% CI 0.51–1.63) and dysentery (OR = 0.80, 95% CI 0.55–1.17).

Table 2. Diarrhea episodes, length of illness, and days ill with diarrhea.

Health Condition	Class or Parameter	n	Control	n	Intervention
Diarrhoea illness overview		**Children**		**Children**	
Days under observation	Median (Q1, 03)	349	263 (213, 274)	376	263 (222, 273)
Days at risk	Median (Q1, 03)	349	246 (192, 265)	376	247 (202, 265)
n Episodes	Median (Q1, 03)	349	1 (0, 3)	376	1 (0, 3)
n Dysentery episodes	Median (Q1, 03)	349	1 (0, 2)	376	1 (0, 2)
Days spent ill	Median (Q1, 03)	349	4 (0, 11)	376	4 (0, 12)
Episode length (d)	Median (Q1, 03)	349	3 (1, 5)	376	3 (2, 5)
Days under observation	Total		79,829		87,140
Days at risk	Total		75,077		82,682
n Episodes	Total		887		808
n Dysentery episodes	Total		460		431
Days spent ill	Total		3,111		3,038
Diarrhoea incidence	**Age class**	**Children**	**IR**	**Children**	**IR**
n Episodes/(child x year at risk)	< 1	16	7.8	15	11.1
	1-2	67	7.1	70	5.5
	2-3	67	4.3	82	3.8
	3-4	77	3.2	75	2.8
	4-5	71	3.4	80	2.1
	5-6	50	2.7	53	2.5
	Total[a]	349	4.3	376	3.6

Table 2. *(Continued)*

Health Condition	Class or Parameter	n	Control	n	Intervention
Diarrhoea prevalence	**Age class**	**Children**	**Mean (50)**	**Children**	**Mean (50)**
n Days ill/(child x year)	< 1	16	27.4 (28.3)	15	42.3 (40.7)
	1-2	67	31.4 (42.2)	70	23.0 (26.1)
	2-3	67	19.0 (47.5)	82	16.4 (28.4)
	3-4	77	11.7 (24.5)	75	7.3 (9.7)
	4-5	71	9.5 (15.1)	80	6.2 (12.4)
	5-6	50	6.9 (11.8)	53	7.7 (10.4)
	Total[a]	349	16.5 (32.8)	376	13.5 (22.4)
Diarrhoea illness	**Days spent ill**	**Children**	**Percent**	**Children**	**Percent**
	0 d	97	27.8	126	33.5
	1-2 d	50	14.3	42	11.2
	3-7 d	91	26.1	80	21.3
	8-14 d	49	14.0	59	15.7
	15-21 d	27	7.7	33	8.8
	22-40 d	18	5.2	21	5.6
	> 40 d	17	4.9	15	4.0
	Total	349	100	376	100
Diarrhoea illness duration	**Episode duration**	**Episodes**	**Percent**	**Episodes**	**Percent**
	1 day	250	28.2	191	23.6
	2-3 d	303	34.2	292	36.1
	4-7 d	258	29.1	250	30.9
	8-13 d	54	6.1	59	7.3
	> 13 d	22	2.5	16	1.9
	Total	887	100	808	100
Prevalence of other symptoms (d/(child xyear])		**Children**	**Mean (50)**	**Children**	**Mean (50)**
Vomit		349	5.5 (13.2)	376	4.0 (8.9)
Fever		349	21.0 (33.0)	376	15.1 (19.8)
Cough		349	41.9 (48.3)	376	30.9 (39.4)
Eyes irritation		349	12.8 (29.8)	376	8.3 (19.5)

' Includes one child per treatment arm with unknown age. SO, standard deviation. doi:1 0.1371/journal.pmed.1 000125.t002

A multivariable model adjusting for age, sex, baseline-existing water treatment practices, and child hand washing was consistent in its estimate of effect (RR = 0.74, 95% CI 0.50–1.11). We repeated the analysis by including confounding covariates in the order of occurrence of the variables in Table 3 to confirm that the conclusions were not sensitive to the choice of covariates. None of the models yielded significant results for the effect of SODIS (all p-values>0.1) or resulted in meaningful changes in estimates of ORs. Figure 2 shows the relationship between study time and diarrhea in the control and intervention arm. We found no statistically significant effect of the interaction of time and intervention in a time-dependent model.

Table 3. Effect of SODIS on diarrhea episodes, longitudinal prevalence, severe diarrhea, and dysentery episodes.

Outcome	Model	n Children	Parameter	RR/OR	95%CI	p-Value
n Episodes (RR)	Unadjusted	725	Intervention	0.81	(0.59-1.12)	0.19
	Adjusted	644	Intervention	0.74	(0.50-1.11)	0.14
			Age	0.75	(0.70-0.81)	< 0.001
			Sex	1.03	(0.84-1.26)	0.80
			Water treatment	1.05	(0.81-1.36)	0.69
			Hand washing	0.93	(0.85-1.02)	0.13
Prevalence (OR)	Unadjusted	725	Intervention	0.92	(0.66-1.29)	0.62
	Adjusted	644	Intervention	0.91	(0.64- 1.30)	0.60
			Age	0.67	(0.61-0.73)	< 0.001
			Sex	1.05	(0.84-1.31)	0.68
			Water treatment	1.00	(0.76-1.33)	0.97
			Hand washing	0.94	(0.84-1.04)	0.23
Severe diarrhoea (OR)	Unadjusted	643	Intervention	0.91	(0.51-1.63)	0.75
	Adjusted	589	Intervention	1.02	(0.52-2.01)	0.95
			Age	0.52	(0.40-0.67)	< 0.001
			Sex	1.12	(0.63-2.01)	0.69
			Water treatment	1.59	(0.81-3.12)	0.18
			Hand washing	0.94	(0.75-1.19)	0.62
Dysentery (OR)	Unadjusted	725	Intervention	0.80	(0.55-1.17)	0.23
	Adjusted	644	Intervention	0.75	(0.47-1.18)	0.20
			Age	0.73	(0.67-0.80)	< 0.001
			Sex	1.00	(0.80-1.26)	0.97
			Water treatment	1.15	(0.87-1.53)	0.33
			Hand washing	0.91	(0.82-1.01)	0.06

Number of episodes, *n* of episodes per days at risk; prevalence, *n* of days ill per days under observation; severe diarrhoea, diarrhoea during > 10% of all days (only children with more than 100 d of observation are included); unadjusted, general linear mixed models, only design factors and treatment are included; adjusted, effects oftreatment and covariates; sex: 0, female; 1, male; water treatment: water treatment at baseline, 0, no treatment; 1, treatment (chlorination or boiling or SODIS); hand washing, reported number of child's hand washing per day at baseline.
doi:1 0.1371 /journal.pmed.1 000125.t003

The ICC was estimated as 0.0009 with a 95% posterior credible region between (0.0001, 0.0025); k was estimated to be 0.27 with a 95% confidence region of (0.11, 0.46).

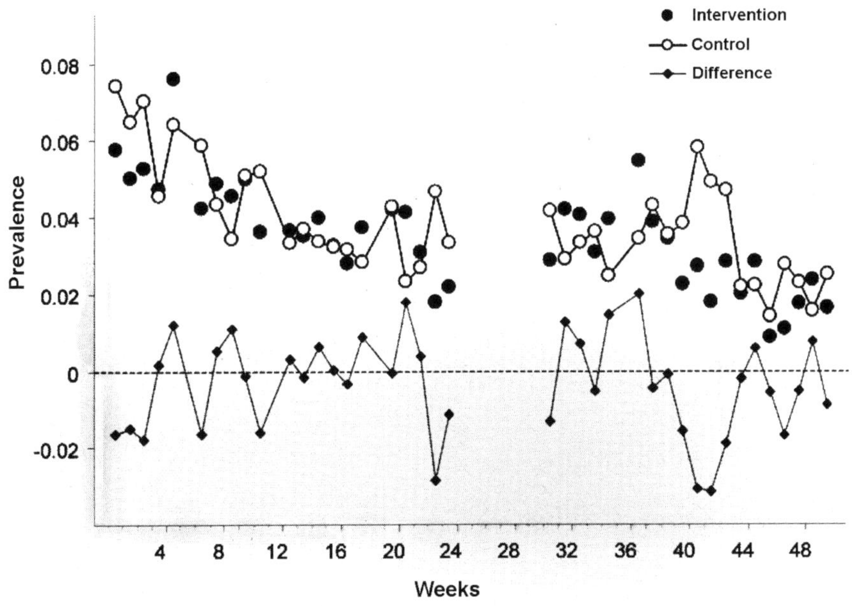

Figure 2. Weekly prevalence of child diarrheal illness. Weekly points are derived from daily prevalence data of each participating child.

Compliance

Community-based field workers who were living in the communities throughout the study observed a mean SODIS-user rate of 32.1% in the intervention arm (minimum 13.5%, maximum 46.8%, based on their personal judgment) (Figure 3). The mean proportion of households with SODIS-bottles exposed to the sun was 5 percentage points higher than the assessment by community-based field workers. In contrast, almost 80% of the households reported using SODIS at the beginning and end of the follow-up. About 14% of the households used the method more than two-thirds (>66%) of the weeks during observation, and 43% of the households applied SODIS in more than 33% of the observed weeks (Table 4).

Diarrheal Illness by Compliance

No positive effect of compliance (proportion of weeks of observed SODIS use) on the IRs in the intervention arm was observed. The incidence did not decline with the increase of weeks using SODIS (Figure 4). Seasonal variation in compliance was observed. The proportion of SODIS-practicing households was consistently below average during weeks 4–16 (January, 2005–April, 2006), which corresponded to the labor intensive cultivating period from November to May.

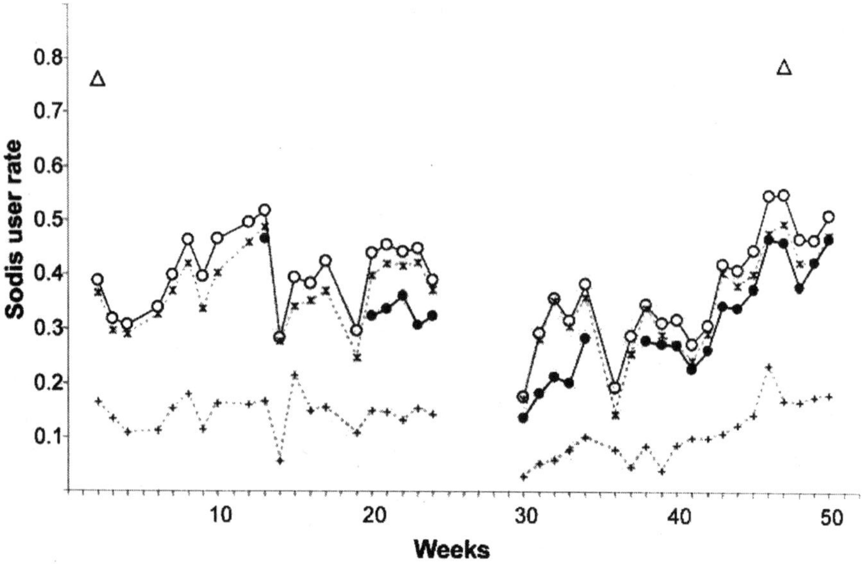

Figure 3. Weekly observed proportion of households using SODIS as point-of-use drinking water purification method. Open triangles, self-reported SODIS use at the beginning (after 3 month of initial SODIS promotion) and at the end of follow-up; filled dots, SODIS use observed by project staff living in the community (see Materials and Methods for definition); open circles, SODIS bottles observed on the roof and/or in the kitchen; stars, SODIS-bottles on the roof; crosses, SODIS-bottles in the kitchen.

Table 4. Climatic conditions and SODIS use of a cluster-randomized trial involving 22 rural communities of Totora District, Bolivia.

Category	Description	Control (n= 11 Clusters)	Intervention (n= 11 Clusters)
Climate	Percentage of sunny days (> 6 h sunshine) [median of clusters (min, max)]	70 (57, 78)	67 (44, 77)
	Average duration of sunshine [median of clusters (min, max)]	7.0 (6.3, 8.0)	7.1 (4.5, 8.3)
S0D1S-use	**Observed level of S0D1S use•**	**Percentage of house-holds**	**Percentage of households**
	0.66-1	0%	14%
	0.33-0.66	0.5%	29%
	0-0.33	99.5%	57%

• Proportion of weeks in which SODIS was used, as estimated by community-based project staff at the end of study. Households with < 10 wk of observation are excluded.

doi:1 0.1371 /journal.pmed.1 000125.t004

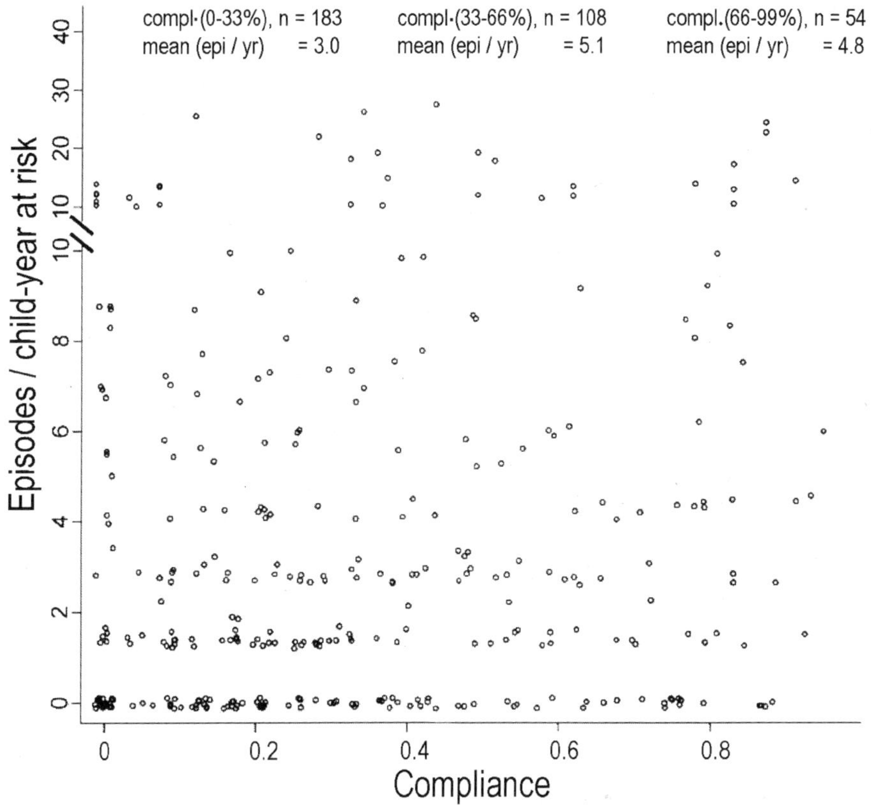

Figure 4. Compliance of using SODIS and child diarrhea in rural Bolivia. Compliance of SODIS use is estimated as the proportion of weeks a family has been classified as a SODIS user by community-based project staff. Dots, number of episodes per child-year at risk. Small random noise was added to the dots to avoid over plotting. Only children with at least 110 d under observation are included.

The median proportion of sunny days with more than 6 hr of sunshine was 70.2% and 67.2% in intervention and control communities, respectively, consistent with the technical and climatic conditions necessary for the proper functioning of the ultraviolet SODIS purification process [29] during the study (Table 4).

DISCUSSION

We conducted a community-randomized trial within the operations of an ongoing national SODIS-dissemination program, which provided an intensive training and repeated reinforcement of the SODIS intervention throughout the study period. In this context of a "natural experiment" we found a RR of 0.81 for the IR of diarrhea episodes among children assigned to SODIS compared to controls. However, the CI was broad and included unity (RR = 0.81, 95% CI 0.59–1.12) and, therefore, we conclude that there is no strong evidence for a substantive reduction in diarrhea among children in this setting. Subsequently, we discuss the primary outcome in the context of other study findings, and explain why we hypothesize that the true effect—if there is any—might be smaller.

First, the estimate for the longitudinal prevalence of diarrhea was substantially smaller (OR = 0.92, 95% CI 0.66–1.29) than the estimate for incidence and there is some evidence that prevalence is a better predictor in terms of mortality and weight gain than incidence [23]. The absence of a time-intervention interaction in our time-dependent analysis suggested no increased health benefits with the ongoing intervention. Furthermore, within the intervention arm, there was no evidence that increased compliance was associated with a lower incidence of diarrhea (Figure 4). However, we interpret this post hoc subgroup analysis cautiously because compliant SODIS users might differ in important ways from noncompliant users. A compliant SODIS user might be more accurately keeping morbidity diaries, whereas less compliant families may tend to underreport diarrheal illness. Or, households with a high burden of morbidity might be more likely to be compliant with the intervention. Both of these scenarios could lead to an underestimation of the effectiveness of SODIS.

Further, analyzing the laboratory results from 197 randomly selected stool specimens also did not provide convincing evidence for an intervention effect: the proportion of *C. parvum* was lower in the intervention children (5/94 vs. 2/103), but other pathogens were found at similar proportions in intervention and control children (*G. lamblia*, 39/94 vs. 40/103; *Salmonella* sp., 2/94 vs. 3/104; *Shigella* sp., 3/94 vs. 3/104). In further exploring the occurrence of other illness symptoms we found the prevalence of eye irritations and cough to be lower in the intervention group compared to the control group. This difference could be the result of the hygiene component in the intervention that increased hygiene awareness among the treatment communities. An alternative explanation is that the lack of blinding led to biased (increased) health outcome reporting in the intervention group.

Due to the nature of the intervention neither participants nor personnel were blinded to treatment assignment. Ideally, blinding to the intervention allocation should apply to the NGO staff administering the SODIS intervention and our enumerators assessing outcomes [30]. Although the former could not be blinded in our study (for obvious reasons), the latter would inevitably be able to identify the intervention status of the cluster through the visible display of bottles to sunlight in the village or directly at the study home during home visits. These problems are consistent with nearly all household water treatment interventions [5] and other public health cluster randomized trials [31, 32]. Schmidt and Cairncross [33] recently argued that reporting bias may have been the dominant problem in unblinded studies included in a meta-analysis reporting a pooled estimate of a 49% reduction of diarrhea in trials investigating the effects of drinking water quality interventions [5]. However, their review of only four available blinded trials showing no effect demonstrates weak support for contrast. In addition, all of the blinded trials exhibited analytical shortcomings or had very broad CIs suggesting very low power. In the absence of blinding—unavoidable in many behavioral change interventions or household water treatment studies—we believe that data collection independent from the implementation is a crucial factor. Future reviews should include reporting on such additional quality parameters.

In our study the lack of blinding may have reduced motivation in the control communities. However, the number of households lost during follow-up and the number of days under observation were almost identical in both arms. Additionally, the control communities knew that they would receive the intervention after study end. Finally, a reduction of diarrhea frequency of 20% might be insufficient to be well perceived, that is, have a noticeable impact in a population with a high burden of child diarrhea and will, thus, not result in a sustainable behavioral change. Faecal contamination in about 60% of the yards indicates a highly contaminated environment with presumably a large potential for transmission pathways other than consuming contaminated water. This simultaneous exposure to a multiplicity of transmission pathways may explain why we found no significant diarrhea reduction due to SODIS.

On the other hand, our result of a 19% reduction in diarrheal episodes appears to be roughly consistent with results of the two other SODIS trials both from Maasai cultural settings conducted by Conroy and colleagues among children <6 year and 5–16 year of age. They report a 16% reduction (in <6 year olds, 2-week prevalence of 48.8% in intervention, and 58.1% in control group) [8] and a 10.3% reduction in the 2-week diarrhea prevalence (in 5–16 year olds) [7]. However, these randomized controlled trials were undertaken in a socio-cultural setting assuring a 100% compliance (as stated by the authors) in water treatment behavior through social control by Maasai elders who promoted the method [7, 8]. In the results presented in these studies adjusted models with post hoc selected covariates were presented (i.e., no unadjusted models were provided). These trials were carried out in conditions of heavily contaminated drinking water and very high diarrhea rates—important considerations when attempting to generalize these results. The only other—quasi-randomized—trial to estimate the effect of solar water disinfection was carried out in the urban slum in Vellore and resulted in a remarkable reduction of diarrhea among children <5 year (IR ratio, 0.64; 95% CI 0.48–0.86) despite 86% of SODIS users also drinking untreated water [9].

To our knowledge this is the first community-randomized trial and the largest study so far to assess the effectiveness of the SODIS method under typical social and environmental conditions in a general rural population setting where children drink untreated water.

Our study was sufficiently powered to detect a 33% reduction in the effectiveness of the SODIS intervention, and we accounted for clustered design in our analysis. On the basis of a post hoc sample size calculations using the model-based estimate for the between-cluster variability ($k = 0.27$), we would have needed a study 2.5 times larger for a 20% difference to be significant.

The implementing NGO, which had global experience in disseminating SODIS, adapted a campaign to the local and cultural needs and also involved the public health and educational system in the roll-out. This comprehensive SODIS campaign resulted in a mean SODIS usage of 32% on any given study day. In using the SODIS-use indicator on the basis of the personal judgment of community-based staff, we intended to measure actual use in combining objective, visible signs of use (e.g., bottles exposed to sunlight) with proxies more responsive to actual treatment behavior (e.g., SODIS water can be offered to drink upon request). We consider this a restrictive, more conservative definition of SODIS use compared to that in other studies, which recorded reported use [9] or the number of bottles exposed to sunlight [34]. Both are indicators that can easily and reliably be measured, but which are prone to over-reporting due to low specificity for actual use. Further studies will need to validate different compliance indicators and formally assess the dimension of reporting bias.

It is possible that respondents would like to please field staff and over-report use out of courtesy. Also, observing exposed bottles on the roof may overestimate use (Figure 3), because some households were noted anecdotally to have placed bottles on the roof to avoid discussions with the SODIS-implementing NGO staff. Figure 3 is indicative of this phenomenon, as reported use at the beginning and reported use and satisfaction with the method at the end of the study reached the 80% mark—a usage figure consistent with other studies relying on reported compliance [9] and evaluation reports from gray literature. We conclude that self-reported SODIS use may overestimate compliance and a combination of reported and objectively measurable indicators provides more accurate SODIS-compliance data.

There are limitations to our study. As in other studies [24, 35], we observed a decline in the reporting of child diarrhea during the observational period in both arms (Figure 2). If true, seasonal variation of diarrhea could be one possible cause; increased awareness leading to more attention to basic hygiene and hence to illness reduction may be another reason. Alternatively, the pattern could be due to survey fatigue.

Despite a comprehensive and intensive intervention promotion campaign, we detected no strong evidence for a significant reduction in the IR of diarrhea in children <5 year in families using SODIS in our trial in a typical setting in rural Bolivia. We believe that clearer understandings of the discrepancy between laboratory and field results (obtained under typical environmental and cultural conditions), the role of compliance in effectiveness, and a direct comparison of SODIS to alternate drinking water treatment methods are needed before further global promotion of SODIS.

EDITORS' SUMMARY

Background

Thirsty? Well, turn on the tap and have a drink of refreshing, clean, safe water. Unfortunately, more than 1 billion people around the world do not have this option. Instead of the endless supply of safe drinking water that people living in affluent, developed countries take for granted, more than a third of people living in developing countries only have contaminated water from rivers, lakes, or wells to drink. Because of limited access to safe drinking water, poor sanitation, and poor personal hygiene, 1.8 million people (mainly children under 5 years old) die every year from diarrheal diseases. This death toll could be greatly reduced by lowering the numbers of disease-causing microbes in household drinking water. One promising simple, low-cost, point-of-use water purification method is SODIS. In SODIS, recycled transparent plastic drinks bottles containing contaminated water are exposed to full sunlight for 6 hr. During this exposure, ultraviolet radiation from the sun, together with an increase in temperature, inactivates the disease-causing organisms in the water.

Why was This Study Done?

SODIS has been promoted as an effective method to purify household water since 1999, and about 2 million people now use the approach (www.SODIS.ch). However, although SODIS works well under laboratory conditions, very few studies have investigated its ability to reduce the number of cases of diarrhea occurring in a population over a specific time period (the incidence of diarrhea) in the real world. Before any more resources are used to promote SODIS—its effective implementation requires intensive and on-going education—it is important to be sure that SODIS really does reduce the burden of diarrhea in communities in the developing world. In this study, therefore, the researchers undertake a cluster-randomized controlled trial (a study in which groups of people are randomly assigned to receive an intervention or to act as controls) in 22 rural communities in Bolivia to evaluate the ability of SODIS to reduce diarrhea in children under 5 years old.

What Did the Researchers Do and Find?

For their trial, the researchers enrolled 22 rural Bolivian communities that included at least 30 children under 5 years old and that relied on drinking water resources that were contaminated with disease-causing organisms. They randomly assigned 11 communities (225 households, 376 children) to receive the intervention—a standardized, interactive SODIS promotion campaign conducted by Project Concern International (a NGO)—and 11 communities (200 households, 349 children) to act as controls. Households in the intervention arm were trained to expose water-filled plastic bottles for at least 6 hr to sunlight using demonstrations, role play, and videos. Mothers in both arms of the trial completed a daily child health diary for a year. Almost 80% of the households self-reported using SODIS at the beginning and end of the study. However, community-based field workers estimated that only 32.1% of households on average used SODIS. Data collected in the child health diaries, which were completed on more than three-quarters of days in both arms of the trial, indicated that the children in the intervention arm had 3.6 episodes of diarrhea per year whereas the children in the control arm had 4.3 episodes of diarrhea per year. The difference in episode numbers was not statistically significant, however. That is, the small difference in the incidence of diarrhea between the arms of the trial may have occurred by chance and may not be related to the intervention.

What do these Findings Mean?

These findings indicate that, despite an intensive campaign to promote SODIS, less than a third of households in the trial routinely treated their water in the recommended manner. Moreover, these findings fail to provide strong evidence of a marked reduction of the incidence of diarrhea among children following implementation of SODIS although some aspects of the study design may have resulted in the efficacy of SODIS being underestimated. Thus, until additional studies of the effectiveness of SODIS in various real world settings have been completed, it may be unwise to extend the global promotion of SODIS for general use any further.

KEYWORDS

- **Generalized linear mixed models**
- **Intracluster correlation coefficient**
- **Nongovernmental organization**
- **Relative rate**
- **Solar drinking water disinfection**

AUTHORS' CONTRIBUTIONS

ICMJE criteria for authorship read and met: Daniel Mäusezahl, Andri Christen, Gonzalo Duran Pacheco, Fidel Alvarez Tellez, Mercedes Iriarte, Maria E. Zapata, Myriam Cevallos, Jan Hattendorf, Monica Daigl Cattaneo, Benjamin Arnold, Thomas A. Smith, and John M. Colford, Jr. Agree with the manuscript's results and conclusions: Daniel Mäusezahl, Andri Christen, Gonzalo Duran Pacheco, Fidel Alvarez Tellez, Mercedes Iriarte, Maria E. Zapata, Myriam Cevallos, Jan Hattendorf, Monica Daigl Cattaneo, Benjamin Arnold, Thomas A. Smith, and John M. Colford, Jr. Designed the experiments/the study: Daniel Mäusezahl, Thomas A. Smith, and John M. Colford, Jr. Analyzed the data: Daniel Mäusezahl, Andri Christen, Gonzalo Duran Pacheco, Jan Hattendorf, Monica Daigl Cattaneo, Benjamin Arnold, and John M. Colford, Jr. Collected data/did experiments for the study: Andri Christen, Myriam Cevallos, and Mercedes Iriarte. Enrolled patients: Andri Christen and Myriam Cevallos. Wrote the first draft of the chapter: Daniel Mäusezahl, Andri Christen, and Jan Hattendorf. Contributed to the writing of the chapter: Daniel Mäusezahl, Andri Christen, Gonzalo Duran Pacheco, Myriam Cevallos, Jan Hattendorf, Monica Daigl Cattaneo, Benjamin Arnold, Thomas A. Smith, and John M. Colford, Jr. Responsible on site for the overall study coordination and supervision: Andri Christen. Contributed to the laboratory studies specifically the microbiological monitoring of water quality: Mercedes Iriarte. Conducted analysis of stool specimen: Maria E. Zapata. Responsible for the coordination and supervision of the field activities and field data collection team: Myriam Cevallos. Administrative and technical support: Jan Hattendorf. Advised on data analysis: Thomas A. Smith.

ACKNOWLEDGMENTS

The authors greatly acknowledge the families who participated in the study, the dedication of our study communities, and the support of the authorities of Cochabamba and Totora District that made this project possible. Project Concern International (PCI) authorized embedding this evaluation trial in their local community development plans—we specifically thank the PCI implementation teams of Carlos Morante and Luciano Cespedes and their field staff for their relentless efforts to intertwine research and development approaches. We are grateful to the study team: Roy Cordova (administrator Bolivia site), Freddy Arauco (data entry, supervisor), Abrahan Cuevas, Fernando Salvatierra, David Villaroel, Dora Claros, Elmer Garvizu, Alfonso Claure (field supervisors and sample collectors), Claudia Lazarte (medical practitioner), and the entire MMS team (Morbidity Monitoring Staff). We thank Sonia Peredo, Gabriela Almanza, and Gonzalo Fillips (laboratory staff); Jenny Rojas, Edgar Sejas, Ana Maria Romero, and Mirjam Mäusezahl for their personal and institutional support. Lee Riley (UCB), Alan Hubbard (UCB), and Joseph Eisenberg, University of Michigan and formerly at UCB, contributed to the development of the research plan for this trial. We greatly appreciate the manifold administrative support of Catherine Wright (UCB) and Ulrich Wasser (STI, Basel). Marcel Tanner (STI) kindly reviewed the manuscript. Markus Niggli, Tim Haley, Michael Hobbins, and Stephan Indergand contributed to the study implementation and initial analyses. The contents of the chapter are solely the responsibility of the authors and do not necessarily represent the official view of the National Institutes of Health (NIH). Portions of this manuscript were presented at the 12th International Congress on Infectious Diseases in Lisbon, Portugal, on June 16, 2006.

Chapter 15

Zinc Treatment for Childhood Diarrhea in Bangladesh

Charles P. Larson, Unnati Rani Saha, and Hazera Nazrul

INTRODUCTION

Zinc treatment of childhood diarrhea has the potential to save 400,000 under-five lives per year in lesser developed countries. In 2004 the World Health Organization (WHO)/UNICEF revised their clinical management of childhood diarrhea guidelines to include zinc. The aim of this study was to monitor the impact of the first national campaign to scale up zinc treatment of childhood diarrhea in Bangladesh.

Between September, 2006 and October, 2008 seven repeated ecologic surveys were carried out in four representative population strata: mega-city urban slum and urban nonslum, municipal, and rural. Households of approximately 3,200 children with an active or recent case of diarrhea were enrolled in each survey round. Caretaker awareness of zinc as a treatment for childhood diarrhea by 10 months following the mass media launch was attained in 90%, 74%, 66%, and 50% of urban nonslum, municipal, urban slum, and rural populations, respectively. By 23 months into the campaign, approximately 25% of urban nonslum, 20% of municipal and urban slum, and 10% of rural under-five children were receiving zinc for the treatment of diarrhea. The scale-up campaign had no adverse effect on the use of oral rehydration salt (ORS).

Long-term monitoring of scale-up programs identifies important gaps in coverage and provides the information necessary to document that intended outcomes are being attained and unintended consequences avoided. The scale-up of zinc treatment of childhood diarrhea rapidly attained widespread awareness, but actual use has lagged behind. Disparities in zinc coverage favoring higher income, urban households were identified, but these were gradually diminished over the 2 years of follow-up monitoring. The scale up campaign has not had any adverse effect on the use of ORS.

About 1.9 million children under the age of 5 year annually die from diarrhea, accounting for 19% of all under-five mortality [1]. Clinical trials of zinc treatment in children 6 months to 5 year of age have consistently demonstrated its ability to reduce disease duration and severity as well as the likelihood of a repeat episode [2-4]. It has been estimated that the successful scaling up of zinc treatment for childhood diarrhea could potentially save 400,000 under-five deaths per year [5]. In response to the curative and preventive evidence in support of zinc treatment, in 2004 WHO/UNICEF revised their clinical management of childhood diarrhea guidelines to include zinc treatment of any episode [6]. The present challenge is the scale up of zinc treatment and other life-saving interventions within resource-deprived health systems with limited capacity to absorb additional services. As has been pointed out by others, efforts to bring zinc treatment to scale have the potential to significantly benefit children, but may also have harmful consequences through their impact on other health behaviors or services [7, 8].

The "Scaling Up of Zinc for Young Children" (SUZY) Project was established in 2003 with the aim of setting Bangladesh on the path to covering all under-five children with zinc treatment of any diarrheal illness episode. A partnership was created that included public, private, nongovernmental organization, and multinational sector agencies. Over a period of 3 year activities in support of preparing for the national scale up included formative and operational research, product registration and technology transfer, awareness building and orientation of health professionals, and preparation of mass media messages. In December, 2006 a national mass media campaign to promote a dispersible tablet zinc formulation, "Baby Zinc," for the treatment of childhood diarrhea was launched. All media messages linked zinc treatment to the continued use of ORS.

In order to monitor the success of the project in achieving its intended aims, as well as other unintended consequences, continuous nationally representative zinc coverage surveys have been carried out in Bangladesh that coincide with the national launch of a mass media zinc treatment promotion campaign. This article describes the results of the national scale-up campaign over the initial 2 years of its conduct. The primary outcomes of interest were the documentation of the proportion of children receiving zinc treatment for a diarrheal illness, identifying disparities in zinc coverage, and monitoring for potential unintended outcomes, in particular decreased use of ORS.

MATERIALS AND METHODS

This study was reviewed and approved by the Research Review and Ethical Review Committees of the International Centre for Diarrheal Disease Research, Bangladesh (ICDDR,B). Given the high levels of suspicion created when asked to sign a document they cannot read and the minimal potential for harm, the Ethics Review Committee of ICDDR,B approved an informed verbal consent. A verbal consent was obtained from all participating interviewees, with the interviewer required to sign that consent when obtained.

Study Design

Repeat ecologic surveys were carried out in four representative population strata: urban Dhaka (mega-city) slum and nonslum, municipal (small city), and rural. Prior to the launch of the planned mass media campaign a baseline survey was completed between September and November, 2006. Thereafter, postlaunch surveys were repeated over the following dates; 12/2006 to 02/2007 (1–3 months), 03 to 05/2007 (4–6 months), 06 to 09/2007 (7–10 months), 10/2007 to 01/2008 (11–14 months), 02 to 05/2008 (15–18 months), and 06 to 10/2008 (19–23 months).

Population

Source Population

There are two large "mega-cities" in Bangladesh, one of which, Dhaka, was arbitrarily chosen (Figure 1). In Dhaka all districts were stratified as either predominantly urban slum or nonslum populations and two districts from each category were randomly selected. Rural sub-districts were purposively selected on the basis of their proximity to ICDDR,B field sites in the west (Abhonagar), southeast (Mirsarai), and northeast (Hobiganj) of Bangladesh. The municipal district capitals representing each rural site were also selected (Khuna, Camila, and Sylhet). The under-five population of these ten sites is approximately 1.5 million children.

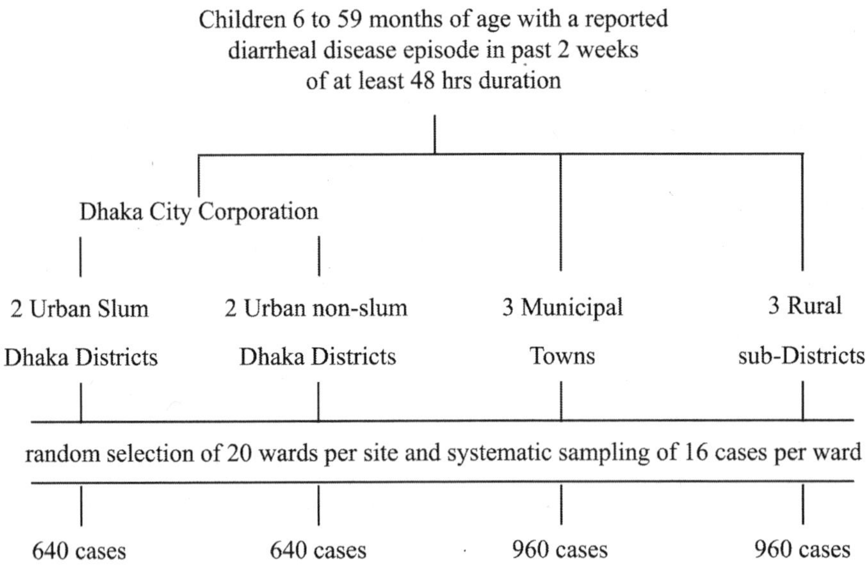

Figure 1. Summary of the cluster sampling framework for the surveys.

Study Population

Within the selected sites, all wards (defined as urban census tracts or an administrative grouping of two to three villages) were listed and 20 randomly selected prior to the start of each survey round. Standard WHO EPI (Expanded Program for Immunization) cluster methods were applied [9]. These methods eliminate the need to enumerate and then randomly select households. In rural settings a site central to the village was chosen and a wheel board spun to determine the direction of households to be visited following a more or less straight line. Upon reaching the peripheral boundary of the village the interviewers returned to the central point and spun the wheel again. Within urban wards a central starting point was chosen and then a systematic, door-to-door survey of households was carried out. Within each rural or urban ward the survey was stopped once 16 children, 6–59 months of age with an active or recent (within the past 2 weeks) episode of diarrhea of at least 48-hour duration had been identified, and informed verbal consent from either parent obtained. If more than one child in the household was eligible, one was randomly chosen.

Sample Size Estimation

Setting the level of confidence at 0.95 and a minimal detectable error of 0.05 around a prevalence estimate, assuming the overall prevalence of zinc use to be 0.20 and adjusting by 2.5 for design effect (DE), a minimum of 615 cases per population strata were estimated to be required. The DE = 1+roh (k–1), where roh (rate of homogeneity) was estimated to be 0.10 and k was arbitrarily set at 16 participants per cluster [9]. The choice of 16 individuals per cluster represents a compromise between data collection efficiency (larger number of participants per cluster) and reducing the DE (smaller number of participants per cluster). Piloting indicated that our research team could, on average, collect 16 cases in a day, but this would vary by time of year and the occurrence of diarrhea outbreaks.

Zinc Treatment Scale Up

Under the title "SUZY Project," preparations and the eventual implementation of a national campaign in support of zinc treatment for childhood diarrhea was carried out over a period of 3 years (2003–2006). The campaign targeted children 6 months to 5 years of age. This was a national effort, involving several public and private agencies, with coordination and financial support provided through the SUZY Project. There were no parallel campaigns occurring. Within the Bangladesh Ministry of Health and Family Welfare (MOHFW), a National Advisory Committee (NAC), chaired by the Health Secretary, was created. This committee was represented by the primary health care division of the MOHFW, the Bangladesh Pediatric Association, and country representatives from WHO and UNICEF. An implementation subcommittee was created to make specific recommendations to the NAC, including a zinc treatment policy statement, over-the-counter and advertising permits, and private sector participation. Private sector involvement included the selection of a pharmaceutical laboratory to produce and distribute a dispersible zinc tablet formulation, a marketing agency, and private provider groups, such as the Bangladesh Pediatric Association and the Bangladesh Village Doctors Association. Inclusion of these professional associations was critical given considerable doubt existed in the beginning about zinc safety and over-the-counter sales.

Following the selection of a pharmaceutical laboratory (ACME Pharmaceuticals, Ltd) a technology transfer from the French nutrition firm holding the patent for the dispersible zinc formulation (Nutriset Ltd) was arranged and eventually completed. Following the technology transfer, this included yearly scheduled quality control visits. The ACME Pharmaceuticals then applied and obtained from the MOHFW Drug Administration for the following in the sequence listed: approval of the dispersible zinc tablet formulation, brand name (Baby Zinc) and packaging design, pricing (18 taka or approximately US $0.25 for a 10-tablet blister pack), an over-the-counter waiver, and a permit to advertise nationally on TV and radio.

Prior to initiating the national scale up several gaps in knowledge were identified and protocols prepared to provide the needed information. These included: (1) a phase IV safety and side effects studies carried out among children attending the ICDDR,B Dhaka hospital [10, 11]; (2) formative studies to better understand household diarrhea management decisions, treatment recommendations made by providers, the influence of drug salesmen, and knowledge of zinc and other micronutrients; (3) the acceptability of the dispersible tablets and adherence to preparation instructions [12]; and (4) a national survey in rural, urban slum, and nonslum populations to determine current childhood diarrhea management practices, health seeking behaviors, and expenditures [13]. These studies enabled us to reassure stakeholders that zinc treatment is safe, but associated with a low risk of nausea and vomiting. The results from the formative studies guided the preparation of messages for caregivers and providers. On the basis of these interviews and focus group discussions we created a frequently asked questions data bank and a uniform set of responses (available at www.icddrb.org). From the surveys we learned that over 90% of parents sought help from private sector providers and over 70% of the time it was with an unregulated provider (village doctor or drug vendor). It was this information that led to the decision to focus the campaign on caretaker decision making and availability of the zinc tablets in the private sector. Nevertheless, all public sector district health and family welfare centers were provided zinc tablets free of charge.

Prior to and following the national launch of the scale-up campaign, training sessions in diarrhea management in line with the revised WHO/UNICEF guidelines were conducted. These were tailored for pediatricians, general MBBS physicians, medical schools, and unregulated providers. For the latter, a 30-min training video was prepared. The ACME drug salesmen provided verbal information and distributed a specially prepared pamphlet to private providers.

Because the large majority of parents interviewed identified TV as their primary source of information, the marketing campaign focused on this medium, but also prepared messages for radio, newspapers, billboards, and buses. In rural settings zinc promotion additionally included the sponsoring of cultural events and courtyard meetings. Four 30-sec TV advertisements were prepared, the first being a "teaser" and the other three informing listeners of Baby Zinc treatment of diarrhea. These advertisements were broadcasted on Bangladesh National Television, which reaches all parts of Bangladesh. The messages included in the mass media promotions included awareness of zinc treatment for childhood diarrhea and its sanctioning by health providers in Bangladesh. All promotional activities linked zinc treatment to the continued use of ORS. Furthermore, as ORS is stopped once the diarrhea subsides, an additional message to continue the zinc for a full 10 day was included. Children 6 months to 5 years of age were targeted.

Survey Interviews

Twelve trained field research assistants divided into two teams carried out the household interviews and completed a 36–40 item questionnaire, depending upon the survey round. Specific household management practices were documented first, followed by knowledge questions, including the question "Prior to this illness were you aware that zinc can be used to treat diarrhea in your child?" The recommended zinc treatment is 20 mg/day for 10 days, either as a dispersible tablet or 5-cc syrup formulation. In addition to asking whether zinc was used, we asked for how many days. Interviewers carried a laminated chart with photos of all syrup and tablet formulations being sold in Bangladesh to assist interviewees to identify the product used. In households where the purchased blister pack or bottle was available, we directly visualized them to confirm the reported number of days of use. Credit was given for any zinc formulation given.

Each of the ten selected survey sites required 7–10 days to complete and each round therefore required 3–4 months. Timing was affected by seasonal floods, however the surveys remained on schedule for the 23 months of follow-up monitoring. The interviews addressed the diarrhea illness history, health seeking behaviors, home management practices, illness-related expenditures, and sociodemographic characteristics of the households.

Analysis

Data were entered and verified using SPSS-PC version 12. These files were then converted to STATA-PC version 10.0 for all analyses. Data files were checked for outliers and reduced categorical variables were generated. The analyses were stratified by location of residence into urban (Dhaka city corporation slum and nonslum districts), municipal, and rural households. To assess differences in categorical outcomes crude relative risks, 95% confidence intervals, and chi-square statistical comparisons of proportions were calculated. Of particular interest was the identification of disparities in the use of zinc by gender, geographic location, and income status of the household. Income status was estimated by determination of a household asset score based upon ownership of consumer items, dwelling characteristics, toilet facilities used, and other household characteristics that are related to wealth status [14, 15]. Each asset is assigned a weight generated through principal components analysis and then standardized scores assigned. The hypotheses we tested were that significant ($p<0.05$) differences in the likelihood of receiving zinc treatment for childhood diarrhea would be found favoring males, higher income, and urban nonslum households. Disparities were tested for significance applying Pearson chi-square statistics. To asses the magnitude and trends in income disparities, concentration index curves were determined for each follow-up period [16].

DISCUSSION

This 2-year follow-up monitoring of the national campaign to scale up zinc treatment of childhood diarrhea in Bangladesh has resulted in several observations of relevance to future scale-up efforts. This campaign largely focused on the promotion of zinc treatment among private sector providers, but with the strong support of the public health sector. The TV and radio promotion of zinc treatment through an electronic mass media campaign was able to rapidly attain high levels of awareness throughout the country. Progress in the actual use of zinc has been slower, with the early adoption primarily observed among urban, higher income households. Among rural and urban slum households, whose children stand to benefit the most from zinc, zinc treatment coverage steadily increased over time and the magnitude of disparity based upon income status was observed to have been reduced. Importantly, the scale-up campaign has not negatively impacted the use of ORS. An important, unmet challenge has been the failure to adhere to a 10-days course of treatment as evidenced by the fact over 50% of caregivers were sold seven or fewer days of zinc treatment.

A potential source of bias and limitation of this study is the populations surveyed, which may not be representative of hard to reach, more remote sites in Bangladesh. These sites were chosen because ICDDR,B researchers were known in these communities, support structures were in place, and local approval to conduct the surveys could be more rapidly obtained. The purpose of these surveys was to document trends in the use of zinc and changes in other practices within the stratified populations described. The findings may not, with confidence, be extrapolated to accurately estimate zinc coverage in all districts of Bangladesh. The sites chosen within each population strata are, nevertheless, typical Bangladeshi communities and we are confident that the observed trends in zinc coverage and reductions in disparities are indicative of what is occurring in Bangladesh as a whole. Bias in estimates of zinc treatment awareness may also have been introduced by repeatedly surveying in the same sub-districts. Those caretakers interviewed may have discussed the experience with relatives or neighbors, including the mention of zinc treatment. Households with a repeated case of diarrhea were not replaced because it was concluded their exclusion would lead to a biased selection of healthier children in subsequent surveys. The sites surveyed contain an estimated population of nearly 1.5 million children under 5 years of age. This was felt to be a large enough population base to minimize biased estimates of zinc awareness and would not have affected zinc coverage estimates. A strength of the surveys was the selection of households where a child had an active or recent diarrhea episode of at least 2-days duration. This eliminated transient, less important episodes and responses were based upon actual practices.

Promotion among health providers followed two strategies: half-day diarrhea training workshops and product promotion by drug salesmen. There are estimated to be over 200,000 health providers in Bangladesh, thus the challenge of reaching them all through workshops is not realistic. We therefore placed an early emphasis on sensitizing and training recognized leaders, such as pediatricians and educators. Less well-trained providers tend to look up to pediatricians and copy their practices. Among the unlicensed providers a training of trainers approach was used. While this set of surveys cannot document the proportion reached, it is likely the majority of health providers remain poorly informed about zinc treatment. Reliance on drug salesmen also has limitations. These individuals and the systems within which they work are profit oriented and based upon prescription medications. Zinc is cheap and it is an over-the-counter product. Not surprisingly, drug salesmen will be more inclined to promote higher priced products. Promotion and distribution of zinc through alternative systems, for example bottled water distribution networks, would reach a far greater number of outlets and increase its availability within rural or urban communities,

Rogers' diffusion of innovation theory is useful for understanding progress with scaling up health interventions in the general population over time. The theory describes the adoption of new innovations as passing through five stages of decision making—awareness, interest, evaluation, trial, and adoption [17]. At any stage a consumer, in our case providers or caretakers, can choose to reject the innovation. The mass media campaign was able to achieve high levels of awareness and probably interest among all segments of the Bangladeshi population. Where it has fallen short is in the transition from awareness to practice (trial and adoption). This gap highlights an important limitation of electronic media, which does not benefit from interpersonal communication, thus showing the need to link mass media messages with personal messages coming from health providers or other influential members of a community. The content of the initial commercials aired in this campaign repeatedly focused on awareness and health provider sanctioning of zinc treatment and not on household decision making. Towards the end of the second year of the campaign the media messages were altered to encourage household level decision making and enhancing self-efficacy to try zinc [18]. This strategy and interpersonal communication with early adopters of zinc treatment are expected to further increase coverage. The choice of electronic media in this scale-up campaign was, in part, based on its previous success in promoting the use of oral rehydration therapies (ORT) [19]. Nonetheless, ORTs and zinc share several characteristics that make them amenable to scaling up through mass media promotion. Both are fairly simple interventions that are easily learned and applied in the home. They are also relatively inexpensive and within the range of typical household diarrheal illness expenditures among Bangladeshi households [13].

Early adopters of new innovations are known to be better educated, of higher income status, and have greater access to mass media [17, 20]. We were able to monitor adoption of zinc treatment by household wealth asset quintiles. Throughout the 2 years of follow up children from higher wealth asset households were more likely to receive zinc treatment. At the outset of the campaign children in the highest quintile, when compared to the lowest quintile households, were seven times more likely to receive zinc. At the end of the second year this disparity had been reduced to less than three times as likely. This reduction in income disparity is further illustrated by the change in the concentration index curves from the beginning to the end of the follow-up monitoring.

Given the preventive effects of zinc are likely to require 8–10 days of treatment, the observation that over half of the children are receiving less than the required amount remains an important, unmet challenge. The mass media messages did include a parent directed reminder to give zinc for 10 days. Unfortunately, in Bangladesh drug vendors commonly sell antibiotics and other curative medications to cover only a few days. If a child remains ill, they return to purchase additional, often alternative, medication. Parents have little or no experience with continuing medications once their child appears to be cured. This behavioral change challenge currently lacks adequate scientific guidance.

All mass media messages in this campaign linked zinc to the use of oral saline. This connection is important, because zinc is an adjunct to and not a replacement for oral saline or other approved rehydration therapies. Future studies need to clarify whether linking zinc to oral saline may lead to caregiver misunderstanding, given they are instructed to discontinue the latter once the diarrhea subsides. Adherence would also be improved if it could be demonstrated that shorter duration zinc treatment schedules have equivalent clinical efficacies.

In summary, the national scale up of zinc treatment for childhood diarrhea in Bangladesh, the first national campaign to be undertaken, has met with modest success over the initial 2 years of promotion and has not had a detrimental effect on the use of ORT. While the relative benefits of provider versus caretaker mass media–focused promotion cannot be separated in this study, it is concluded that both strategies are important to undertake. To improve upon the overall coverage attained of about 20% of childhood diarrhea episodes will require greater, more consistent support from health providers and strengthened demand for zinc treatment from caregivers. Both of these objectives would benefit from an improved understanding of health system constraints as well as societal, household, and provider characteristics facilitating sustained behavioral change leading to the successful scale up of zinc and other life-saving interventions.

RESULTS

Seven survey rounds were completed with the range in the 2-weeks prevalence (p) of diarrhea and the number of cases (n) surveyed by location as follows; urban slums p = 0.20–0.25 and n = 642–646; urban nonslums p = 0.17–0.23 and n = 641–658; municipalities p = 0.15–0.19 and n = 965–979; rural p = 0.19–0.23 and n = 962–976. The mean age of the children at each survey ranged from 26.3 ± 0.3 months to 27.4 ± 0.3 months and the percentage of cases being female ranged from 44.2% to 46.6%. Of all cases identified, 98.5% were enrolled in the surveys.

Figure 2 summarizes caretaker awareness of zinc as a treatment for childhood diarrhea over time stratified by the location of the household. At baseline, prior to the launch of the mass media campaign, but following multiple workshops with licensed pediatricians, awareness was under 5% for all but urban, nonslum caregivers (99% mothers). In all locations awareness rapidly increased following the onset of mass media zinc promotion, reaching peak levels by 10 months into the campaign. Among the urban slum and rural populations surveyed, zinc treatment awareness reached 65% and 55%, respectively, by 23 months.

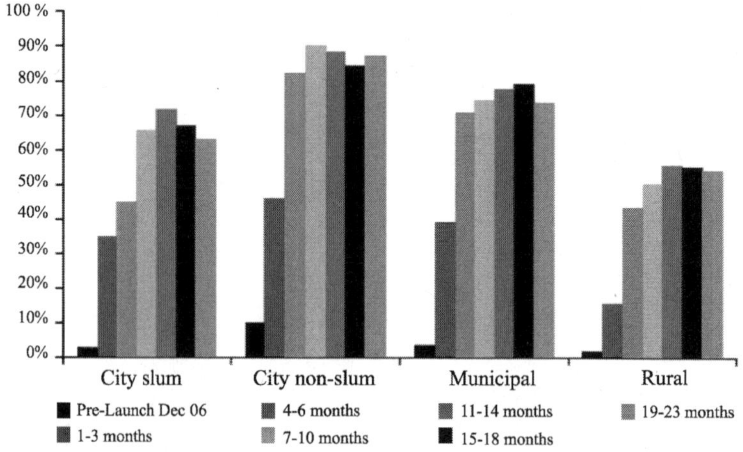

Figure 2. Caretaker awareness of zinc treatment for childhood diarrhea before and up to 23 months following the onset of the mass media campaign.

As illustrated in Figure 3, the actual use of zinc falls far short of awareness. With the exception of urban nonslum children, few received zinc prior to the national mass media launch. By the second year of the national scale-up campaign approximately 25–30% of urban nonslum, 15–20% of urban slum or municipal, and 9–13% of rural children were receiving zinc for their diarrheal illness episode. In urban nonslum and municipal households the use of zinc levelled off by the end of the first year, while a steady increase in zinc coverage has been observed in rural and urban slum areas. Table 1 summarizes the proportion of children who received zinc stratified by the type of provider seen during the final, 19- to 23-months survey round; this is further subgrouped by whether a tablet or syrup formulation was received. Children seen by a private, MBBS provider were the most likely to receive zinc, at nearly 40%. Dispersible zinc tablets accounted for 60% of all zinc purchased.

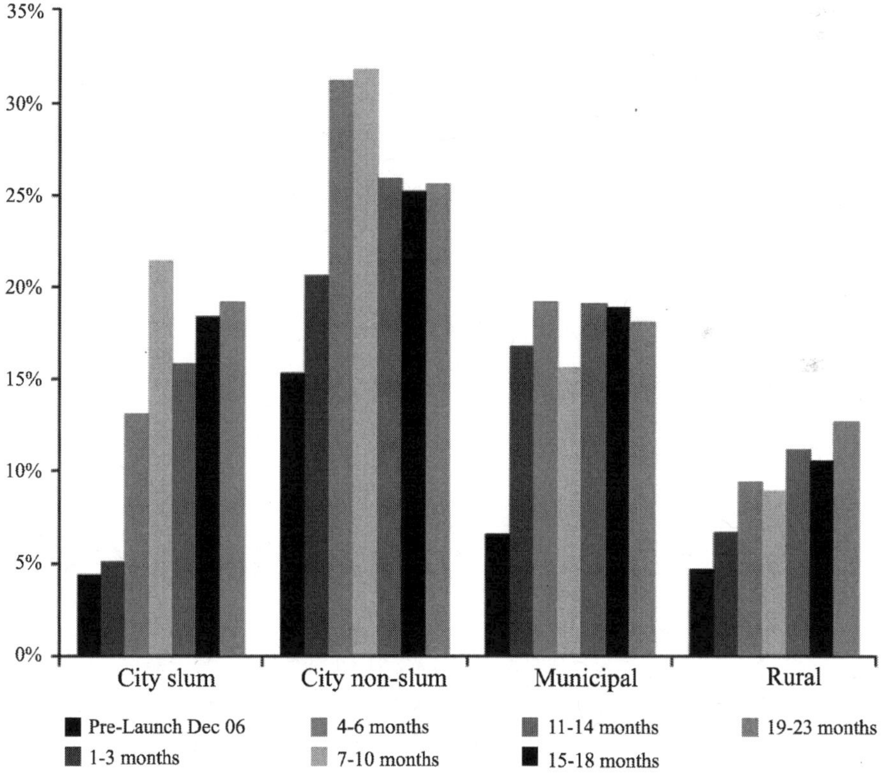

Figure 3. The proportion of children receiving zinc treatment by location of household before and up to 23 months following the onset of the mass media campaign.

Table 1. The proportion of children receiving zinc treatment and among these children the formulation taken by type of provider seen at 19–23 months post-launch of the scale-up campaign.

Provider	Child		Formulation Taken	
	n Children Seen	Received Zinc (%)	Tablet (%)	Syrup (%)
Private, unlicensed village doctors or drug vendors	1,670	18.1	73.5	26.5
Private, licensed MBBS doctors	372	39.2	43.2	56.8
Public sector, MOHFW health providers	220	25.9	49.1	50.9
Overall	2,262	20.3	60.9	39.1

72% of the children surveyed were seen by a provider for their diarrheal Illness.

doi:10.1371/journal.pmed.1000175.t001

In each survey period significant differences ($p<0.001$) in the use of zinc were observed favoring higher quintile wealth asset households (Figure 4). At 18 months significant disparities in the likelihood of receiving zinc treatment on the basis of gender were limited to municipal households and favored males (21% versus 16%, $p = 0.024$). No gender bias at any time interval was observed in urban slum and rural poor households. As can be seen from the concentration index curves summarized in Figure 5, income disparities in the use of zinc decreased over time. Referring to the figure, if there were no disparity in the use of zinc on the basis of household income status (asset score) then the poorest 60% of children would have accounted for 60% of the total zinc treatments received. At the outset of the mass media campaign (1–3 months) this lower 60% accounted for only 28% of the zinc treatments received, but by the end of the second year (19–23 months) this had risen to 46%.

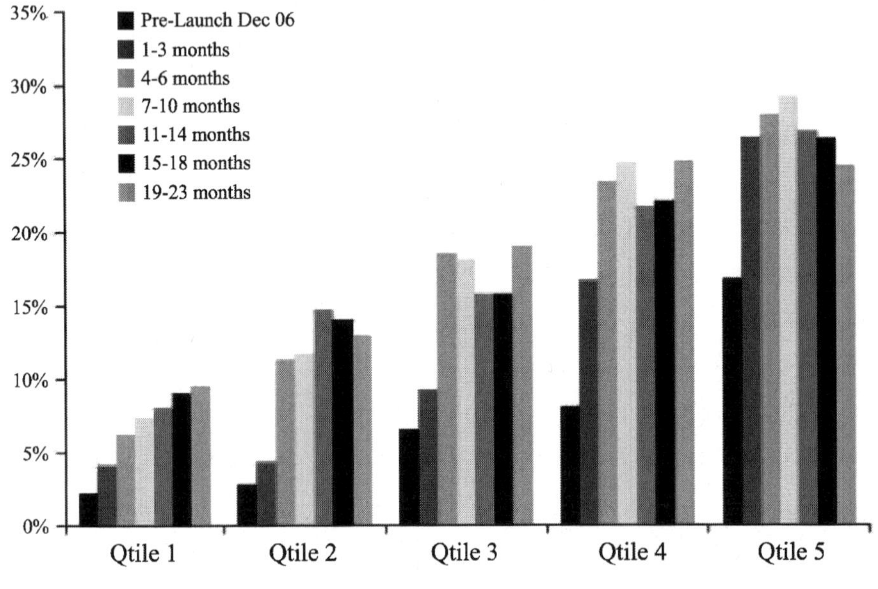

Figure 4. Changes in zinc coverage over time by household wealth asset quintile (1 lowest, 5 highest).

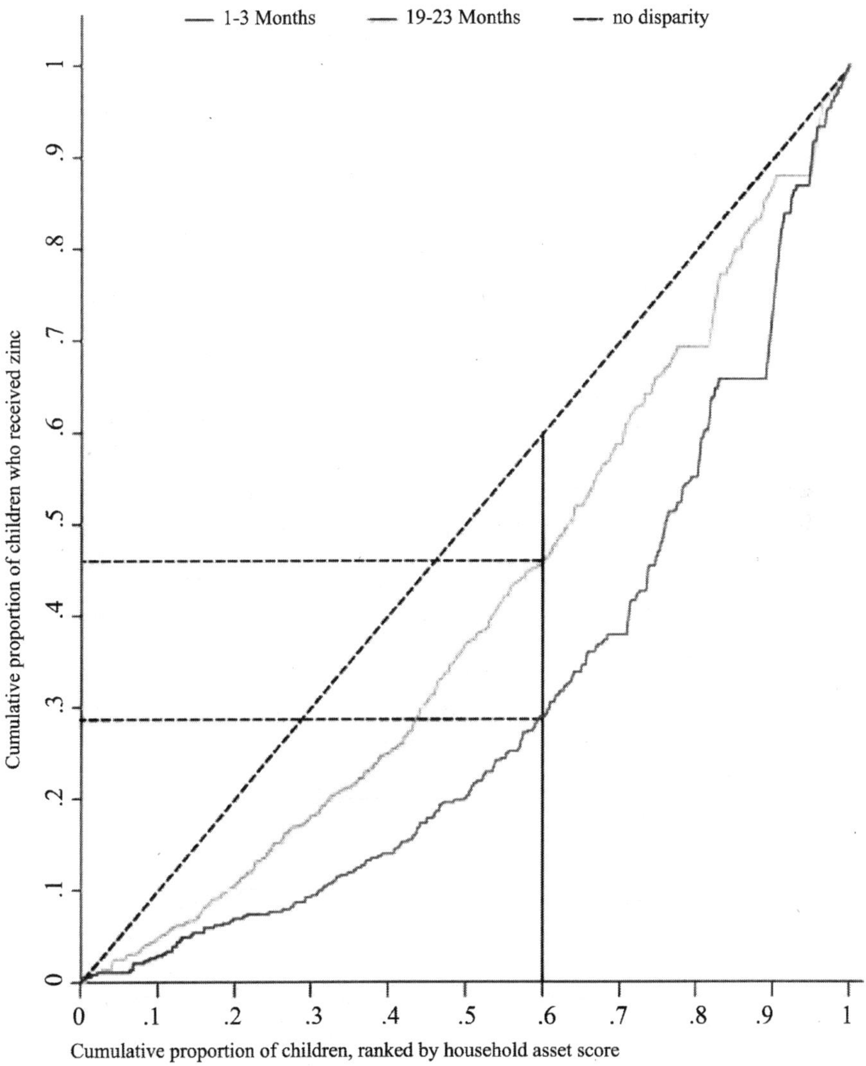

Figure 5. Income disparities in receiving zinc treatment as measured by concentration index curves 1–3 months and 19–23 months following the onset of the mass media campaign. Distribution of use of zinc is equal among income levels if the concentration curve coincides with diagonal. During the 1–3 months interval the 60% of poorest households accounted for 28% of the zinc treatments received, while from 19–23 months they accounted for 46%.

The cross-sectional design of these surveys does not permit a determination of the number of days the children received zinc treatment. As a proxy for duration of treatment, caretakers were asked the number of zinc tablets they purchased (Figure 6). By 19–23 months following the launch of the mass media campaign in each of the population strata nearly 50% or more of zinc tablet purchases were for <8 days.

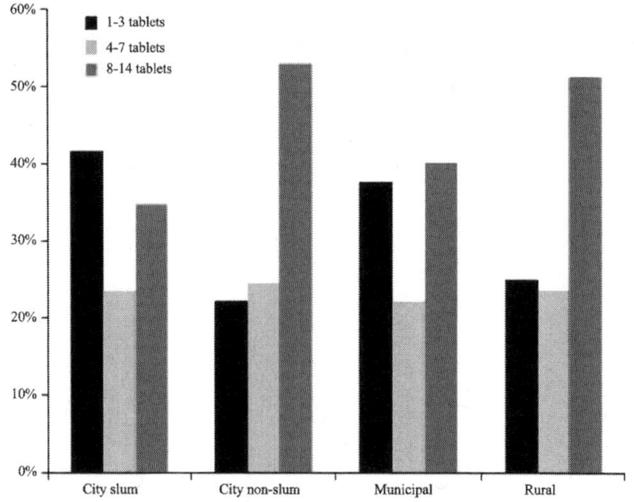

Figure 6. Number of zinc tablets purchased by caregivers during the final survey interval 19–23 months following the mass media launch.

Figure 7 summarizes changes in the use of ORS during the zinc scale-up campaign by household location. Variations in the estimated proportion of children receiving ORS occurred within each of the population strata over time, but no significant trend, upwards or downwards, in the use of ORS was observed. There was no significant change documented in the use of antidiarrheals. A significant reduction over the 2 years of follow-up in the use of antibiotics from 34.7 to 27.6% ($p<0.001$),were observed among urban nonslum households, but not in the other populations surveyed.

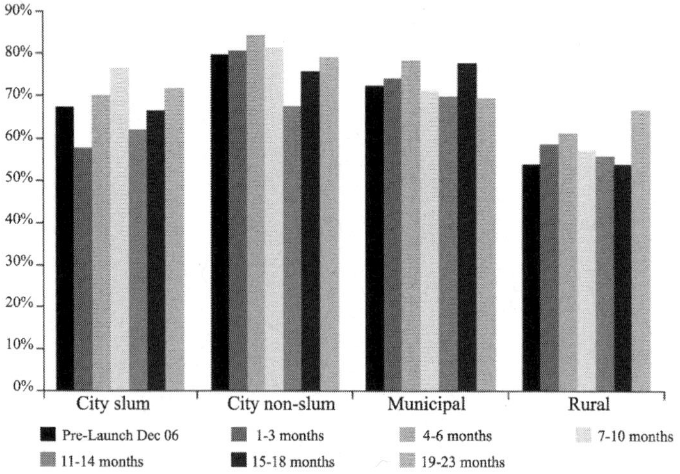

Figure 7. ORS utilization prior to and over the first 23 months of the mass media zinc scale-up campaign.

EDITORS' SUMMARY

Background

Diarrheal disease is a significant global health problem with approximately 4 billion cases and 2.5 million deaths annually. The overwhelming majority of cases are in developing countries where there is a particularly high death rate among children under 5 years of age. Diarrhea is caused by bacterial, parasitic, or viral pathogens, which often spread in contaminated water. Poor hygiene and sanitation, malnutrition, and lack of medical care all contribute to the burden of this disease. Replacing lost fluids and salts is a cheap and effective method to rehydrate people following dehydration caused by diarrhea. Clinical trials show that zinc, as part of a treatment for childhood diarrhea, not only helps to reduce the severity and duration of diarrhea but also reduces the likelihood of a repeat episode in the future. Zinc is now included in the guidelines by the WHO/UNICEF for treatment of childhood diarrhea.

Why was This Study Done?

Zinc treatment together with traditional ORS therapy following episodes of diarrhea could potentially benefit millions of children in areas where diarrheal disease is prevalent. The "Scaling Up of Zinc for Young Children" (SUZY) project was established in 2003 to provide zinc treatment for diarrhea in all children under 5 years of age in Bangladesh. The project was supported by a partnership of public, private, nongovernmental organization, and multinational sector agencies during its scale up to a national campaign across Bangladesh. The partners helped to develop the scale-up campaign, produce and distribute zinc tablets, train health professionals to provide zinc treatment, and create media campaigns (such as advertisements in TV, radio, and newspapers) to raise awareness and promote the use of zinc for diarrhea. The researchers wanted to monitor how effective and successful the national campaign was at promoting zinc treatment for childhood diarrhea. Also, they wanted to highlight any potential problems during the implementation of health care initiatives in areas with deprived health systems.

What Did the Researchers Do and Find?

The researchers set up survey sites to monitor results from the first 2 years of the SUZY campaign. Four areas, each representing different segments of the population across Bangladesh were surveyed; urban slums, urban nonslums, municipal (small city), and rural. There are approximately 1.5 million children under the age of five across these sites. Households in each survey site were selected at random, and seven surveys were conducted at each site between September, 2006 and October, 2008—about 3,200 children with diarrhea for each survey. Over 90% of parents used private sector providers of drug treatment so the campaign focused on distribution of zinc tablets in the private sector. They were also available free of charge in the public health sector. The TV and radio campaigns for zinc treatment rapidly raised awareness across Bangladesh. Awareness was less than 10% in all communities prelaunch and peaked 10 months later at 90%, 74%, 66%, and 50% in urban nonslum, municipal, urban slum, and rural sites, respectively. However, after 23 months only 25% of urban nonslum,

20% of municipal and urban slum, and 10% of rural children under 5 years of age were actually using zinc for childhood diarrhea. Use of zinc was shown to be safe, with few side-effects, and did not affect the use of traditional treatments for diarrhea. Researchers also found that many children were not given the correct 10-days course of treatment; 50% of parents were sold seven or fewer zinc tablets.

What Do These Findings Mean?

These findings show that the first national campaign promoting zinc treatment for childhood diarrhea in Bangladesh has had some success. Addition of zinc tablets for diarrhea treatment did not interfere with existing therapies. Mass media campaigns, using TV and radio, were useful for promoting health care initiatives nationwide alongside the education of health care providers and care givers. The study also identified areas where more work is needed. Surveys in more remote, hard to reach sites in Bangladesh would provide better representation of the country as a whole. High awareness of zinc did not translate into high use. Repeated surveying in the same subdistricts may have overestimated actual awareness levels. Furthermore, mass media messages must link with messages from health care providers to help to reinforce and promote understanding of the use of zinc. A change in focus of media messages from awareness to promoting household decision-making may aid the adoption of zinc treatment for childhood diarrhea and improve adherence.

KEYWORDS

- **Childhood diarrhea**
- **Oral rehydration salt**
- **Oral rehydration therapies**
- **World Health Organization**
- **Zinc treatment**

AUTHORS' CONTRIBUTIONS

ICMJE criteria for authorship read and met: Charles P. Larson, Unnati Rani Saha, and Hazera Nazrul. Agree with the manuscript's results and conclusions: Charles P. Larson, Unnati Rani Saha, and Hazera Nazrul. Designed the experiments/the study: Charles P. Larson and Hazera Nazrul. Analyzed the data: Charles P. Larson, Unnati Rani Saha, and Hazera Nazrul. Enrolled patients: Hazera Nazrul. Wrote the first draft of the chapter: Charles P. Larson. Contributed to the writing of the chapter: Unnati Rani Saha and Hazera Nazrul . Participated in planning and writing the protocol, oversaw data management:Unnati Rani Saha, . Participated in the planning and writing the protocol, managed field site data collection:Hazera Nazrul.

ACKNOWLEDGMENTS

ICDDR,B acknowledges with gratitude the commitment of the Bill & Melinda Gates Foundation to the Centre's research efforts.

Permissions

Chapter 1: Ecological Equivalence: Niche Theory as a Testable Alternative to Neutral Theory was originally published as "Ecological Equivalence: Niche Theory as a Testable Alternative to Neutral Theory" in *PLoS ONE 10:14, 2009*. Reprinted with permission under the Creative Commons Attribution License or equivalent.

Chapter 2: Energy Sprawl or Energy Efficiency: Climate Policy Impacts on Natural Habitat was originally published as "Energy Sprawl or Energy Efficiency: Climate Policy Impacts on Natural Habitat for the United States of America" in *PLoS ONE 8:26, 2009*. Reprinted with permission under the Creative Commons Attribution License or equivalent.

Chapter 3: Mediterranean Ecosystems Worldwide: Climate Adaptation was originally published as "Climate Change, Habitat Loss, Protected Areas and the Climate Adaptation Potential of Species in Mediterranean Ecosystems Worldwide" in *PLoS ONE 7:29, 2009*. Reprinted with permission under the Creative Commons Attribution License or equivalent.

Chapter 4: Environmental Knowledge Management was originally published as "The impact of industrial knowledge management and environmental strategy on corporate performance of iso-14000 companies in Taiwan: The application of structural equation modeling" in *African Journal of Business Management Vol. 4 (1), pp. 021-030, January 2010 ISSN 1993-8233 © 2010 Academic Journals*. Reprinted with permission under the Creative Commons Attribution License or equivalent.

Chapter 5: Electromagnetic Radiation Effect on Foraging Bats was originally published as "The Aversive Effect of Electromagnetic Radiation on Foraging Bats—A Possible Means of Discouraging Bats from Approaching Wind Turbines" in *PLoS ONE 7:16, 2009*. Reprinted with permission under the Creative Commons Attribution License or equivalent.

Chapter 6: Phylogeographic Analysis of Tick-borne Pathogen was originally published as "Phylogeographic analysis reveals association of tick-borne pathogen, *Anaplasma marginale*, MSP1a sequences with ecological traits affecting tick vector performance" in *BioMed Central 9:1, 2009*. Reprinted with permission under the Creative Commons Attribution License or equivalent.

Chapter 7: Cancer Risk Among Residents of Rhineland-Palatinate Winegrowing Communities was originally published as "Cancer risk among residents of Rhineland-Palatinate winegrowing communities: a cancer-registry based ecological study" in *BioMed Central 6:6, 2008*. Reprinted with permission under the Creative Commons Attribution License or equivalent.

Chapter 8: Risk of Congenital Anomalies Around a Municipal Solid Waste Incinerator was originally published as "Risk of congenital anomalies around a municipal solid waste incinerator: a GIS-based case-control study" in in *BioMed Central 2:10, 2009*. Reprinted with permission under the Creative Commons Attribution License or equivalent.

Chapter 9: Spatial Analysis of Plague in California was originally published as "Spatial analysis of plague in California: niche modeling predictions of the current distribution and potential response to climate change" in *BioMed Central 6:28, 2009*. Reprinted with permission under the Creative Commons Attribution License or equivalent.

Chapter 10: Australia's Dengue Risk: Human Adaptation to Climate Change was originally published as "Australia's Dengue Risk Driven by Human Adaptation to Climate Change" in *Beebe NW, Cooper RD, Mottram P, Sweeney AW (2009) Australia's Dengue Risk Driven by Human Adaptation to Climate Change. PLoS Negl Trop Dis 3(5): e429. Doi: 10.1371/journal.pntd.0000429*. Reprinted with permission under the Creative Commons Attribution License or equivalent.

Chapter 11: Threats from Oil and Gas Projects in the Western Amazon was originally published as "Oil and Gas Projects in the Western Amazon: Threats to Wilderness, Biodiversity, and Indigenous

References

1

1. Marschner, H. (1995). *Mineral Nutrition of Higher Plants*. Academic Press Limited, London.
2. Barker, A. V. and Bryson, G. M. (2007). Nitrogen. In *Handbook of Plant Nutrition*. A. V. Barker and D. J. Pilbeam (Eds.). CRC Press, Boca Raton, pp. 21–50.
3. Stitt, M. (1999). Nitrate regulation of metabolism and growth. *Curr. Opin. Plant Biol.* **2**, 178–186.
4. Brouquisse, R., Masclaux, C., Feller, U., and Raymond, P. (2001). Protein hydrolysis and nitrogen remobilisation in plant life and senescence. In *Plant Nitrogen*. P. J. Lea and J. F. Morot-Gaudry (Eds.). Springer-Verlag Berlin Hidelberg, Hidelberg, pp. 275–293.
5. Stitt, M., Müller, C., Matt, P., Gibon, Y., Carillo, P., Morcuende, R., Sheible, W. R., and Krapp, A. (2002). Steps towards an integrated view of nitrogen metabolism. *J. Exp. Bot.* **53**, 959–970.
6. Miller, A. J. and Cramer, M. D. (2004). Root nitrogen acquisition and assimilation. *Plant Soil* **274**, 1–36.
7. Scheible, W. R., Morcuende, R., Czechowski, T., Fritz, C., Osuna, D., Palacios-Rojas, N., Schindelasch, D., Thimm, O., Udvardi, M. K., and Stitt, M. (2004). Genome-wide programming of primary and secondary metabolism, protein synthesis, cellular growth processes, and the regulatory infrastructure of *Arabidopsis* in response to nitrogen. *Plant Physiol.* **136**, 2483–2499.
8. Jackson, L. E., Burger, M., and Cavagnaro, T. R. (2008). Roots, nitrogen transformation and ecosystem services. *Annu. Rev. Plant Biol.* **59**, 341–363.
9. Lawlor, D. W., Lemaire, G., and Gastal, F. (2001). Nitrogen, plant growth and crop yield. In *Plant Nitrogen*. P. J. Lea and J. F. Morot-Gaudry (Eds.). Springer-Verlag Berlin Hidelberg, Hidelberg, 343–367.
10. Hirel, B., Bertin, P., Quillere, I., Bourdoncle, W., Attagnant, C., Dellay, C., Gouy, A., Cadiou, S., Retailliau, C., Falque, M., and Gallais, A. (2001). Towards a better understanding of the genetic and physiological basis for nitrogen use efficiency in maize. *Plant Physiol.* **125**, 1258–1270.
11. Hagedorn, F., Bucher, J. B., and Schleppi, P. (2001). Contrasting dynamics of dissolved inorganic and organic nitrogen in soil and surface waters of forested catchments with Gleysols. *Geoderma* **100**, 173–192.
12. Owen, A. G. and Jones, D. L. (2001). Competition for amino acids between wheat roots and rhizosphere microorganisms and the role of amino acids in plant N acquisition. *Soil Biol. Biochem.* **33**, 651–657.
13. Orsel, M., Filleur, S., Fraisier, V., and Daniel-Vedele, F. (2002). Nitrate transport in plants: Which gene and which control? *J. Exp. Bot.* **53**, 825–833.
14. Ullrich, C. I. and Novacky, A. J. (1990). Extra and intracellular pH and membrane potential changes by K^+, Cl^-, H_2PO_4 and NO_3 uptake and fusicoccin in root hairs of *Limnobium stoloniferum*. *Plant Physiol.* **94**, 1561–1567.
15. McClure, P. R., Kochian, L. V., Spanwick, R. M., and Shaff, J. E. (1990). Evidence for cotransport of nitrate and protons in maize roots. I. Effects of nitrate on the membrane potential. *Plant Physiol.* **93**, 281–289.
16. Meharg, A. A. and Blatt, M. R. (1995). NO_3^- transport across the plasma membrane of *Arabidopsis thaliana* root hairs: Kinetic control by pH and membrane voltage. *J. Membrane Biol.* **145**, 49–66.
17. Crawford, N. M. and Glass, A. D. M. (1998). Molecular and physiological aspects of nitrate uptake in plants. *Trends Plant Sci.* **3**, 389–395.
18. Huang, N. C., Liu, K. H., Lo, H. J., and Tsay, Y. F. (1999). Cloning and functional characterization of an *Arabidopsis* nitrate transporter gene that encodes a constitutive component of low-affinity uptake. *Plant Cell* **11**, 1381–1392.

19. Espen, L., Nocito, F. F., and Cocucci, M. (2004). Effect of NO_3^- transport and reduction on intracellular pH: An *in vivo* NMR study in maize roots. *J. Exp. Bot.* **55**, 2053–2061.
20. Palmgren, M. G. (2001). Plant plasma membrane H^+-ATPases: Powerhouses for nutrient uptake. *Annu. Rev. Plant Physiol. Plant Mol. Biol.* **52**, 817–845.
21. Santi, S., Locci, G., Monte, R., Pinton, R., and Varanini, Z. (2003). Induction of nitrate uptake in maize roots: Expression of putative high affinity nitrate transporter and plasma membrane H^+-ATPase isoforms. *J. Exp. Bot.* **54**, 1851–1864.
22. Sondergaard, T. E., Schulz, A., and Palmgren, M. G. (2004). Energization of transport processes in plants. Roles of the plasma membrane H1-ATPase. *Plant Physiol.* **136**, 2475–2482.
23. Oaks, A. and Hirel, B. (1985). Nitrogen metabolism in roots. *Annu. Rev. Plant Physiol.* **36**, 345–365.
24. Meyer, C. and Stitt, M. (2001). Nitrate reduction and signalling. In *Plant Nitrogen*. P. J. Lea and J. F. Morot-Gaudry (Eds.). Springer-Verlag Berlin Hidelberg, Hidelberg, pp. 37–59.
25. Hirel, B. and Lea, P. J. (2001). Ammonia assimilation. In *Plant Nitrogen*. P. J. Lea and J. F. Morot-Gaudry (Eds.). Springer-Verlag Berlin Hidelberg, Hidelberg, pp. 79–99.
26. Crawford, N. M. (1995). Nitrate: Nutrient and signal for plant growth. *Plant Cell* **7**, 859–868.
27. Paul, M. J. and Foyer, C. H (2001). Sink regulation of photosynthesis. *J. Exp. Bot.* **52**, 1383–1400.
28. Forde, B. G. (2002). Local and long-range signalling pathways regulating plant responses to nitrate. *Annu. Rev. Plant Biol.* **53**, 203–224.
29. Wang, R., Guegler, K., LaBrie, S. T., and Crawford, N. M. (2000) Genomic analysis of a nutrient response in *Arabidopsis* reveals diverse expression patterns and novel metabolic and potential regulatory genes induced by nitrate. *Plant Cell* **12**, 1491–1509.
30. Rossignol, M. (2001). Analysis of the plant proteome. *Curr. Opin. Biotech.* **12**, 131–134.
31. Roberts, J. K. M. (2002). Proteomics and future generation of plant molecular biologists. *Plant Mol. Biol.* **48**, 143–154.
32. Yarmush, M. L. and Jayaraman, A. (2002). Advances in proteomic technologies. *Annu. Rev. Biomed. Eng.* **4**, 349–373.
33. Patterson, S. D. and Aebersold, R. H. (2003). Proteomics: the first decade and beyond. *Nat. Genet. Suppl.* **33**(Suppl.), 311–323.
34. Jorrín-Novo, J. V., Maldonato, A. M., Eche-varríaZomeòo, S., Valledor, L., Castllejo, M. A., Curto, M., Valero, J., Sghaier, B., Donoso, G., and Redonado, I. (2009). Plant Proteomics update (2007–2008): Second-generation proteomic techniques, an appropriate experimental design, and data analysis to fulfil MIAPE standards, increase plant proteome coverage and expand biological knowledge. *J. Proteomics* **72**, 285–314.
35. Lawrence, C. J., Dong, Q., Mary, L., Polacco, M. L., Seigfried, T. E., and Brendel, V. (2004). MaizeGDB, the community database for maize genetics and genomics. *Nucleic Acids Res.* **32**, D393–397.
36. Porubleva, L., Velden, K. V., Kothari, S., David, J., Oliver, D. J., Parag, R., and Chitnis, P. R. (2001). The proteome of maize leaves: Use of gene sequences and expressed sequence tag data for identification of proteins with peptide mass Fingerprints. *Electrophoresis* **22**, 1724–1738.
37. Majeran, W., Cai, Y., Sun, Q., and van Wijk, K. J. (2005). Functional differentiation of bundle sheath and mesophyll maize chloroplasts determined by comparative proteomics. *Plant Cell* **17**, 3111–3140.
38. Dembinsky, D., Woll, K., Saleem, M., Liu, Y., Fu, Y., Borsuk, L. A., Lamkemeyer, T., Fladerer, C., Madlung, J., Barbazuk, B., Nordheim, A., Nettleton, D., Schnable, P. S., and Hochholdinger, F. (2007). Transcriptomic and proteomic analyses of pericycle cells of the maize primary root. *Plant Physiol.* **145**, 575–588.
39. Bahrman, N., Le Gouls, J., Negroni, L., Amilhat, L., Leroy, P., Lainé, A. L., and Jaminon, O. (2004). Differential protein expression assessed by two-dimensional gel electrophoresis for two wheat varieties

grown at four nitrogen levels. *Proteomics* **4**, 709–719.

40. Bahrman, N., Gouy, A., Devienne-Barret, F., Hirel, B., Vedele, F., and Le Gouis, J. (2005). Differential change in root protein pattern of two wheat varieties under high and low nitrogen nutrition levels. *Plant Sci.* **168**, 81–87.
41. Foyer, C. H., Ferrario-Méry, S., and Noctor, G. (2001). Interactions between carbon and nitrogen metabolism. In *Plant Nitrogen*. P. J. Lea and J. F. Morot-Gaudry (Eds.). Springer-Verlag Berlin Hidelberg, Hidelberg, 237–254.
42. Sivasankar, S., Rothstein, S., and Oaks, A. (1997). Regulation of the accumulation and reduction of nitrate by nitrogen and carbon metabolites in maize seedlings. *Plant Physiol.* **114**, 583–589.
43. Klein, D., Morcuende, R., Stitt, M., and Krapp, A. (2000). Regulation of nitrate reductase expression in leaves by nitrate and nitrogen metabolism is completely overridden when sugars fall below a critical level. *Plant Cell Environ.* **23**, 863–871.
44. Huppe, H. C. and Turpin, D. H. (1996). Appearance of novel glucose-6-phosphate dehydrogenase isoforms in *Chlamydomonas reinhardtii* during growth on nitrate. *Plant Physiol.* **110**, 1431–1433.
45. Wang, Y. H., Garvin, D. F., and Kochian, L. V. (2001). Nitrate-induce genes in tomato roots. Array analysis reveals novel genes that may play a role in nitrogen nutrition. *Plant Physiol.* **127**, 345–359.
46. Li, M., Villemur, R., Hussey, P. J., Silflow, C. D., Gantt, J. S., and Snustad, D. P. (1993). Differential expression of six glutamine synthetase genes in *Zea mays*. *Plant Mol. Biol.* **23**, 401–407.
47. Sakakibara, H., Kawabata, S., Hase, T., and Sugiyama, T. (1992). Differential effects of nitrate and light on the expression of glutamine synthetases and ferredoxin-dependent glutamate synthase in maize. *Plant Cell Physiol.* **33**, 1193–1198.
48. Rockel, P., Strube, F., Rockel, A., Wildt, J., and Kaiser, W. M. (2002). Regulation of nitric oxide (NO) production by plant nitrate reductase *in vivo* and *in vitro*. *J. Exp. Bot.* **53**, 103–110.
49. Igamberdiev, A. U., Bycova, N. V., and Hill, R. D. (2006). Nitric oxide scavenging by barley hemoglobin is facilitated by a monodehydroascorbate reductase-mediated ascorbate reduction of methemoglobin. *Planta* **223**, 1033–1040.
50. Lamattina, L., Garcìa-Mata, C., and Pagnussat, G. (2003). Nitric oxide: The versatility of an extensive signal molecule. *Annu. Rev. Plant Biol.* **54**, 109–136.
51. Stöhr, C. and Stremlau, S. (2006). Formation and possible roles of nitric oxide in plant roots. *J. Exp. Bot.* **57**, 463–470.
52. Zhao, D. Y., Tian, Q. Y., Li, L. H., and Zhang, W. H. (2007). Nitric oxide is involved in nitrate-induced inhibition of root elongation in *Zea mays*. *Ann. Bot.* **100**, 497–503.
53. Peschke, V. M. and Sachs, M. M. (1994). Characterization and expression of transcripts induced by oxygen deprivation in maize (*Zea mays* L.). *Plant Physiol.* **104**, 387–394.
54. Igamberdiev, A. U. and Hill, R. D. (2004). Nitrate NO and haemoglobin in plant adaptation to hypoxia: An alternative to classic fermentation pathways. *J. Exp. Bot.* **408**, 2473–2482.
55. Fritz, C., Palacios-Rojas, N., Fell, R., and Stitt, M. (2006). Regulation of secondary metabolism by the carbon-nitrogen status in tobacco: Nitrate inhibits large sectors of phenylpropanoid metabolism. *Plant J.* **46**, 533–548.
56. Kingston-Smith, A. H., Bollard, A. L., and Minchin, F. R. (2005). Stress-induced changes in protease composition are determined by nitrogen supply in non-nodulating white clover. *J. Exp. Bot.* **56**, 745–753.
57. Simões, I. and Faro, C. (2004). Structure and function of plant aspartic proteinases. *Eur. J. Biochem.* **271**, 2067–2075.
58. Askura, T., Watanabe, H., Abe, K., and Arai, S. (1995). Rice aspartic proteinases, oryzasin, expressed during seed ripening and germination, has a gene organization distinct from those of animal and microbial aspartic proteinases. *Eur. J. Biochem.* **232**, 77–83.
59. Raven, J. A. (1986). Biochemical disposal of excess H^+ in growing plants? *New Phytol.* **104**, 175–206.

60. Sakano, K. (1998). Revision of biochemical pH-stat: Involvement of alternative pathway metabolisms. *Plant Cell Physiol.* **39**, 467–473.
61. Britto, D. T. and Kronzucker, H. J. (2005). Nitrogen acquisition, PEP carboxylase, and cellular pH homeostasis: New views on old paradigms. *Plant Cell Environ.* **28**, 1396–1409.
62. Uhrig, R. G., She, Y. M., Leach, C. A., and Plaxton, W. C. (2008). Regulatory monoubiquitination of phosphoenolpyruvate carboxylase in germinating castor oil seeds. *JBC* **283**, 29650–29657.
63. de Vetten, N. C. and Ferl, R. J. (1994). Two genes encoding GF14 (14-3-3) proteins in *Zea mays*. Structure, expression, and potential regulation by G-box-binding complex. *Plant Physiol.* **106**, 1593–1604.
64. Bihn, E. A., Paul, A. L., Wang, S. W., Erdos, G. W., and Ferl, R. J. (1997). Localization of 14-3-3 proteins in the nuclei of *Arabidopsis* and maize. *Plant J.* **12**, 1439–1445.
65. Roberts, M. R. (2000). Regulatory 14-3-3 protein-protein interactions in plant cells. *Curr. Opin. Plant Biol.* **3**, 400–405.
66. Bachmann, M., Huber, J. L., Athwal, G. S., Wu, K., Ferl, R. J., and Huber, S. C. (1996). 14-3-3 proteins associate with the regulatory phosphorylation site of spinach leaf nitrate reductase in an isoform-specific manner and reduce dephosphorylation of Ser-543 by endogenous protein phosphatases. *FEBS Lett.* **398**, 26–30.
67. Ikeda, Y., Koizumi, N., Kusano, T., and Sano, H. (2000). Specific binding of a 14-3-3 protein to autophosphorylated WPK4, an SNF1-related wheat protein kinase, and to WPK-4-phosphorylated nitrate reductase. *JBC* **275**, 31695–31700.
68. Dickson, R., Weiss, C., Howard, R. J., Alldrick, S. P., Ellis, R. J., Lorimer, G., Azem, A., and Viitenen, P. V. (2000). Reconstitution of higher plant chloroplast chaperonin 60 tetradecamers active in protein folding. *JBC* **275**, 11829–11835.
69. Averill, R. H., Bailey-Serres, J., and Kruger, N. J. (1998). Co-operation between cytosolic and plastidic oxidative pentose phosphate pathways revealed by 6-phosphogluconate dehydrogenase-deficient genotypes of maize. *Plant J.* **14**, 449–457.
70. Malkin, R. and Niyogi, K. (2000). Photosynthesis. In *Biochemistry and Molecular Biology of Plants*. B. Buchanan, W. Gruissem, and R. Jones (Eds.). *Am. Soc. Plant Physiol. Rockville* 568–628.
71. Rapala-Kozik, M., Kowalaska, E., and Ostrowska, K. (2008). Modulation of thiamine metabolism in *Zea mays* seedlings under conditions of abiotic stress. *J. Exp. Bot.* **59**, 4133–4143.
72. Edwards, G. E., Franceschi, V. R., and Voznesenskaya, E. V. (2004). Single-cell C_4 phothosynthesis versus the dual-cell (Kranz) paradigm. *Annu. Rev. Plant Biol.* **55**, 173–196.
73. Ueno, Y., Imanari, E., Emura, J., Yoshizawa-Kumagaye, K., Nakajiama, K., Inami, K., Shiba, T., Sakakibara, H., Sugiyama, T., and Izui, K. (2000). Immunological analysis of the phosphorylation state of maize C_4-form phosphoenolpyruvate carboxylase with specific antibodies raised against a synthetic phosphorylated peptide. *Plant J.* **21**, 17–26.
74. Izui, K., Matsumura, H., Furumoto, T., and Kai, Y. (2004). Phosphoenolpyruvate carboxylase: A new era of structural biology. *Annu. Rev. Plant. Biol.* **55**, 69–84.
75. Nemchenko, A., Kunze, S., Feussner, I., and Kolomietes, M. (2006). Duplicate maize 13-lipoxygenase genes are differentially regulated by circadian rhythm, cold stress, wounding, pathogen infection, and hormonal treatments. *J. Exp. Bot.* **57**, 3767–3779.
76. Feussner, I., Bachmann, A., Höhne, M., and Kindl, H. (1998). All three acyl moieties of trilinolein are efficiently oxygenated by recombinant His-tagged lipid body lipoxygenase *in vitro*. *FEBS Lett.* **431**, 433–436.
77. James, H. E. and Robinson, C. (1991). Nucleotide sequence of cDNA encoding the precursor of the 23 kDa protein of the photosynthetic oxygen-evolving complex from wheat. *Plant Mol. Biol.* **17**, 179–182.
78. Yoshiba, Y., Yamaguchi-Shinozaki, K., Shinozaki, K., and Harada, Y. (1995). Characterization of a cDNA clone encoding 23 kDa polypeptide of the oxygen-evolving

complex of photosystem II in rice. *Plant Cell Physiol.* **36**, 1677–1682.

79. Sourosa, M. and Aro, E. M. (2007). Expression, assembly and auxiliary functions of photosystem II oxygen-evolving proteins in higher plants. *Photosynth. Res.* **93**, 89–100.
80. Ifuku, K., Yamamoto, Y., Ono, T., Ishihara, S., and Sato, F. (2005). PsbP protein, but not PsbQ protein, is essential for the regulation and stabilization of photosystem II in higher plants. *Plant Physiol.* **139**, 1175–1184.
81. Gaude, N., Bréhélin, C., Tischendorf, G., Kessler, F., and Dörmann, P. (2007). Nitrogen deficiency in *Arabidopsis* affects galactolipid composition and gene expression and results in accumulation of fatty acid phytyl esters. *Plant J.* **49**, 729–739.
82. Sakurai, I., Mizusawa, N., Wada, H., and Sato, N. (2007). Digalactosyldiacylglycerol is required for stabilization of the oxygen-evolving complex in photosystem II. *Plant Physiol.* **145**, 1361–1370.
83. Cataldo, D. A., Haroon, M., Schrader, L. E., and Youngs, V. L. (1975). Rapid colorimetric determination of nitrate in plant tissue by nitration of salicylic acid. *Commun. Soil Sci. Plant Anal.* **6**, 71–80.
84. Ferrario-Méry, S., Valadier, M. H., and Foyer, C. H. (1998). Overexpression of nitrate reductase in tobacco delays drought-induced decreases in nitrate reductase activity and mRNA. *Plant Physiol.* **117**, 293–302.
85. Nelson, N. A. (1944). A photometric adaptation of the Somogy method for the determination of glucose. *JBC*, **153**, 375–384.
86. Moore, S. and Stein, W. H. (1954). A modified ninhydrin reagent for the photometric determination of amino acids and related compounds. *JBC* **211**, 907–913.
87. Martínez-Garcia, J. F., Monte, E., and Quall, P. H. (1999). A simple, rapid and quantitative method for preparing *Arabidopsis* protein extracts for immunoblot analysis. *Plant J.* **20**, 251–257.
88. Lichtenthaler, H. K. (1987). Chlorophylls and carotenoids: Pigments of photosynthetic biomembranes. *Met. Enzymol.* **148**, 350–382.
89. Heat, R. L. and Packer, K. (1968). Photoperoxidation in isolated chloroplasts. I. Kinetics and stoichiometry of fatty acid peroxidation. *Arch. Biochem. Biophys.* **125**, 189–198.
90. Krause, G. H. and Weis, E. (1991). Chlorophyll fluorescence and photosynthesis: The basics. *Annu. Rev. Plant Physiol. Plant Mol. Biol.* **42**, 313–349.
91. Genty, B., Briantais, J. M., and Baker, N. R. (1989). The relationship between the quantum yield of photosynthetic electron transport and quenching of chlorophyll fluorescence. *BBA* **990**, 87–92.
92. Hurkman, W. J. and Tanaka, C. K. (1986). Solubilization of plant membrane proteins for analysis by two-dimensional gel electrophoresis. *Plant Physiol.* **81**, 802–806.
93. Laemmli, U. K. (1970). Cleavage of structural proteins during the assembly of the head of bacteriophage. T4. *Nature* **227**, 680–685.
94. Neuhoff, V., Arold, N., Taube, D., and Ehrhardt, W. (1988). Improved staining of proteins in polyacrylamide gels including isoelectric focusing gels with clear background at nanogram sensitivity using Coomassie Brilliant Blue G-250 and R-250. *Electrophoresis* **9**, 255–262.
95. Magni, C., Scarafoni, A., Herndl, A., Sessa, F., Prinsi, B., Espen, L., and Duranti, M. (2007). Combined electrophoretic approaches for the study of white lupin mature seed storage proteome. *Phytochemistry* **68**, 997–1007.
96. National Center for Biotechnology Information. Retrieved from [http://www.ncbi.nlm.nih.gov/].
97. Eng, J. K., McCormack, A. L., and Yates, J. R. III (1994). An approach to correlate tandem mass spectral data of peptides with amino acid sequences in a protein database. *J. Am. Soc. Mass Spectrom.* **5**, 976–989.
98. Mackey, A. J., Haystead, T. A. J., and Pearson, W. R. (2002). Getting more from less: Algorithms for rapid protein identification with multiple short peptide sequences. *Mol. Cell Proteomics* **1**, 139–147.
99. ExPASy Proteomics Server. Retrieved from [http://www.expasy.org/].

2

1. Sanborn, M., Kerr, K. J., Sanin, L. H., Cole, D. C., Bassil, K. L., and Vakil, C. (2007). Non-cancer health effects of pesticides: Systematic review and implications for family doctors. *Can. Fam. Physician.* **53**(10), 1712–20.
2. Alavanja, M. C. R., Hoppin, J. A., and Kamel, F. (2004). Health effects of chronic pesticde exposure: Cancer and neurotoxicity. *Ann. Rev. Public Health* **25**, 155–197.
3. Costa, L. G. (2006). Current issues in organophosphate toxicology. *Clinica Chimica Acta* **366**, 1–13.
4. Tahmaz, N., Soutar, A., and Cherrie, J. W. (2003). Chronic fatigue and organophosphate pesticides in sheep farming: A retrospective study amongst people reporting to a UK pharmacovigilance scheme. *Ann. Occup. Hyg.* **47**(4), 261–267.
5. Hospital Episode Statistics. Retrieved from [http://www.hesonline.nhs.uk].
6. Rushton, L. and Mann, V. (2008). Retrieved from [http://www.hse.gov.uk/research/rrhtm/rr608.htm]. *Estimating the Prevalence and Incidence of Pesticide-Related Illness Presented to General Practitioners in Great Britain. Research Report 608, HSE.*
7. General Practice Research Framework. Retrieved from [http://www.gprf.mrc.ac.uk].
8. General Practitioner Workload (2004). Retrieved from [http://www.rcgp.org.uk/pdf/ISS_INFO_03_APRIL04.pdf].
9. Thundyil, J. G., Stober, J., Besbelli, N., and Pronczuk, J. (2008). Acute pesticide poisoning: A proposed classification tool. *Bull. WHO* **86**(3), 206–211.
10. Keifer, M., McConnell, R., Pachero, F., Daniel, W., and Rosenstock, L. (1996). Estimating underreported pesticide poisonings in Nicaragua. *Am. J. Ind. Med.* **30**, 195–201.
11. Das, R., Steege, A., Baron, S., Beckman, J., and Harrison, R. (2001). Pesticide-related illness among Migrant Farm workers in the United States. *Int. J. Occup. Environ. Health* **7**, 303–312.
12. Casey, P. and Vale, J. A. (1994). Deaths from pesticide poisoning in England and Wales: 1945–1989. *Hum. Exper. Toxicol.* **13**, 95–101.
13. Profile of UK General Practitioners (2006). Retrieved from [https://www.rcgp.org.uk/pdf/ISS_INFO_01_JUL06.pdf].
14. Rudent, J., Manegaux, F., Leverger, G., et al. (2007). Household exposure to pesticides and risk of childhood haematopoietic malignancies: The ESCALE study (SFCE). *Env. Health Perspect* **115**, 1787–1793.
15. Ma, X., Buffler, P. A., Gurvier, R. B., et al. (2002). Critical windows of exposure to household pesticides and risk of childhood leukaemia. *Env. Health Perspect.* **110**, 955–960.
16. Grey, C. N. B., Nieuwenhuijsen, M. J., and Golding, J. (2006). Usage and storage of domestic pesticides in the UK. *Sci. Tot. Env.* **368**, 465–470.
17. Weale, V. P. and Goddard, H. (1998). *The Effectiveness of Non-agricultural Pesticide Labelling (Contract Research Report 161/1998)*. HSE Books, Sudbury, UK.
18. Calvert, G. M., Plate, D. K., Das, R., Rosales, R., Sahfey, O., Thomsen, C., et al. (2004). Acute occupational pesticide-related illness in the US, 1998–1999: Surveillance findings from the SENSOR-pesticides program. *Am. J. Industrial. Med.* **45**, 14–23.
19. Health and Safety Executive (HSE) (2004). *Pesticide Incidents Report 1 April 2003–31 March 2004.*
20. Deckers, J. G. M., Paget, W. J., Schellevis, F. G., and Fleming, D. M. (2006). European primary care surveillance networks: Their structure and operation. *Family Prac.* **23**, 151–158.
21. Green, L. A. and Hickner, J. (2006). A short history of primary care practice-based research networks: From concept to essential research laboratories. *JABFM* **19**(1), 1–10.
22. Freedman, D. O., Weld, L. H., Kozarsky, P. E., Fisk, T., Robins, R., von Sonnenburg, F., et al. (2006). The GeoSentinel Surveillance Network: Spectrum of disease and relationship to place of exposure among ill returned travellers. *New Engl. J. Med.* **354**, 119–130.
23. Jelinek, T. and Muhlberger, N. (2005). Surveillance of imported diseases as a window to travel health risks. *Infect. Dis. Clin. North Am.* **19**, 1–13.

24. London, L., Bourne, D., Sayed, R., and Eastman, R. (2004). Guillain-Barre syndrome in a rural farming district in South Africa: A possible relationship to environmental organophosphate exposure. *Arch. Environ. Health* **59**(11), 575–80.
25. Soutar, C. S. (2001). *Frequencies of Disease Presenting to General Practitioners According to Patients' Occupation. Contract Research Report 340*. HSE Books, Sudbury, UK.
26. Kass, D. E., Their, A. L., Leighton, J., Cone, J. E., and Jeffery, N. L. (2004). Developing a comprehensive pesticide health effects tracking system for an urban setting: New York City's approach. *Env. Health Perspec.* **112**(4), 1419–1423.
27. Wakefield, J. (2003). Pesticides initiative: basic training for health care providers. *Env. Health Perspec.* **111**(10), A520–522.
28. Sandiford, P. (1992). What can information systems do for Primary Health Care? An international perspective. *Soc. Sci. Med.* **34**, 1077–1087.
29. London, L. and Bailie, R. (2001). Challenges for improving surveillance for pesticide poisoning: Policy implications for developing countries. *Int. J. Epid.* **30**, 564–570.

3

1. DeLucia, E. H., Hamilton, J. G., Naidu, S. L., Thomas, R. B., Andrews, J. A., et al. (1999). Net primary production of a forest ecosystem with experimental CO_2 enrichment. *Science* **284**, 1177–1179.
2. Allen, A. S., Andrews, J. A., Finzi, A. C., Matamala, R., Richter, D. D., and Schlesinger, W. H. (2000). Effects of free-air CO_2 enrichment (FACE) on belowground processes in a Pinus taeda forest. *Ecol. Appl.* **10**, 437–448.
3. Schlesinger, W. H. and Lichter, J. (2001). Limited carbon storage in soil and litter of experimental forest plots under increased atmospheric CO_2. *Nature* **411**, 466–469.
4. Finzi, A. C., Allen, A. S., DeLucia, E. H., Ellsworth, D. S., and Schlesinger, W. H. (2001). Forest litter production, chemistry, and decomposition following two years of free-air CO_2 enrichment. *Ecology* **82**, 470–484.
5. Zak, D. R., Holmes, W. E., Finzi, A. C., Norby, R. J., and Schlesinger, W. H. (2003). Soil nitrogen cycling under elevated CO_2: A synthesis of forest FACE experiments. *Ecol. Appl.* **13**, 1508–1514.
6. Zhang, X., Zwiers, F. W., Hegerl, G. C., Lambert, H., Gillett, N. P., et al. (2007). Detection of human influence on twentieth-century precipitation trends. *Nature* **448**, 462–465.
7. Raich, J. W., Russell, A. E., Kitayama, K., Parton, W. J., and Vitousek, P. M. (2006). Temperature influences carbon accumulation in moist tropical forests. *Ecology* **87**, 76–87.
8. Bardgett, R. D. and Wardle, D. A. (2003). Herbivore mediated linkages between aboveground and belowground communities. *Ecology* **84**, 2258–2268.
9. Sayer, E. J. (2006). Using experimental manipulation to assess the roles of leaf litter in the functioning of forest ecosystems. *Biol. Rev.* **81**, 1–31.
10. Dixon, R. K., Brown, S., Houghton, R. A., Solomon, A. M., Trexler, M. C., and Wisniewski, J. (1994). Carbon pools and flux of global forest ecosystems. *Science* **263**, 185–190.
11. Bernoux, M., Carvalho, M. S., Volkoff, B., and Cerri, C. C. (2002). Brazil's soil carbon stocks. *Soil. Sci. Soc. Am. J.* **66**, 888–896.
12. Clark, D. A. (2004). Sources or sinks? The responses of tropical forests to current and future climate and atmospheric composition. *Philos. T. Roy. Soc. B.* **359**, 477–491.
13. Jobbagy, E. G. and Jackson, R. B. (2000). The vertical distribution of soil organic carbon and its relation to climate and vegetation. *Ecol. Appl.* **10**, 423–436.
14. Raich, J. W., Potter, C. S., and Bhagawati, D. (2002). Interannual variability in global soil respiration, 1980–1994. *Glob. Change. Biol.* **8**, 800–812.
15. Palmroth, S., Oren, R., McCarthy, H. R., Johnsen, K. H., Finzi, A. C., et al. (2006). Aboveground sink strength in forests controls the allocation of carbon below ground

and its [CO_2]-induced enhancement. *Proc. Nat. Acad. Sci. USA* **103**, 19362–19367.

16. Kirschbaum, M. U. F. (2000). Will changes in soil organic carbon act as a positive or negative feedback on global warming? *Biogeochemistry* **48**, 21–51.
17. Davidson, E. A. and Janssens, I. A. (2006). Temperature sensitivity of soil carbon decomposition and feedbacks to climate change. *Nature* **440**, 165–173.
18. Neff, J. C., Townsend, A. R., Gleixner, G., Lehman, S. J., Turnbull, J., and Bowman, W. D. (2002). Variable effects of nitrogen additions on the stability and turnover of soil carbon. *Nature* **419**, 915–917.
19. Reich, P. B., Hungate, B. A., and Luo, Y. (2006). Carbon-nitrogen interactions in terrestrial ecosystems in response to rising atmospheric carbon dioxid. *Annu. Rev. Ecol. Evol. Syst.* **37**, 611–636.
20. Cleveland, C. C. and Townsend, A. R. (2006). Nutrient additions to a tropical rain forest drive substantial soil carbon dioxide losses to the atmosphere. *Proc. Natl. Acad. Sci. USA* **103**, 10316–10321.
21. Raich, J. W. and Nadelhoffer, K. J. (1989). Belowground carbon allocation in forest ecosystems: Global trends. *Ecology* **70**, 1346–1354.
22. Davidson, E. A., Savage, K., Bolstad, P., Clark, D. A., Curtis, P. S., et al. (2002). Belowground carbon allocation in forests estimated from litterfall and IRGA-based soil respiration measurements. *Agr. Forest. Meteorol.* **113**, 39–51.
23. Cornejo, F. H., Varela, A., and Wright, S. J. (1994). Tropical forest litter decomposition under seasonal drought: Nutrient release, fungi and bacteria. *Oikos* **70**, 183–190.
24. Vasconcelos, S. S., Zarin, D. J., Capanu, M., Littell, R., Davidson, E. A., et al. (2004). Moisture and substrate availability constrain soil trace gas fluxes in an eastern Amazonian regrowth forest. *Global. Biogeochem. Cy.* **18**, GB2009.
25. Li, Y., Xu, M., Sun, O. J., and Cui, W. (2004). Effects of root and litter exclusion on soil CO_2 efflux and microbial biomass in wet tropical forests. *Soil. Biol. Biochem.* **36**, 2111–2114.
26. Priess, J. A. and Folster, H. (2001). Microbial properties and soil respiration in submontane forests of Venezuelan Guyana: characteristics and response to fertilizer treatments. *Soil. Biol. Biochem.* **33**, 503–509.
27. Sayer, E. J., Tanner, E. V. J., and Lacey, A. L. (2006a). Effects of litter manipulation on early-stage decomposition and meso-arthropod abundance in a tropical moist forest. *Forest. Ecol. Manage.* **229**, 285–293.
28. Sayer, E. J., Tanner, E. V. J., and Cheesman, A. W. (2006b). Increased litterfall changes fine root distribution in a moist tropical forest. *Plant. Soil.* **281**, 5–13.
29. Pregitzer, K. S., Laskowski, M. J., and Burton, A. J. (1998). Variation in sugar maple root respiration with root diameter and soil depth. *Tree Physiol.* **18**, 665–670.
30. Desrochers, A., Landhausser, S. M., and Lieffers, V. J. (2002). Coarse and fine root respiration in aspen (Populus tremuloides). *Tree Physiol.* **22**, 725–732.
31. Singh, J. S. and Gupta, S. R. (1977). Plant decomposition and soil respiration in terrestrial ecosystems. *Bot. Rev.* **43**, 449–528.
32. Anderson, J. M., Proctor, J., and Vallack, H. W. (1983). Ecological studies in four contrasting lowland rain forests in Gunung-Mulu National Park, Sarawak. 3. Decomposition processes and nutrient losses from leaf litter. *J. Ecol.* **71**, 503–527.
33. Rout, S. K. and Gupta, S. R. (1989). Soil respiration in relation to abiotic factors, forest floor litter, root biomass and litter quality in forest ecosystems of Siwaliks in northern India. *Acta. Oecol.* **10**, 229–244.
34. Metcalfe, D. B., Meir, P., Aragão, L. E. O. C., Mahli, Y., da Costa, A. C. L., et al. (2007). Factors controlling spatio-temporal variation in carbon dioxide efflux from surface litter, roots, and soil organic matter at four rain forest sites in eastern Amazon. *J. Geophys. Res.* **112**, G04001.
35. Bingeman, C. W., Varner, J. E., and Martin, W. P. (1953). The effect of the addition of organic materials on the decomposition of an organic soil. *Soil. Sci. Soc. Am. Proc.* **29**, 692–696.

36. Kuzyakov, Y., Friedel, J. K., and Stahr, K. (2000). Review of mechanisms and quantification of priming effects. *Soil Biol. Biochem.* **32**, 1485–1498.
37. DeNobili, M., Contin, M., Mondini, C., and Brookes, P. C. (2001). Soil microbial biomass is triggered into activity by trace amounts of substrate. *Soil Biol. Biochem.* **33**, 1163–1170.
38. Hamer, U. and Marschner, B. (2005). Priming effects in soils after combined and repeated substrate additions. *Geoderma* **128**, 38–51.
39. Fontaine, S., Bardoux, G., Abbadie, L., and Mariotti, A. (2004a). Carbon input to soil may decrease soil carbon content. *Ecol. Lett.* **7**, 314–320.
40. Cavalier, J. (1992). Fine-root biomass and soil properties in a semi-deciduous and a lower montane rain forest in Panama. *Plant Soil* **142**, 187–201.
41. Powers, J. S., Treseder, K. K., and Lerdau, M. T. (2005). Fine roots, arbuscular mycorrhizal hyphae and soil nutrients in four neotropical rain forests: Patterns across large geographic distances. *New Phytol.* **165**, 913–921.
42. Leigh, E. G. (1999). *Tropical Forest Ecology*. Oxford University Press, Oxford.
43. Leigh, E. G. and Wright, S. J. (1990). Barro Colorado Island and tropical biology. In. *Four Neotropical Rainforests*. A. H. Gentry (Ed.). Yale University Press, New Haven, pp. 28–47.
44. Brookes, P. C., Landman, A., Pruden, G., and Jenkinson, D. S. (1985). Chloroform fumigation and the release of soil nitrogen: A rapid direct extraction method to measure microbial biomass nitrogen in the soil. *Soil Biol. Biochem.* **17**, 837–842.
45. Beck, T., Joergensen, R. G., Kandeler, E., Makeschin, F., Nuss, E., et al. (1997). An inter-laboratory comparison of ten different ways of measuring soil microbial biomass C. *Soil Biol. Biochem.* **29**, 1023–1032.

4

1. Wallace, A. R. (1867). *Proc. Entomol. Soc. Lond.* (4 March).
2. Merilaita, S. and Ruxton, G. D. (2007). Aposematic signals and the relationship between conspicuousness and distinctiveness. *J. Theor. Biol.* **245**, 268–277.
3. Fisher, R. A. (1930). The genetical theory of natural selection. Clarendon, Oxford.
4. Harvey, P. H., Bull, J. J., Pemberton, M., and Paxton, R. J. (1982). The evolution of aposematic coloration in distasteful prey: A family model. *Am. Nat.* **119**, 710–718.
5. Harvey, P. H. and Paxton, R. J. (1981). The evolution of aposematic coloration. *Oikos* **37**, 391–396.
6. Eisner, T. and Grant, R. P. (1981). Toxicity, Odor Aversion, and Olfactory Aposematism. *Science* **213**, 476–476.
7. Cott, H. B. (1940). Adaptive Coloration in Animals. Methuen, London.
8. Rothschild, M. (1961). Defensive odours and Müllerian mimicry among insects. *Trans. R. Entomol. Soc. Lond.* **113**, 101–121.
9. Prudic, K. L., Noge, K., and Becerra, J. X. (2008). Adults and Nymphs Do Not Smell the Same: The Different Defensive Compounds of the Giant Mesquite Bug (Thasus neocalifornicus: Coreidae). *J. Chem. Ecol.* **34**, 734–741.
10. Eisner, T., Eisner, M., and Seigler, M. (2005). Secret Weapons: Defenses of Insects, Spiders, Scorpions, and Other Many-Legged Creatures. Belknap Press of Harvard University Press, Cambridge, Mass.
11. Speed, M. P. and Ruxton, G. D. (2005). Warning displays in spiny animals: One (more) evolutionary route to aposematism. *Evolution* **59**, 2499–2508.
12. Gamberale-Stille, G. (2000). Decision time and prey gregariousness influence attack probability in naïve and experienced predators. *Anim. Behav.* **60**, 95–99.
13. Merilaita, S. and Ruxton, G. D. (2007). Aposematic signals and the relationship between conspicuousness and distinctiveness. *J. Theor. Biol.* **245**, 268–277.
14. Skelhorn, J. and Rowe, C. (2006). Avian predators taste–reject aposematic prey on the basis of their chemical defence. *Biol. Let.* **2**(3), 348–350.

15. Rowe, C. (1999). Receiver psychology and the evolution of multicomponent signals. *Anim. Behav.* **58**, 921–931.
16. Rowe, C. and Guilford, T. (1999). The evolution of multimodal warning displays. *Evol. Ecol.* **13**, 655–671.
17. Malakoff, D. (1999). Olfaction: Following the scent of avian olfaction. *Science* **286**(5440), 704–705.
18. Steiger, S. S., Fidler, A. E., Valcu, M., and Kempenaers, B. (2008). Avian olfactory receptor gene repertoires: Evidence for a well-developed sense of smell in birds? *Proc. R. Soc. B.* **275**, 2309–2317.

5

1. Yonaha, M., Saito, M., and Sagai, M. (1983). Stimulation of lipid peroxidation by methyl mercury in rats. *Life Sci.* **32**, 1507–1514.
2. Sarafian, T. and Verity, M. A. (1991). Oxidative mechanisms underlying methyl mercury neurotoxicity. *Int. J. Dev. Neurosci.* **9**, 147–153.
3. Shanker, G. and Aschner, M. (2003). Methylmercury-induced reactive oxygen species formation in neonatal cerebral astrocytic cultures is attenuated by antioxidants. *Mol. Brain. Res.* **110**, 85–91.
4. Shanker, G., Aschner, J. L., Syversen, T., and Aschner, M. (2004). Free radical formation in cerebral cortical astrocytes in culture induced by methylmercury. *Mol. Brain. Res.* **128**, 48–57.
5. Ali, S. F., LeBel, C. P., and Bondy, S. C. (1992). Reactive oxygen species formation as a biomarker of methylmercury and trimethyltin neurotoxicity. *Neurotoxicology* **13**, 637–648.
6. Thompson, S. A., White, C. C., Krejsa, C. M., Eaton, D. L., and Kavanagh, T. J. (2000). Modulation of glutathione and glutamate-L-cysteine ligase by methylmercury during mouse development. *Toxicol. Sci.* **57**, 141–146.
7. Ding, Y., Gonick, H. C., and Vaziri, N. D. (2000). Lead promotes hydroxyl radical generation and lipid peroxidation in cultured aortic endothelial cells. *Am. J. Hypertens.* **13**, 552–555.
8. Hsu, P., Liu, M., Hsu, C., Chen, L., and Guo, Y. (1997). Lead exposure causes generation of reactive oxygen species and functional impairment in rat sperm. *Toxicology* **122**, 133–143.
9. Ercal, N., Treratphan, P., Hammond, T. C., Mathews, R. H., Grannemann, N. H., et al. (1996). *In vivo* indices of oxidative stress in lead exposed C57BL/6 mice are reduced by treatment with meso-2,3-dimercaptosuccinic acid or N-acetyl cysteine. *Free Radic Biol. Med.* **21**, 157–161.
10. Stahnke, T. and Richter-Landsberg, C. (2004). Triethyltin-induced stress responses and apoptotic cell death in cultured oligodendrocytes. *Glia* **46**, 334–344.
11. Jenkins, S. M. and Barone, S. (2004). The neurotoxicant trimethyltin induces apoptosis via caspase activation, p38 protein kinase, and oxidative stress in PC12 cells. *Toxicol. Lett.* **147**, 63–72.
12. Fowler, B. A., Whittaker, M. H., Lipsky, M., Wang, G., and Chen, X. Q. (2004). Oxidative stress induced by lead, cadmium and arsenic mixtures: 30-day, 90-day, and 180-day drinking water studies in rats: An overview. *Biometals* **17**, 567–568.
13. Souza, V., Escobar Mdel, C., Bucio, L., Hernandez, E., Gutierrez-Ruiz, M. C. (2004). Zinc pretreatment prevents hepatic stellate cells from cadmium-produced oxidative damage. *Cell Biol. Toxicol.* **20**, 241–251.
14. Hei, T. K. and Filipic, M. (2004). Role of oxidative damage in the genotoxicity of arsenic. *Free Radic Biol. Med.* **37**, 574–581.
15. McDonough, K. H. (2003). Antioxidant nutrients and alcohol. *Toxicology* **189**, 89–97.
16. Abdollahi, M., Ranjbar, A., Shadnia, S., Nikfar, S., and Rezale, A. (2004). Pesticides and oxidative stress: A review. *Med. Sci. Monit.* **10**, RA141–147.
17. Suntres, Z. E. (2002). Role of antioxidants in paraquat toxicity. *Toxicology* **180**, 65–77.
18. Smith, L. L., Rose, M. S., and Wyatt, I. (1978). The pathology and biochemistry of paraquat. *Ciba. Found. Symp.* **65**, 321–341.
19. Giray, B. (2001). Cypermethrin-induced oxidative stress in rat brain and liver is

prevented by vitamin E or allopurinol. *Toxicol. Lett.* **118**, 139–146.

20. Gupta, A. (1999). Effect of pyrethroid-based liquid mosquito repellent inhalation on the blood-brain barrier function and oxidative damage in selected organs of developing rats. *J. Appl. Toxicol.* **19**, 67–72.
21. Kale, M., Rathore, N., John, S., and Bhathagar, D. (1999). Lipid peroxidative damage on pyrethroid exposure and alteration in antioxidant status in rat erythrocytes. A possible involvement of reactive oxygen species. *Toxicol. Lett.* **105**, 197–205.
22. Gultekin, F. (2000). The effect of organophosphate insecticide chlorpyrifos-ethyl on lipid peroxidation and antioxidant enzymes (*in vitro*). *Arch. Toxicol.* **74**, 533–538.
23. Gupta, R. C. (2001). Depletion of energy metabolites following acetylcholinesterase inhibitor-induced status epilepticus: Protection by antioxidants. *Neurotoxicology* **22**, 271–282.
24. Akhgari, M., Abdollahi, M., and Kebryaeezadeh, A., (2003). Biochemical evidence for free radical-induced lipid peroxidation as a mechanism for subchronic toxicity of malathion in blood and liver of rats. *Hum. Exp. Toxicol.* **22**, 205–211.
25. Banerjee, B. D., Seth, V., Bhattacharya, A., Pasha, S. T., and Chakraborty, A. K. (1999). Biochemical effects of some pesticides on lipid peroxidation and freeradical scavengers. *Toxicol. Lett.* **107**, 33–47.
26. Ranjbar, A., Pasalar, P., and Abdollahi, M. (2002). Induction of oxidative stress and acetylcholinesterase inhibition in organophosphorous pesticide manufacturing workers. *Hum. Exp. Toxicol.* **21**, 179–182.
27. Noble, M., Mayer-Proschel, M., and Proschel, C. (2005). Redox regulation of precursor cell function: Insights and paradoxes. *Antioxid. Redox. Signal.* **7**, 1456–1467.
28. Nathan, C. (2003). Specificity of a third kind: Reactive oxygen and nitrogen intermediates in cell signaling. *J. Clin. Invest.* **111**, 769–778.
29. Droge, W. (2006). Redox regulation in anabolic and catabolic processes. *Curr. Opin. Clin. Nutr. Metab. Care* **9**, 190–195.
30. Cerdan, S., Rodrigues, T. B., Sierra, A., Benito, M., Fonseca, L. L., et al. (2006). The redox switch/redox coupling hypothesis. *Neurochem. Int.* **48**, 523–530.
31. Squier, T. C. (2006). Redox modulation of cellular metabolism through targeted degradation of signaling proteins by the proteasome. *Antioxid. Redox. Signal* **8**, 217–228.
32. Sager, P. R., Doherty, R. A., and Olmsted, J. B. (1983). Interaction of methylmercury with microtubules in cultured cells and *in vitro*. *Exp. Cell. Res.* **146**, 127–137.
33. Lopachin, R. M. and Barber, D. S. (2006). Synaptic cysteine sulfhydryl groups as targets of electrophilic neurotoxicants. *Toxicol. Sci.* **94**, 240–255.
34. Denny, M. F. and Atchison, W. D. (1996). Mercurial-induced alterations in neuronal divalent cation homeostasis. *Neurotoxicology* **17**, 47–61.
35. Goldstein, G. W. (1993). Evidence that lead acts as a calcium substitute in second messenger metabolism. *Neurotoxicology* **14**, 97–102.
36. Simons, T. J. B. (1993). Lead-calcium interactions in cellular lead toxicity. *Neurotoxicology* **14**, 77–86.
37. Costa, L. G., Guizzetti, M., Lu, H., Bordi, F., Vitalone, A., et al. (2001). Intracellular signal transduction pathways as targets for neurotoxicants. *Toxicology* **160**, 19–26.
38. Deng, W. and Poretz, R. D. (2002). Protein kinase C activation is required for the lead-induced inhibition of proliferation and differentiation of cultured oligodendroglial progenitor cells. *Brain. Res.* **929**, 87–95.
39. Choi, B. H., Yee, S., and Robles, M. (1996). The effects of glutathione glycoside in methylmercury poisoning. *Toxicol. Appl. Pharmacol.* **141**, 357–364.
40. Shenker, B. J., Guo, T. L. O. I., and Shapiro, I. M. (1999). Induction of apoptosis in human T-cells by methyl mercury: Temporal relationship between mitochondrial dysfunction and loss of reductive reserve. *Toxicol. Appl. Pharmacol.* **157**, 23–35.
41. Anderson, A. C., Puerschel, S. M., and Linakis, J. G. (1996). Pathophysiology of lead poisoning. In *Lead Poisoning In Children*. S. M. Pueschel, J. G. Linakis, and A. C. Anderson (Eds.). P.H. Brookes, Baltimore, pp. 75–96.

42. He, L., Poblenz, A. T., Medrano, C. J., and Fox, D. A. (2000). Lead and calcium produce rod photoreceptor cell apoptosis by opening the mitochondrial permeability transition pore. *J. Biol. Chem.* **275**, 12175–12184.
43. Tiffany-Castiglioni, E., Sierra, E. M., Wu, J. N, and Rowles, T. K. (1989). Lead toxicity in neuroglia. *Neurotoxicol* **10**, 417–443.
44. Bressler, J. P. and Goldstein, G. W. (1991). Mechanisms of lead neurotoxicity. *Biochem. Pharmacol.* **41**, 479–484.
45. Pounds, J. G. (1984). Effect of lead intoxication on calcium homeostasis and calcium-mediated cell function: A review. *Neuro.Toxicology.* **5**, 295–332.
46. Trasande, L., Landrigan, P. J., and Schechter, C. (2005). Public health and economic consequences of methyl mercury toxicity to the developing brain. *Environ. Health Perspect.* **113**, 590–596.
47. Raff, M. C., Miller, R. H., and Noble, M. (1983). A glial progenitor cell that develops *in vitro* into an astrocyte or an oligodendrocyte depending on the culture medium. *Nature* **303**, 390–396.
48. Barres, B. A., Hart, I. K., Coles, H. S., Burne, J. F., Voyvodic, J. T., et al. (1992). Cell death in the oligodendrocyte lineage. *J. Neurobiol.* **23**, 1221–1230.
49. Noble, M., Mayer-Proschel, M., and Miller, R. H. (2005). The oligodendrocyte. In *Developmental Neurobiology*. M. S. Rao and M. Jacobson (Eds.). Kluwer Academic/Plenum, New York.
50. Noble, M., Pröschel, C., and Mayer-Proschel, M. (2004). Getting a GR(i)P on oligodendrocyte development. *Dev. Biol.* **265**, 33–52.
51. Levine, J. M., Reynolds, R., and Fawcett, J. W. (2001). The oligodendrocyte precursor cell in health and disease. *TINS* **24**, 39–47.
52. Miller, R. H. (2002). Regulation of oligodendrocyte development in the vertebrate CNS. *Prog. Neurobiol.* **67**, 451–467.
53. Deng, W., McKinnon, R. D., and Poretz, R. D. (2001). Lead exposure delays the differentiation of oligodendroglial progenitors *in vitro*, and at higher doses induces cell death. *Toxicol. Appl. Pharmacol.* **174**, 235–244.
54. Bichenkov, E. and Ellingson, J. S. (2001). Ethanol exerts different effects on myelin basic protein and 2′,3′-cyclic nucleotide 3′-phosphodiesterase expression in differentiating CG-4 oligodendrocytes. *Brain. Res. Dev. Brain. Res.* **128**, 9–16.
55. Zoeller, R. T., Butnariu, O. V., Fletcher, D. L., and Riley, E. P. (1994). Limited postnatal ethanol exposure permanently alters the expression of mRNAS encoding myelin basic protein and myelin-associated glycoprotein in cerebellum. *Alcohol. Clin. Exp. Res.* **18**, 909–916.
56. Harris, S. J., Wilce, P., and Bedi, K. S. (2000). Exposure of rats to a high but not low dose of ethanol during early postnatal life increases the rate of loss of optic nerve axons and decreases the rate of myelination. *J. Anat.* **197**(Prt. 3), 477–485.
57. Özer, E., Saraioglu, S., and Güre, A. (2000). Effect of prenatal ethanol exposure on neuronal migration, neurogenesis and brain myelination in the mice brain. *Clin. Neuropathol.* **19**, 21–25.
58. O'Callaghan, J. P. and Miller, D. B. (1983). Acute postnatal exposure to triethyltin in the rat: Effects on specific protein composition of subcellular fractions from developing and adult brain. *J. Pharmacol. Exp. Ther.* **224**, 466–472.
59. Smith, J., Ladi, E., Mayer-Pröschel, M., and Noble, M. (2000). Redox state is a central modulator of the balance between self-renewal and differentiation in a dividing glial precursor cell. *Proc. Natl. Acad. Sci. USA* **97**, 10032–10037.
60. Noble, M., Murray, K., Stroobant, P., Waterfield, M. D., and Riddle, P. (1988). Platelet-derived growth factor promotes division and motility and inhibits premature differentiation of the oligodendrocyte/type-2 astrocyte progenitor cell. *Nature* **333**, 560–562.
61. Richardson, W. D., Pringle, N., Mosley, M., Westermark, B., and Dubois-Dalcq, M. (1988). A role for platelet-derived growth factor in normal gliogenesis in the central nervous system. *Cell* **53**, 309–319.
62. Calver, A., Hall, A., Yu, W., Walsh, F., Heath, J., et al. (1998). Oligodendrocyte

population dynamics and the role of PDGF *in vivo*. *Neuron* **20**, 869–882.

63. Barres, B. A., Lazar, M. A., and Raff, M. C. (1994). A novel role for thyroid hormone, glucocorticoids and retinoic acid in timing oligodendrocyte development. *Development* **120**, 1097–1108.
64. Ibarrola, N., Mayer-Proschel, M., Rodriguez-Pena, A., and Noble, M. (1996). Evidence for the existence of at least two timing mechanisms that contribute to oligodendrocyte generation *in vitro*. *Dev. Biol.* **180**, 1–21.
65. Grinspan, J. B., Edell, E., Carpio, D. F., Beesley, J. S., Lavy, L., et al. (2000). Stage-specific effects of bone morphogenetic proteins on the oligodendrocyte lineage. *J. Neurobiol.* **43**, 1–17.
66. Mabie, P., Mehler, M., Marmur, R., Papavasiliou, A., Song, Q., et al. (1997). Bone morphogenetic proteins induce astroglial differentiation of oligodendroglial-astroglial progenitor cells. *Neurosci* **17**, 4112–4120.
67. Castoldi, A. F., Barni, S., Turin, I., Gandini, C., Manzo, L. (2000). Early acute necrosis, delayed apoptosis and cytoskeletal breakdown in cultured cerebellar granule neurons exposed to methylmercury. *J. Neurosci. Res.* **60**, 775–787.
68. Park, S. T., Lim, K. T., Chung, Y. T., and Kim, S. U. (1996). Methylmercury induced neurotoxicity in cerebral neuron culture is blocked by antioxidants and NMDA receptor antagonists. *Neurotoxicology* **17**, 37–46.
69. Aschner, M., Yao, C. P., Allen, J. W., and Tan, K. H. (2000). Methylmercury alters glutamate transport in astrocytes. *Neurochem. Int.* **37**, 199–206.
70. Markowski, V. P., Flaugher, C. B., Baggs, R. B., Rawleigh, R. C., Cox, C., et al. (1998). Prenatal and lactational exposure to methylmercury affects select parameters of mouse cerebellar development. *Neurotoxicology* **19**, 879–892.
71. Peckham, N. H. and Choi, B. H. (1988). Abnormal neuronal distribution within the cerebral cortex after prenatal methylmercury intoxication. *Acta. Neuropathol.* **76**, 222–226.
72. Kakita, A., Inenaga, C., Sakamoto, M., and Takahashi, H. (2002). Neuronal migration disturbance and consequent cytoarchitecture in the cerebral cortex following transplacental administration of methylmercury. *Acta. Neuropathol. (Berl)* **104**, 409–417.
73. Faustman, E. M., Ponce, R. A., Ou, Y. C., Mendoza, M. A., Lewandowski, T., et al. (2002). Investigations of methylmercury-induced alterations in neurogenesis. *Environ. Health Perspect.* **110**, 859–864.
74. Choi, B. H. (1986). Methylmercury poisoning of the developing nervous system: I. Pattern of neuronal migration in the cerebral cortex. *Neurotoxicology* **7**, 591–600.
75. Murata, K., Budtz-Jorgensen, E., and Grandjean, P. (2002). Benchmark dose calculations for methylmercury-associated delays on evoked potential latencies in two cohorts of children. *Risk. Anal.* **22**, 465–474.
76. Murata, K., Weihe, P., Araki, S., Budtz-Jorgensen, E., and Grandjean, P. (1999). Evoked potentials in Faroese children prenatally exposed to methylmercury. *Neurotoxicol. Teratol.* 471–472.
77. Murata, K., Weihe, P., Budtz-Jorgensen, E., Jorgensen, P. J., and Grandjean, P. (2004). Delayed brainstem auditory evoked potential latencies in 14-year-old children exposed to methylmercury. *J. Pediatr.* **144**, 177–183.
78. Hamada, R., Yoshida, Y., Kuwano, A., Mishima, I., and Igata, A. (1982). Auditory brainstem responses in fetal organic mercury poisoning (in Japanese). *Shinkei-Naika* **16**, 282–285.
79. Nakamura, K., Houzawa, J., and Uemura, T. (1986). Auditory brainstem responses in rats with methylmercury poisoning. *Audiol. Jpn.* **29**, 445–446.
80. Algarin, C., Peirano, P., Garrido, M., Pizarro. F., and Lozoff, B. (2003). Iron deficiency anemia in infancy: Long–lasting effects on auditory and visual system functioning. *Pediatr. Res.* **53**, 217–223.
81. Roncagliolo, M., Garrido, M., Walter, T., Peirano, P., and Lozoff, B. (1998). Evidence of altered central nervous system development in infants with iron deficiency anemia at 6 mo: Delayed maturation of

auditory brainstem responses. *Am. J. Clin. Nutr.* **68**, 683–690.

82. Heldin, C. H., Ostman, A., and Ronnstrand, L. (1998). Signal transduction via platelet-derived growth factor receptors. *Biochim. Biophys. Acta.* **1378**, F79–113.
83. Rupprecht, H. D., Sukhatme, V. P., Lacy, J., Sterzel, R. B., and Coleman, D. L. (1993). PDGF-induced Egr-1 expression in rat mesangial cells is mediated through upstream serum response elements. *Am. J. Physiol.* **265**, F351–360.
84. Franke, T. F., Yang, S. I., Chan, T. O., Datta, K., Kazlauskas, A., et al. (1995). The protein kinase encoded by the Akt proto-oncogene is a target of the PDGF-activated phosphatidylinositol 3-kinase. *Cell* **81**, 727–736.
85. Choudhury, G. G. (2001). Akt serine threonine kinase regulates platelet-derived growth factor-induced DNA synthesis in glomerular mesangial cells: Regulation of c-fos AND p27(kip1) gene expression. *J. Biol. Chem.* **276**, 35636–35643.
86. Raff, M. C., Lillien, L. E., Richardson, W. D., Burne, J. F., and Noble, M. D. (1988). Platelet-derived growth factor from astrocytes drives the clock that times oligodendrocyte development in culture. *Nature* **333**, 562–565.
87. Lamballe, F., Klein, R., Barbacid, M. (1991). trkC, a new member of the trk family of tyrosine protein kinases, is a receptor for neurotrophin-3. *Cell* **66**, 967–979.
88. Miyake, S., Mullane-Robinson, K. P., Lill, N. L., Douillard, P., and Band, H. (1999). Cbl-mediated negative regulation of platelet-derived growth factor receptor-dependent cell proliferation. A critical role for Cbl tyrosine kinase-binding domain. *J. Biol. Chem.* **274**, 16619–16628.
89. Miyake, S., Lupher, M. L. J., Druke, B., and Band, H. (1998). The tyrosine kinase regulator Cbl enhances the ubiquitination and degradation of the platelet-derived growth factor receptor alpha. *Proc. Natl. Acad. Sci. USA* **95**, 7927–7932.
90. Duan, L., Miura, Y., Dimri, M., Majumder, B., Dodge, I. L., et al. (2003). Cbl-mediated ubiquitinylation is required for lysosomal sorting of epidermal growth factor receptor but is dispensable for endocytosis. *J. Biol. Chem.* **278**, 28950–28960.
91. Rosenkranz, S., Ikuno, Y., Leong, F. L., Klinghoffer, R. A., Miyake, S., et al. (2000). Src family kinases negatively regulate platelet-derived growth factor alpha receptor-dependent signaling and disease progression. *J. Biol. Chem.* **275**, 9620–9627.
92. Schmidt, M. H. and Dikic, I. (2005). The Cbl interactome and its functions. *Nat. Rev. Mol. Cell. Biol.* **6**, 907–919.
93. Tsygankov, A. Y., Mahajan, S., Fincke, J. E., and Bolen, J. B. (1996). Specific association of tyrosine-phosphorylated c-Cbl with Fyn tyrosine kinase in T cells. *J. Biol. Chem.* **271**, 27130–27137.
94. Hunter, S., Burton, E. A., Wu, S. C., and Anderson, S. M. (1999). Fyn associates with Cbl and phosphorylates tyrosine 731 in Cbl, a binding site for phosphatidylinositol 3-kinase. *J. Biol. Chem.* **274**, 2097–2106.
95. Feshchenko, E. A., Langdon, W. Y., and Tsygankov, A. Y. (1998). Fyn, Yes, and Syk phosphorylation sites in c-Cbl map to the same tyrosine residues that become phosphorylated in activated T cells. *J. Biol. Chem.* **273**, 8223–8331.
96. Kassenbrock, C. K., Hunter, S. F., Garl, P., Johnson, G. L., and Anderson, S. M. (2002). Inhibition of Src family kinases blocks epidermal growth factor (EGF)-induced activation of Akt, phosphorylation of c-Cbl, and ubiquitination of the EGF receptor. *J. Biol. Chem.* **277**, 24967–24975.
97. Abe, J. and Berk, B. C. (1999). Fyn and JAK2 mediate ras activation by reactive oxygen species. *J. Biol. Chem.* **274**, 21003–21010.
98. Abe, J., Okuda, M., Huang, Q., Yoshizumi, M., and Berk, B. C. (2000). Reactive oxygen species activate p90 ribosomal S6 kinase via Fyn and Ras. *J. Biol. Chem.* **275** 1739–1748.
99. Sanguinetti, A. R., Cao, H., and Corley Mastick, C. (2003). Fyn is required for oxidative- and hyperosmotic-stress-induced tyrosine phosphorylation of caveolin-1. *Biochem. J.* **376**, 159–168.

100. Hehner, S. P., Breitfreutz, R., Shubinsky, G., Unsoeld, H., Schulze-Osthoff, K., et al. (2000) Enhancement of T cell receptor signaling by a mild oxidative shift in the intracellular thiol pool. *J. Immunol.* **165**, 4319–4328.

101. Osterhout, D. J., Wolven, A., Wolf, R. M., Resh, M. D., and Chao, M. V. (1999). Morphological differentiation of oligodendrocytes requires activation of Fyn tyrosine kinase. *J. Cell. Biol.* **145**, 1209–1218.

102. Wolf, R. M., Wilkes, J. J., Chao, M. V., and Resh, M. D. (2001). Tyrosine phosphorylation of p190 RhoGAP by Fyn regulates oligodendrocyte differentiation. *J. Neurobiol.* **49**, 62–78.

103. Poole, B. and Ohkuma, S. (1981). Effect of weak bases on the intralysosomal pH in mouse peritoneal macrophages. *J. Cell. Biol.* **90**, 665–669.

104. Brown, W. J., Goodhouse, J., and Farquhar, M. G. (1986). Mannose-6-phosphate receptors for lysosomal enzymes cycle between the Golgi complex and endosomes. *J. Cell. Biol.* **103**, 1235–1247.

105. Laing, J. G., Tadros, P. N., Green, K., Saffitz, J. E., and Beyer, E. C. (1998). Proteolysis of connexin43-containing gap junctions in normal and heat-stressed cardiac myocytes. *Cardiovasc. Res.* **38**, 711–718.

106. Taher, T. E., Tjin, E. P., Beuling, E. A., Borst, J., Spaargaren, M., et al. (2002). c-Cbl is involved in Met signaling in B cells and mediates hepatocyte growth factor-induced receptor ubiquitination. *J. Immunol.* **169**, 3793–3780.

107. Thien, C. B. and Langdon, W. Y. (2005). Negative regulation of PTK signalling by Cbl proteins. *Growth Factors* **23**, 161–167.

108. van Leeuwen, J. E., Paik, P. K., and Samelson, L. E. (1999). The oncogenic 70Z Cbl mutation blocks the phosphotyrosine binding domain-dependent negative regulation of ZAP-70 by c-Cbl in Jurkat T cells. *Mol. Cell Biol.* **19**, 6652–6664.

109. Deng, W. and Poretz, R. D. (2003). Oliogodendroglia in developmental neurotoxicity. *Neurotoxicol.* **24**, 161–178.

110. Hausburg, M. A., Dekrey, G. K., Salmen, J. J., Palic, M. R., and Gardiner, C. S. (2005). Effects of paraquat on development of preimplantation embryos *in vivo* and *in vitro*. *Reprod. Toxicol.* **20**, 239–246.

111. McCarthy, S., Somayajulu, M., Sikorska, M., Borowy-Borowski, H., and Pandey, S. (2004). Paraquat induces oxidative stress and neuronal death; neuroprotection by water-soluble Coenzyme Q10. *Toxicol. Appl. Pharmacol.* **201**, 21–31.

112. Matsuda, S., Gomi, F., Katayama, T., Koyama, Y., Tohyama, M., et al. (2006). Induction of connective tissue growth factor in retinal pigment epithelium cells by oxidative stress. *Jpn. J. Ophthalmol.* **50**, 229–234.

113. Kim, S. J., Kim, J. E., and Moon, I. S. (2004). Paraquat induces apoptosis of cultured rat cortical cells. *Mol. Cells* 17

114. Shimizu, K., Matsubara, K., Ohtaki, K., and Shiono, H. (2003). Paraquat leads to dopaminergic neural vulnerability in organotypic midbrain culture. *Neurosci. Res.* **46**, 523–532.

115. Aruoma, O. I., Halliwell, B., Hoey, B. M., and Butler, J. (1989). The antioxidant action of N-acetylcysteine: Its reaction with hydrogen peroxide, hydroxyl radical, superoxide and hypochlorous acid. *Free Radic Biol. Med.* **6**, 593–597.

116. Meister, A., Anderson, M. E., and Hwang, O. (1986). Intracellular cysteine and glutathione delivery systems. *J. Am. Coll. Nutr.* **5**, 137–151.

117. Hoffer, E., Avidor, I., Benjaminov, O., Shenker, L., Tabak, A., et al. (1993). N-acetylcysteine delays the infiltration of inflammatory cells into the lungs of paraquat-intoxicated rats. *Toxicol. Appl. Pharmacol.* **120**, 8–12.

118. Mayer, M. and Noble, M. (1994). N-acetyl-L-cysteine is a pluripotent protector against cell death and enhancer of trophic factor-mediated cell survival *in vitro*. *Proc. Natl. Acad. Sci. USA* **91**, 7496–7500.

119. Chen, Y. W., Huang, C. F., Tsai, K. S., Yang, R. S., Yen, C. C., et al. (2006). The role of phosphoinositide 3-kinase/Akt signaling in low-dose mercury-induced mouse pancreatic {beta}-cell dysfunction *in vitro* and *in vivo*. *Diabetes* **55**, 1614–16124.

120. Ballatori, N., Lieberman, M. W., and Wang, W. (1998). N-acetylcysteine as an

antidote in methylmercury poisoning. *Environ. Health Perspect.* **106**, 267–271.

121. Shanker, G., Syversen, T., and Aschner, M. (2005). Modulatory effect of glutathione status and antioxidants on methylmercury-induced free radical formation in primary cultures of cerebral astrocytes. *Brain Res. Mol. Brain Res.* **137**, 11–22.
122. Nehru, B. and Kanwar, S. S. (2004). N-acetylcysteine exposure on lead-induced lipid peroxidative damage and oxidative defense system in brain regions of rats. *Biol. Trace. Elem. Res.* **101**, 257–264.
123. Neal, R., Copper, K., Gurer, H., and Ercal, N. (1998). Effects of N-acetyl cysteine and 2,3-dimercaptosuccinic acid on lead induced oxidative stress in rat lenses. *Toxicology* **130**, 167–174.
124. Yeh, S. T., Guo, H. R., Su, Y. S., Lin, H. J., Hou, C. C., et al. (2006). Protective effects of N-acetylcysteine treatment post acute paraquat intoxication in rats and in human lung epithelial cells. *Toxicology* **223**, 181–190.
125. Satoh, E., Okada, M., Takadera, T., and Ohyashiki, T. (2005). Glutathione depletion promotes aluminum-mediated cell death of PC12 cells. *Biol. Pharm. Bull.* **28**, 941–946.
126. Tandon, S. K., Singh, S., Prasad, S., Khandekar, K., Dwivedi, V. K., et al. (2003). Reversal of cadmium induced oxidative stress by chelating agent, antioxidant or their combination in rat. *Toxicol. Lett.* **145**, 211–217.
127. Flora, S. J. (1999). Arsenic-induced oxidative stress and its reversibility following combined administration of N-acetylcysteine and meso 2,3-dimercaptosuccinic acid in rats. *Clin. Exp. Pharmacol. Physiol.* **26**, 865–869.
128. Zaragoza, A., Diez-Fernandez, C., Alvarez, A. M., Andres, D., and Cascales, M. (2001). Mitochondrial involvement in cocaine-treated rat hepatocytes: Effect of N-acetylcysteine and deferoxamine. *Br. J. Pharmacol.* **132**, 1063–1070.
129. Roberts, J., Nagasawa, H., Zera, R., Fricke, R., and Goon, D. (1987). Prodrugs of L-cysteine as protective agents against acetaminophen-induced hepatotoxicity. 2-(Polyhydroxyalkyl)- and 2-(polyacetoxyalkyl)thiazolidine-4(R)-carboxylic acids. *J. Med. Chem.* **30**, 1891–1896.
130. Yan, H. and Rivkees, S. A. (2002). Hepatocyte growth factor stimulates the proliferation and migration of oligodendrocyte progenitor cells. *J. Neurosci. Res.* **69**, 597–606.
131. Bottaro, D. P., Rubin, J. S., Faletto, D. L., Chan, A. M., Kmiecik, T. E., et al. (1991). Identification of the hepatocyte growth factor receptor as the c-met proto-oncogene product. *Science* **251**, 802–804.
132. Naldini, L., Vigna, E., Narsimhan, R. P., Gaudino, G., Zarnegar, R., et al. (1991). Hepatocyte growth factor (HGF) stimulates the tyrosine kinase activity of the receptor encoded by the proto-oncogene c-MET. *Oncogene* **6**, 501–504.
133. Knapp, P. E. and Adams, M. H. (2004). Epidermal growth factor promotes oligodendrocyte process formation and regrowth after injury. *Exp. Cell Res.* **296**, 135–144.
134. Levkowitz, G., Klapper, L. N., Tzahar, E., Freywald, A., Sela, M., et al. (1996). Coupling of the c-Cbl protooncogene product to ErbB-1/EGF-receptor but not to other ErbB proteins. *Oncogene* **12**, 1117–1125.
135. Rubin, C., Gur, G., and Yarden, Y. (2005). Negative regulation of receptor tyrosine kinases: Unexpected links to c-Cbl and receptor ubiquitylation. *Cell Res.* **15**, 66–71.
136. de Melker, A. A., van der Horst, G., and Borst, J. (2004). c-Cbl directs EGF receptors into an endocytic pathway that involves the ubiquitin-interacting motif of Eps15. *J. Cell Sci.* **117**, 5001–5012.
137. Ravid, T., Heidinger, J. M., Gee, P., Khan, E. M., and Goldkorn, T. (2004). c-Cbl-mediated ubiquitinylation is required for epidermal growth factor receptor exit from the early endosomes. *J. Biol. Chem.* **279**, 37153–37162.
138. Garcia-Guzman, M., Larsen, E., and Vuori, K. (2000). The proto-oncogene c-Cbl is a positive regulator of Met-induced MAP kinase activation: A role for the adaptor protein Crk. *J. Immunol.* **19**, 4058–4065.

139. Tiffany-Castiglioni, E. (1993). Cell culture models for lead toxicity in neuronal and glial cells. *Neurotoxicol.* **14**, 513–536.

140. Krigman, M. R., Druse, M. J, Traylor, T. D., Wilson, M. H., Newell, L. R., et al. (1974). Lead encephalopathy in the developing rat: Effect on myelination. *J. Neuropathol. Exp. Neurol.* **33**, 58–73.

141. Dabrowska-Bouta, B., Sulkowski. G., Bartosz. G., Walski. M., and Rafalowska. U. (1999). Chronic lead intoxication affects the myelin membrane status in the central nervous system of adult rats. *J. Mol. Neurosci.* **13**, 127–139.

142. Deng, W. and Poretz, R. D. (2001). Chronic dietary lead exposure affects galactolipid metabolic enzymes in the developing rat brain. *Toxicol. Appl. Pharmacol.* **172**, 98–107.

143. Weiss, B., Stern, S., Cox, C., and Balys, M. (2005). Perinatal and lifetime exposure to methylmercury in the mouse: Behavioral effects. *Neurotoxicology* **26**, 675–690.

144. Stern, S., Cox, C., Cernichiari, E., Balys, M., and Weiss, B. (2001). Perinatal and lifetime exposure to methylmercury in the mouse: Blood and brain concentrations of mercury to 26 months of age. *Neurotoxicology* **22**, 467–477.

145. Goulet, S., Dore, F. Y., and Mirault, M. E. (2003). Neurobehavioral changes in mice chronically exposed to methylmercury during fetal and early post-natal development. *Neurotoxicol. Teratol.* **25**, 335–347.

146. Sakamoto, M., Kakita, A., de Oliveira, R. B., Pan, H. S., and Takahashi, H. (2004). Dose-dependent effects of methylmercury administered during neonatal brain spurt in rats. *Dev. Brain. Res.* **152**, 171–176.

147. Barone, S. Jr., Haykal-Coates, N., Parran, D. K., and Tilson, H. A. (1998). Gestational exposure to methylmercury alters the developmental pattern of trk-like immunoreactivity in the rat brain and results in cortical dysmorphology. *Brain. Res. Dev. Brain. Res.* **109**, 13–31.

148. Dietrich, J., Han, R., Yang, Y., Mayer-Pröschel, M., and Noble, M. (2006). CNS progenitor cells and oligodendrocytes are targets of chemotherapeutic agents *in vitro* and *in vivo*. *J. Biol.* **5**, 22.

149. Rowitch, D. H., Lu, R. Q., Kessaris, N., and Richardson, W. D. (2002). An "oligarchy" rules neural development. *Trends Neurosci.* **25**, 417–422.

150. Takebayashi, H., Nabeshima, Y., Yoshida, S., Chisaka, O., Ikenaka, K., et al. (2002). The basic helix-loop-helix factor olig2 is essential for the development of motoneuron and oligodendrocyte lineages. *Curr. Biol.* **12**, 1157–1163.

151. Zhou, Q., Choi, G., and Anderson, D. J. (2001). The bHLH transcription factor Olig2 promotes oligodendrocyte differentiation in collaboration with nkx2.2. *Neuron* **31**, 791–807.

152. Mukouyama, Y. S., Deneen, B., Lukaszewicz, A., Novitch, B. G., Wichterle, H., et al. (2006). $Olig2^+$ neuroepithelial motoneuron progenitors are not multipotent stem cells *in vivo*. *Proc. Natl. Acad. Sci. USA* **103**, 1551–1556.

153. Fancy, S. P., Zhao, C., and Franklin, R. J. (2004). Increased expression of Nkx2.2 and Olig2 identifies reactive oligodendrocyte progenitor cells responding to demyelination in the adult CNS. *Mol. Cell Neurosci.* **27**, 247–254.

154. Talbott, J. F., Loy, D. N., Liu, Y., Qiu, M. S., Bunge, M. B., et al. (2005). Endogenous Nkx2.2+/$Olig2^+$ oligodendrocyte precursor cells fail to remyelinate the demyelinated adult rat spinal cord in the absence of astrocytes. *Exp. Neurol.* **192**, 11–24.

155. Homolya, L., Varadi, A., and Sarkadi, B. (2003). Multidrug resistance-associated proteins: Export pumps for conjugates with glutathione, glucuronate or sulfate. *Biofactors* **17**, 103–114.

156. Leslie, E. M., Deeley, R. G., and Cole, S. P. (2001). Toxicological relevance of the multidrug resistance protein 1, MRP1 (ABCC1) and related transporters. *Toxicology* **167**, 3–23.

157. Liang, X., Draghi, N. A., and Resh, M. D. (2004). Signaling from integrins to Fyn to Rho Family GTPases regulates morphologic differentiation of oligodendrocytes. *J. Neurosci.* **24**, 7140–7149.

158. Tsatmali, M., Walcott, E. C., and Crossin, K. L. (2005). Newborn neurons acquire high levels of reactive oxygen species and

increased mitochondrial proteins upon differentiation from progenitors. *Brain Res.* **1040**, 137–150.

159. Goldsmit. Y., Erlich, S., and Pinkas-Kramarski, R. (2001). Neuregulin induces sustained reactive oxygen species generation to mediate neuronal differentiation. *Cell Mol. Neurobiol.* **211**, 753–769.

160. Puceat, M. (2005). Role of Rac-GTPase and reactive oxygen species in cardiac differentiation of stem cells. *Antioxid Redox Signal* **7**, 1435–1439.

161. McGrath, S. A. (1998). Induction of p21WAF/CIP1 during hyperoxia. *Am. J. Respir. Cell Mol. Biol.* **18**, 179–187.

162. McGrath-Morrow, S. A., Cho, C., Soutiere, S., Mitzner, W., and Tuder, R. (2004). The effect of neonatal hyperoxia on the lung of p21Waf1/Cip1/Sdi1-deficient mice. *Am. J. Respir. Cell Mol. Biol.* **30**, 635–640.

163. Seomun, Y., Kim, J. T., Kim, H. S., Park, J. Y., Joo, and C. K. (2005). Induction of p21Cip1-mediated G2/M arrest in H2O2-treated lens epithelial cells. *Mol. Vis.* **11**, 764–774.

164. Esposito, F., Russo, L., Chirico, G., Ammendola, R., Russo, T., et al. (2001). Regulation of p21waf1/cip1 expression by intracellular redox conditions. *IUBMB Life* **52**, 67–70.

165. Barnouin, K., Dubuisson, M. L., Child, E. S., Fernandez De Mattos, S., Glassford, J., et al. (2002). H2O2 induces a transient multi-phase cell cycle arrest in mouse fibroblasts through modulating cyclin D and p21Cip1 expression. *J. Biol. Chem.* **277**, 13761–13770.

166. Hu, Y., Wang, X., Zeng, L., Cai, D. Y., Sabapathy, K., et al. (2005). ERK phosphorylates p66shcA on Ser36 and subsequently regulates p27kip1 expression via the Akt-FOXO3a pathway: Implication of p27kip1 in cell response to oxidative stress. *Mol. Biol. Cell* **16**, 3705–3718.

167. WHO (1990). Environmental health criteria 101: *Methylmercury*. World Health Organization, Geneva. Retrieved from [http://www.inchem.org/documents/ehc/ehc/ehc101.htm]. Accessed December 15, 2006.

168. Cernichiari, E., Brewer, R., Myers, G. J., Marsh, D. O., Lapham, L. W., et al. (1995). Monitoring methylmercury during pregnancy: Maternal hair predicts fetal brain exposure. *Neurotoxicology* **16**, 705–710.

169. Goyer, R. A. (1993). Lead toxicity: Current concerns. *Environ. Health Perspect* **100**, 177–187.

170. Banks, E. C., Ferretti, L. E., and Shucard, D. W. (1997). Effects of low level lead exposure on cognitive function in children: A review of behavioral, neuropsychological and biological evidence. *Neurotoxicology* **18**, 237–282.

171. Lidsky, T. I. and Schneider, J. S. (2003). Lead neurotoxicity in children: Basic mechanisms and clinical correlates. *Brain* **126**, 5–19.

172. Needleman, H. L. and Gatsonis, C. A. (1990). Low level lead exposure and the IQ of children. A meta-analysis of modern studies. *JAMA* **263**, 673–678.

173. Finkelstein, Y., Markowitz, M. E., and Rosen, J. F. (1998). Low-level lead-induced neurotoxicity in children: An update on central nervous system effects. *Brain Res. Brain Res. Rev.* **27**, 168–176.

174. Ballinger, D., Leviton, A., Waternoux, C., Needleman, H., and Rabinowitz, P. (1987). Longitudinal analysis of prenatal and postnatal lead exposure and early cognitive development. *N. Engl. J. Med.* **316**, 1037–1043.

175. Winneke, G., Brockhaus, A., Ewers, U., Krämer, U., and Neuf, M. (1990). Results from the European Multicenter Study on lead neurotoxicity in children: Implications for risk assessment. *Neurotox. Teratol.* **12**, 553–559.

176. Bellinger, D. and Needleman, H. L. (1992). Neurodevelopmental effects of low-level lead exposure in children. In *Human Lead Exposure*. H. Needleman (Ed.). CRC Press, Boca Raton (Florida), pp. 191–208.

177. WHO (1995) Environmental health criteria 165: Inorganic lead. Geneva: World Health Organization. Available: [http://www.inchem.org/documents/ehc/ehc/ehc165.htm.] Accessed December 15, 2006.

178. Kaiser, R., Henderson, A. K., Daley, W. R., Naughton, M., Khan, M. H., et al. (2001). Blood lead levels of primary school children in Dhaka, Bangladesh. *Environ. Health Perspect* **109**, 563–566.

179. Yakovlev, A. Y., Boucher, K., Mayer-Pröschel, M., and Noble, M. (1998). Quantitative insight into proliferation and differentiation of O-2A progenitor cells *in vitro*: The clock model revisited. *Proc. Natl. Acad. Sci. USA* **95**, 14164–14167.

180. Hyrien, O., Mayer-Proschel, M., Noble, M., and Yakovlev, A. (2005). Estimating the life-span of oligodendrocytes from clonal data on their development in cell culture. *Math. Biosci.* **193**, 255–274.

181. Hyrien, O., Mayer-Proschel, M., Noble, M., and Yakovlev, A. (2005). A stochastic model to analyze clonal data on multi-type cell populations. *Biometrics* **61**, 199–207.

182. Tamm, C., Duckworth, J., Hermanson, O., and Ceccatelli, S. (2006). High susceptibility of neural stem cells to methylmercury toxicity: Effects on cell survival and neuronal differentiation. *J. Neurochem.* **97**, 69–78.

183. Rothenberg, S. J., Poblano, A., and Schnaas, L. (2000). Brainstem auditory evoked response at five years and prenatal and postnatal blood lead. *Neurotoxicol. Teratol.* **22**, 503–510.

184. Bleecker, M. L., Ford, D. P., Lindgren, K. N., Scheetz, K., and Tiburzi, M. J. (2003). Association of chronic and current measures of lead exposure with different components of brainstem auditory evoked potentials. *Neurotoxicology* **24**, 625–631.

185. Lester, B. M., Lagasse, L., Seifer, R., Tronick, E. Z., Bauer, C. R., et al. (2003). The Maternal Lifestyle Study (MLS): Effects of prenatal cocaine and/or opiate exposure on auditory brain response at one month. *J. Pediatr.* **142**, 279–285.

186. Tan-Laxa, M. A., Sison-Switala, C., Rintelman, W., and Ostrea, E. M. J. (2004). Abnormal auditory brainstem response among infants with prenatal cocaine exposure. *Pediatrics* **113**, 357–360.

187. Poblano, A., Belmont, A., Sosa, J., Ibarra, J., Rosas, Y., et al. (2002). Effects of prenatal exposure to carbamazepine on brainstem auditory evoked potentials in infants of epileptic mothers. *J. Child Neurol.* **17**, 364–368.

188. Fruttiger, M., Karlsson, L., Hall, A., Abramsson, A., Calver, A., et al. (1999). Defective oligodendrocyte development and severe hypomyelination in PDGF-A knockout mice. *Development* **126**, 457–467.

189. Hoch, R. V. and Soriano, P. (2003). Roles of PDGF in animal development. *Development* **130**, 4769–4784.

190. Betsholtz, C. (2004). Insight into the physiological functions of PDGF through genetic studies in mice. *Cytokine. Growth Factor Rev.* **15**, 215–228.

191. Wong, R. W. C. and Guillaud, L. (2004). The role of epidermal growth factor and its receptors in mammalian CNS. *Cytokine Growth Factor Rev.* **15**, 147–156.

192. Xian, C. J. and Zhou, X. F. (2004). EGF family of growth factors: Essential roles and functional redundancy in the nerve system. *Front. Biosci.* **9**, 85–92.

193. Holbro, T. and Hynes, N. E. (2004). ErbB receptors: Directing key signaling networks throughout life. *Annu. Rev. Pharmacol. Toxicol.* **44**, 195–217.

194. Gutierrez, H., Dolcet, C., Tolcos, M., and Davies, A. (2004). HGF regulates the development of cortical pyramidal dendrites. *Development* **131**, 3717–3726.

195. Birchmeier, C. and Gherardi, E. (1998). Developmental role of HGF/SF and its receptor, the c-Met tyrosine kinase. *Trends Cell Biol.* **8**, 404–410.

196. Morita, A., Yamashita, N., Sasaki, Y., Uchida, Y., Nakajima, O., et al. (2006). Regulation of dendritic branching and spine maturation by semaphorin3A-Fyn signaling. *J. Neurosci.* **26**, 2971–2980.

197. He, J., Nixon, K., Shetty, A. K., and Crews, F. T. (2005). Chronic alcohol exposure reduces hippocampal neurogenesis and dendritic growth of newborn neurons. *Eur. J. Neurosci.* **21**, 2711–2720.

198. Newey, S. E., Velamoor, V., Govek, E. E., and Van Aeist, L. (2005). Rho GTPases, dendritic structure, and mental retardation. *J. Neurobiol.* **64**, 58–74.

199. Power, J., Mayer-Proschel, M., Smith, J., and Noble, M. (2002). Oligodendrocyte precursor cells from different brain regions express divergent properties consistent with the differing time courses of myelination in these regions. *Dev. Biol.* **245**, 362–375.

200. Sakamoto, M., Kakita, A., Wakabayashi, K., Takahashi, H., Nakano, A., et al. (2002). Evaluation of changes in methylmercury accumulation in the developing rat brain and its effects: A study with consecutive and moderate dose exposure throughout gestation and lactation periods. *Brain Res.* **949**, 51–59.

6

1. Garner, C. E., Jefferson, W. N., Burka, L. T., Matthews, H. B., and Newbold, R. R. (1999). *In vitro* estrogenicity of the catechol metabolites of selected polychlorinated biphenyls. *Toxicol. Appl. Pharmacol.* **154**, 188–197.
2. Moore, M., Mustain, M., Daniel, K., Chen, I., Safe, S., Zacharewski, T., Gillesby, B., Joyeux, A., and Balaguer, P. (1997). Antiestrogenic activity of hydroxylated polychlorinated biphenyl congeners identified in human serum. *Toxicol. Appl. Pharmacol.* **142**, 160–168.
3. Salama, J., Chakraborty, T. R., Ng, L., and Gore, A. C. (2003). Effects of polychlorinated biphenyls on estrogen receptor-beta expression in the anteroventral periventricular nucleus. *Environ. Health Perspect.* **111**, 1278–1282.
4. United States Environmental Protection Agency, Office of Water Update (2003). National Listing of Fish and Wildlife Advisories. EPA-823-F-03-003.
5. Fernandez, M. A., Gomara, B., Bordajandi, L. R., Herrero, L., Abad, E., Abalos, M., Rivera, J., and Gonzalez, M. J. (2004). Dietary intakes of polychlorinated dibenzo-p-dioxins, dibenzofurans and dioxin-like polychlorinated biphenyls in Spain. *Food Addit. Contam.* **21**, 983–991.
6. Smith, A. G. and Gangolli, S. D. (2002). Organochlorine chemicals in seafood: Occurrence and health concerns. *Food Chem. Toxicol.* **40**, 767–779.
7. Antunes, P. and Gil, O. (2004). PCB and DDT contamination in cultivated and wild Sea bass from Ria de Aveiro, Portugal. *Chemosphere* **54**, 1503–1507.
8. Svensson, B. G., Hallberg, T., Schultz, A., and Hagmar, L. (1994). Parameters of immunological competence in subjects with high consumption of fish contaminated with persistent organochlorine compounds. *Arch. Occup. Environ. Health* **65**, 351–358.
9. Faroon, O., Keith, M. S., Jones, D., and de Rosa, C. (2001). Effects of polychlorinated biphenyls on development and reproduction. *Toxicol. Ind. Health* **17**, 63–93.
10. Brouwer, A., Longnecker, M. P., Birnbaum, L. S., Cogliano, J., Kostyniak, P., Moore, J., Schantz, S., and Winneke, G. (1999). Characterization of potential endocrine-related health effects at low dose levels of exposure to PCBs. *Environ. Health Perspect.* **107**(Suppl. 4), 639–649.
11. Judd, N., Griffith, W. C., and Faustman, E. M. (2004). Contribution of PCB exposure from fish consumption to total dioxin-like dietary exposure. *Regul. Toxicol. Pharmacol.* **40**, 125–135.
12. Schantz, S. L., Gasior, D. M., Polverejan, E., McCaffrey, R. J., Sweeney, A. M., Humphrey, H. E., and Gardiner, J. C. (2001). Impairments of memory and learning in older adults exposed to polychlorinated biphenyls via consumption of Great Lakes fish. *Environ. Health Perspect.* **109**, 605–611.
13. Istituto Superiore di Sanità (2002). Istisan Report 02/38 ISSN 1123-3117, ISS, Roma.
14. Liu, J. W. and Picard, D. (1998). Bioactive steroids as contaminants of the common carbon source galactose. *FEMS Microbiol. Lett.* **159**, 167–171.
15. Liu, J. W., Jeannin, E., and Picard, D. (1999). The anti-estrogen hydroxytamoxifen is a potent antagonist in a novel yeast system. *Biol. Chem.* **380**, 1341–1345.
16. Pinto, B., Picard, D., and Reali, D. (2004). A recombinant yeast strain as a short term bioassay to assess estrogen-like activity of xenobiotics. *Ann. Ig* **16**, 579–585.

17. Pinto, B., Garritano, S., and Reali, D. (2005). Occurrence of estrogen-like substances in the marine environment of the Northern Mediterranean Sea. *Mar. Poll. Bull.* **50**, 1681–1685.

18. Miller, J. H. (1972). *Experiments in Molecular Genetics*. Cold Spring Harbour Laboratory Press, New York.

19. Gong, Y., Chin, H. S., Lim, L. S., Loy, C. J., Obbard, J. P., and Yong, E. L. (2003). Clustering of sex hormone disruptors in Singapore's marine environment. *Environ. Health Perspect.* **111**, 1448–1453.

20. Jackson, J. E. (1991). *A User's Guide to Principal Components*. John Wiley & Sons, Inc, New York, NY.

21. Jolliffe, I. T. (1986). *Principal Components Analysis*. Springer-Verlag, New York Inc., NY.

22. Martens, H. and Naes, T. (1989). *Multivariate Calibration*. John Wiley & Sons, Chichester, UK.

23. Storelli, M. M., Giacominelli-Stuffler, R., D'Addabbo, R., and Marcotrigiano, G. O. (2003). Health risk of coplanar polychlorinated biphenyl congeners in edible fish from the Mediterranean Sea. *J. Food Prot.* **66**, 2176–2179.

24. Perugini, M., Cavaliere, M., Giammarino, A., Mazzone, P., Olivieri, V., and Amorena, M. (2004). Levels of polychlorinated biphenyls and organochlorine pesticides in some edible marine organisms from the Central Adriatic Sea. *Chemosphere* **57**, 391–400.

25. Layton, A. C., Sanseverino, J., Gregory, B. W., Easter, J. P., Sayler, G. S., and Schultz, T. W. (2002). *In vitro* estrogen receptor binding of PCBs: Measured activity and detection of hydroxylated metabolites in a recombinant yeast assay. *Toxicol. Appl. Pharmacol.* **180**, 157–163.

26. Bonefeld-Jørgensen, E. C., Andersen, H. R., Rasmussen, T. H., and Vinggaard, A. M. (2001). Effect of highly bioaccumulated polychlorinated biphenyl congeners on estrogen and androgen receptor activity. *Toxicology* **158**, 141–153.

27. Rivas, A., Fernandez, M. F., Cerrillo, I., Ibarluzea, J., Olea-Serrano, M. F., Pedraza, V., and Olea, N. (2001). Human exposure to endocrine disrupters: Standardisation of a marker of estrogenic exposure in adipose tissue. *APMIS* **109**, 185–197.

28. Miao, X. -S., Swenson, C., Woodward, L. A., and Li, Q. X. (2000). Distribution of polychlorinated biphenyls in marine species from French Frigate Shoals, North Pacific Ocean. *Sci. Tot. Environ.* **257**, 17–28.

29. Bayarri, S., Baldassarri, L. T., Iacovella, N., Ferrara, F., and di Domenico, A. (2001). PCDDs, PCDFs, PCBs and DDE in edible marine species from the Adriatic Sea. *Chemosphere* **43**, 601–610.

30. Llobet, J. M., Bocio, A., Domingo, J. L., Teixido, A., Casas, C., and Muller, L. (2003). Levels of polychlorinated biphenyls in food from Catalonia, Spain: Estimated dietary intake. *J. Food Prot.* **66**, 479–484.

31. Oberdörster, E. and Cheek, A. O. (2000). Gender benders at the beach: Endocrine disruption in marine and estuarine organisms. *Environ. Toxicol. Chem.* **20**, 23–36.

32. De Metrio, G., Corriero, A., Desantis, S., Zubani, D., Cirillo, F., Deflorio, M., Bridges, C. R., Eicker, J., de la Serna, J. M., Megalofonou, P., and Kime, D. E. (2003). Evidence of a high percentage of intersex in the Mediterranean swordfish (*Xiphias gladius* L.). *Mar. Poll. Bull.* **46**, 358–361.

33. Jobling, S., Casey, D., Rogers-Gray, T., Oehlmann, J., Schulte-Oehlmann, U., Pawlowski, S., Baunbeck, T., Turner, A. P., and Tyler, C. R. (2004). Comparative responses of molluscs and fish to environmental estrogens and an estrogenic effluent. *Aquat. Toxicol.* **66**, 207–222.

34. Nash, J. P., Kime, D. E., Van der Ven, L. T., Wester, P. W., Brion. F., Maack, G., Stahlschmidt-Allner, P., and Tyler, C. R. (2004). Long-term exposure to environmental concentrations of the pharmaceutical ethynylestradiol causes reproductive failure in fish. *Environ. Health Perspect.* **112**, 1725–1733.

35. Tyler, C. R., Spary, C., Gibson, R., Santos, E. M., Shears, J., and Hill, E. M. (2005). Accounting for differences in estrogenic responses in rainbow trout (*Oncorhynchus mykiss*: salmonidae) and roach (*Rutilus rutilus*: Cyprinidae) exposed to effluents from wastewater treatment works. *Environ. Sci. Technol.* **39**, 2599–2607.

36. Grimvall, E., Rylander, L., Nilsson-Ehle, P., Nilsson, U., Strömberg, U., Hagmar, L., and Östman, C. (1997). Monitoring of polychlorinated biphenyls in human blood plasma: Methodological developments and influence of age, lactation, and fish consumption. *Arch. Environ. Contam. Toxicol.* **32**, 329–336.
37. Grandjean, P., Weihe, P., Burse, V. W., Needham, L. L., Storr-Hansen, E., Heinzow, B., Debes, F., Murata, K., Simonsen, H., Ellefsen, P., Budtz-Jorgensen, E., Keiding, N., and White, R. F. (2001). Neurobehavioural deficits associated with PCB in 7-year-old children prenatally exposed to seafood neurotoxicants. *Neurotoxicol. Teratol.* **23**, 305–317.
38. Fossi, M. C., Casini, S., Marsili, L., Neri, G., Mori, G., Ancora, S., Moscatelli, A., Ausili, A., and Notarbartolo-di-Sciara, G. (2002). Biomarkers for endocrine disruptors in three species of Mediterranean large pelagic fish. *Marine Environ. Res.* **54**, 667–671.
39. Dewailly, E. and Weihe, P. (2003). The effect of Arctic pollution on population health. [http://www.amap.no] AMAP 2003. *AMAP Assessment 2002: Human health in the Arctic, Chapter 9* Arctic Monitoring and Assessment Programme (AMAP), Oslo, Norway, xiv+137. HH_CO9.pdf ISBN 82-7971-016-7.
40. Bonefeld-Jørgensen, E. C. and Ayotte, P. (2003). Toxicological properties of persistent organic pollutants and related health effects of concern for the Arctic populations. [http://www.amap.no] *AMAP 2003. AMAP Assessment 2002: Human health in the Arctic, Chapter 6* Arctic Monitoring and Assessment Programme (AMAP), Oslo, Norway, xiv+137. HH_CO6.pdf ISBN 82-7971-016-7.
41. Dallinga, J. W., Moonen, E. J., Dumoulin, J. C., Evers, J. L., Geraedts, J. P., and Kleinjans, J. C. (2002). Decreased human semen quality and organochlorine compounds in blood. *Hum. Reprod.* **17**, 1973–1979.

7

1. International Boundary and Water Commission (1998). *Second phase of the binational study regarding the presence of toxic substances in the Rio Grande/Rio Bravo and its tributaries along the boundary portion between the United States and Mexico.* Final Report, United States and Mexico, I.
2. Texas Natural Resource Conservation Commission, Watershed Management Division (1994). *Regional assessment of water quality in the Rio Grande Basin. Austin, TX.*
3. Singh, A. (1992) Detection methods for waterborne pathogens. In *Environmental Microbiology*, R. Mitchell (Ed.). A John Wiley & Sons, Inc, New York, pp. 125156.
4. Craun, G. F., Berger, P. S., and Calderon, R. L. (1997). Coliform bacteria and waterborne disease outbreaks. *J. Am. Water Works Ass.* **89**(3), 96104.
5. International Boundary and Water Commission (1997). *Second phase of the binational study regarding the presence of toxic substances in the Rio Grande/Rio Bravo and its tributaries along the boundary portion between the United States and Mexico. Final Report. United States and Mexico*, II.
6. Owen, R. J. (1995). Bacteriology of Helicobacter pylori. *Baillieres Clin. Gastroenterol.* **9**(3), 415446.
7. Peura, D. A. (1997). The report of the international update conference on Helicobacter pylori. *Digestive Disease Week.* Washington, DC May 14, 1997.
8. Megraud, F. (1995). Transmission of Helicobacter pylori: Fecal-oral versus oraloral route. *Aliment. Pharmacol. Ther.* **9**(2), 8591.
9. Hulten, K., Han, S. W., Enroth, H., Klein, P. D., Opejun, A. R., Gilman, R. H., Evans, D. G., Engstrand, L., Graham, D. Y., and El-Zaatari, F. A. (1996). Helicobacter pylori in the drinking water in Peru. *Gastroenterology* **110**(4), 10311035.
10. Redlinger, T., O'Rourke, K., and Goodman, K. J. (1999). Age distribution of Helicobacter pylori seroprevalence among young children in a United States/Mexico border community: Evidence for transitory infection. *Am. J. Epidemiol.* **150**, 225230.
11. Rozak, D. B. and Colwell, R. R. (1987). Survival strategies of bacteria in the natural environment. *Microbiol. Rev.* **51**, 365379.

12. Shahamat, M., u Mai, Paszko-Kova, C., Sessel, M., and Colwell, R. R. (1993). Use of autoradiography to assess viability of Helicobacter pylori in water. *Appl. Environ. Microbiol.* **59**(4), 12311235.

13. Alvarez, M. E., Aguilar, M., Fountain, A., Gonzalez, N.,. Rascon, O., and Saenz, D. (2000). Inactivation of MS2 phage and poliovirus in groundwater. *Can. J. Microbiol.* **46**, 159165.

14. Botsford, J. L. (1998). A simple assay for toxic chemicals using a bacterial indicator. *World J. Microb. Biot.* **14**, 369376.

15. American Public Health Association (1995). Standard Methods for the Examination of Water and Wastewater 19 edition, Washington DC.

16. Elmund, G. K., Allen, M. J., and Rice, E. W. (1999). Comparison of Escherichia coli, total coliform, and fecal coliform populations as indicators of wastewater treatment efficiency. *Water Environ. Res.* **71**(3), 332339.

8

1. Berg, G., Knaape, C., Ballin, G., and Seidel, D. (1994). Biological control of Verticillium dahliae KLEB by naturally occurring rhizosphere bacteria. *Arch. Phytopathol. Dis. Prot.* **29**, 249–262.

2. Debette, J. and Blondeau, R. (1980). Presence of Pseudomonas maltophilia in the rhizosphere of several cultivated plants. *Can. J. Microbiol.* **26**, 460–463.

3. Heuer, H. and Smalla, K. (1999). Bacterial phyllosphere communities of Solanum tuberosum L and T4-lysozyme producing genetic variants. *FEMS Microbiol. Ecol.* **28**, 357–371.

4. Lambert, B. and Joos, H. (1989). Fundamental aspects of rhizobacterial plant growth promotion research. *Trends Biotechnol.* **7**, 215–219.

5. Whipps, J. (2001). Microbial interactions and biocontrol in the rhizosphere. *J. Exp. Bot.* **52**, 487–511.

6. Binks, P. R., Nicklin, S., and Bruce, N. C. (1995). Degradation of RDX by *Stenotrophomonas maltophilia* PB1. *Appl. Environ. Microbiol.* **61**, 1813–1822.

7. Lee, E. Y., Jun, Y. S., Cho, K. S., and Ryu, H. W. (2002). Degradation characteristics of toluene, benzene, ethylbenzene, and xylene by *Stenotrophomonas maltophilia* T3-c. *J. Air Waste Manag. Assoc.* **52**, 400–406.

8. Juhasz, A. L., Stanley, G. A., and Britz, M. L. (2000). Microbial degradation and detoxification of high molecular weight polycyclic aromatic hydrocarbons by *Stenotrophomonas maltophilia* strain VUN 10,003. *Lett. Appl. Microbiol.* **30**, 396–401.

9. Quinn, J. P. (1998). Clinical problems posed by multiresistant nonfermenting gram-negative pathogens. *Clin. Infect. Dis.* **27**, 117–124.

10. Valdezate, S., Vindel, A., Loza, E., Baquero, F., and Canton, R. (2001). Antimicrobial susceptibilities of unique *Stenotrophomonas maltophilia* clinical strains. *Antimicrob. Agents Chemother.* **45**, 1581–1584.

11. Dignani, M. C., Grazziutti, M., and Anaissie, E. (2003). *Stenotrophomonas maltophilia* infections. *Semin. Respir. Crit. Care Med.* **24**, 89–98.

12. Yamazaki, E., Ishii, J., Sato, K., and Nakae, T. (1989). The barrier function of the outer membrane of Pseudomonas maltophilia in the diffusion of saccharides and beta-lactam antibiotics. *FEMS Microbiol. Lett.* **51**, 85–88.

13. Li, X. Z., Zhang, L., and Poole, K. (2002). SmeC, an outer membrane multidrug efflux protein of Stenotrophomonas maltophilia. *Antimicrob. Agents Chemother.* **46**, 333–343.

14. Alonso, A. and Martinez, J. L. (2000). Cloning and characterization of SmeDEF, a novel multidrug efflux pump from Stenotrophomonas maltophilia. *Antimicrob. Agents Chemother.* **44**, 3079–3086.

15. Zhang, L., Li, X. Z., and Poole, K. (2001). SmeDEF multidrug efflux pump contributes to intrinsic multidrug resistance in Stenotrophomonas maltophilia. *Antimicrob. Agents Chemother.* **45**, 3497–3503.

16. Berg, G., Eberl, L., and Hartmann, A. (2005). The rhizosphere as a reservoir for opportunistic human pathogenic bacteria. *Environ. Microbiol.* **7**, 1673–85.

17. Knudsen, G. R., Walter, M. V., Porteous, L. A., Prince, V. J., Amstrong, J. L., et al.

(1988). Predictive model of conjugated plasmid transfer in the rhizosphere and phyllosphere. *Appl. Environ. Microbiol.* **54**, 343–347.

18. Dungan, R. S., Yates, S. R., Frankenberger, W. T. Jr. (2003). Transformations of selenate and selenite by *Stenotrophomonas maltophilia* isolated from a seleniferous agricultural drainage pond sediment. *Environ. Microbiol.* **5**, 287–295.
19. Sauge-Merle, S., Cuine, S., Carrier, P., Lecomte-Pradines, C., Luu, D. T., et al. (2003). Enhanced toxic metal accumulation in engineered bacterial cells expressing Arabidopsis thaliana phytochelatin synthase. *Appl. Environ. Microbiol.* **69**, 490–494.
20. Alonso, A., Sanchez, P., and Martinez, J. L. (2000). *Stenotrophomonas maltophilia* D457R contains a cluster of genes from gram-positive bacteria involved in antibiotic and heavy metal resistance. *Antimicrob. Agents Chemother.* **44**, 1778–1782.
21. Fauchon, M., Lagniel, G., Aude, J. C., Lombardia, L., Soularue, P., et al. (2002). Sulfur sparing in the yeast proteome in response to sulfur demand. *Mol. Cell* **9**, 713–723.
22. Park, S. and Imlay, J. A. (2003). High levels of intracellular cysteine promote oxidative DNA damage by driving the Fenton reaction. *J. Bacteriol.* **185**, 1942–50.
23. Holmes, J. D., Richardson, D. J., Saed, S., Evans-Gowing, R., Russell, D. A., et al. (1997). Cadmium-specific formation of metal sulfide "Q-particles" by Klebsiella pneumoniae. *Microbiology* **143**, 2521–2530.
24. Sharma, P. K., Balkwill, D. L., Frenkel, A., and Vairavamurthy, M. A. (2000). A new Klebsiella planticola strain (Cd-1) grows anaerobically at high cadmium concentrations and precipitates cadmium sulfide. *Appl. Environ. Microbiol.* **66**, 3083–3087.
25. Kredich, N. M., Foote, L. J., and Keenan, B. S. (1973). The stoichiometry and kinetics of the inducible cysteine desulfhydrase from Salmonella typhimurium. *J. Biol. Chem.* **218**, 6187–6196.
26. Wang, C. L., Lum, A. M., Ozuna, S. C., Clark, D. S., and Keasling, J. D. (2001). Aerobic sulfide production and cadmium precipitation by Escherichia coli expressing the Treponema denticola cysteine desulfhydrase gene. *Appl. Microbiol. Biotechnol.* **56**, 425–430.
27. Pagès, D., Sanchez, L., Conrod, S., Gidrol, X., Fekete, A., et al. (2007). Exploration of intraclonal strategies of Pseudomonas brassicacearum facing Cd toxicity. *Environ. Microbiol.* **9**, 2820–35.
28. Rijstenbil, J. W. and Wijnholds, J. A. (1996). HPLC analysis of nonprotein thiols in planktonic diatoms: pool size, redox state and response to copper and cadmium exposure. *Mar. Biol.* **127**, 45–54.
29. Michalowicz, A. (1991). *Logiciels pour la Chimie*. Société Francaise de Chimie, Paris.
30. Teo, B. K. (1986). *Inorganic Chemistry Concepts*. Springer-Verlag, Berlin.
31. Rehr, J. J. and Albers, R. C. (1990). Scattering-matrix formulation of curved-wave multiple-scattering theory: Application to x-ray-absorption fine structure. *Phys. Rev. B Condens. Matter* **41**, 8139–8149.
32. Zabinsky, S. I., Rehr, J. J., Ankudinov, A., Albers, R. C., and Eller, M. J. (1995). Multiple-scattering calculations of x-ray-absorption spectra. *Phys. Rev. B* **52**, 2995–3009.

10

1. Berenbaum, M. R. (1995). The chemistry of defense—Theory and practice. *Proc. Natl. Acad. Sci. USA* **92**, 2–8.
2. Hay, M. E. and Fenical, W. (1988). Marine plant-herbivore interactions: The ecology of chemical defense. *Ann. Rev. Ecol. Syst.* **19**, 111–145.
3. Pawlik, J. R. (1993). Marine invertebrate chemical defenses. *Chem. Rev.* **93** 1911–1922.
4. Rosenthal, G. A. and Janzen, D. H. (1979). *Herbivores: Their Interaction with Secondary Plant Metabolites*. Academic Press, Orlando, Florida.
5. Jensen, P. R. and Fenical, W. (1994). Strategies for the discovery of secondary metabolites from marine bacteria: Ecological perspectives. *Ann. Rev. Microbiol.* **48** 559–584.
6. Costerton, J. W., Stewart, P. S., and Greenberg, E. P. (1999). Bacterial biofilms: A

common cause of persistent infections. *Science* **284**, 1318–1322.

7. Hall-Stoodley, L., Costerton, J. W., and Stoodley, P. (2004). Bacterial biofilms: From the natural environment to infectious diseases. *Nat. Rev. Microbiol.* **2**, 95–108.
8. Darby, C., Hsu, J. W., Ghori, N., and Falkow, S. (2002). *Caenorhabditis elegans*: Plague bacteria biofilm blocks food intake. *Nature* **417**, 243–244.
9. Matz, C. and Kjelleberg, S. (2005). Off the hook—How bacteria survive protozoan grazing. *Trends Microbiol.* **13**, 302–307.
10. Fenchel, T. (1987). *Ecology of Protozoa: The Biology of Free-living Phagotrophic Protists.* Science Tech Publishers, Madison, WI.
11. Sherr, E. B. and Sherr, B. F. (2002). Significance of predation by protists in aquatic microbial food webs. *Antonie Leeuwenhoek* **81**, 293–308.
12. Parry, J. D. (2004). Protozoan grazing of freshwater biofilms. *Adv. Appl. Microbiol.* **54**, 167–196.
13. Fenchel, T. and Blackburn, N. (1999). Motile chemosensory behaviour of phagotrophic protists: Mechanisms for and efficiency in congregating at food patches. *Protist* **150**, 325–336.
14. Kiørboe, T., Tang, K., Grossart, H. P., and Ploug, H. (2003). Dynamics of microbial communities on marine snow aggregates: Colonization, growth, detachment, and grazing mortality of attached bacteria. *Appl. Environ. Microbiol.* **69**, 3036–3047.
15. Jürgens, K. and Matz, C. (2002) Predation as a shaping force for the phenotypic and genotypic composition of planktonic bacteria. *Antonie Leeuwenhoek* **81**, 413–434.
16. Chrzanowski, T. H. and Šimek, K. (1990). Prey-size selection by freshwater flagellated protozoa. *Limnol. Oceanogr.* **35**, 1429–1436.
17. Hahn, M. W., Moore, E. R. B, and Höfle, M. G. (1999). Bacterial filament formation, a defense mechanism against flagellate grazing, is growth rate controlled in bacteria of different phyla. *Appl. Environ. Microbiol.* **65**, 25–35.
18. Pernthaler, J., Sattler, B., Šimek, K., Schwarzenbacher, A., and Psenner, R. (1996). Top-down effects on the size-biomass distribution of a freshwater bacterioplankton community. *Aquat. Microb. Ecol.* **10**, 255–263.
19. Matz, C. and Jürgens, K. (2005). High motilityreduces grazing mortality of planktonic bacteria. *Appl. Environ. Microbiol.* **71**, 921–929.
20. Hay, M. E. and Kubanek, J. (2002). Community and ecosystem level consequences of chemical cues in the plankton. *J. Chem. Ecol.* **28**, 2001–2016.
21. Egan, S., Thomas, T., Holmström, C., and Kjelleberg, S. (2000). Phylogenetic relationship and antifouling activity of bacterial epiphytes from the marine alga *Ulva lactuca*. *Environ. Microbiol.* **2**, 343–347.
22. Burmølle, M., Webb, J. S., Rao, D., Hansen, L. H., Sørensen, S. J., et al. (2006). Enhanced biofilm formation and increased resistance to antimicrobial agents and bacterial invasion are caused by synergistic interactions in multispecies biofilms. *Appl. Environ. Microbiol.* **72**, 3916–3923.
23. Rao, D., Webb, J. S., and Kjelleberg, S. (2006). Microbial colonization and competition on the marine alga *Ulva australis*. *Appl. Environ. Microbiol.* **72**, 5547–5555.
24. Rao, D., Webb, J. S., Holmström, C., Case, R., Low, A., et al. (2007). Low densities of epiphytic bacteria from the marine alga *Ulva australis* inhibit settlement of fouling organisms. *Appl. Environ. Microbiol.* **73**, 7844–7852.
25. Longford, S. R., Tujula, N. A., Crocetti, G. R., Holmes, A. J., Holmström, C., et al. (2007). Comparison of diversity of bacterial communities associated with three sessile marine eukaryotes. *Aquat. Microb. Ecol.* **48**, 217–229.
26. Matz, C., McDougald, D., Moreno, A. M., Yung, P. Y., Yildiz, F. H., et al. (2005). Biofilm formation and phenotypic variation enhance predation-driven persistence of *Vibrio cholerae*. *Proc. Natl. Acad. Sci. USA* **102**, 16819–16824.
27. Patterson, D. J. and Lee, W. Y. (2000). The flagellates—Unity, diversity and evolution. B. S. C. Leadbeater and J. C. Green (Eds.). Taylor & Francis, London, pp. 269–287.

28. Boenigk, J. and Arndt, H. (2000). Comparative studies on the feeding behavior of two heterotrophic nanoflagellates: The filter-feeding choanoflagellate *Monosiga ovata* and the raptorial-feeding kinetoplastid *Rhynchomonas nasuta*. *Aquat. Microb. Ecol.* **22**, 243–249.
29. Matz, C., Bergfeld, T., Rice, S. A., and Kjelleberg, S. (2004). Microcolonies, quorum sensing and cytotoxicity determine the survival of *Pseudomonas aeruginosa* biofilms exposed to protozoan grazing. *Environ. Microbiol.* **6**, 218–226.
30. Weitere, M., Bergfeld, T., Rice, S. A., Matz, C., and Kjelleberg, S. (2005). Grazing resistance of *Pseudomonas aeruginosa* biofilms depends on type of protective mechanism, developmental stage and protozoan feeding mode. *Environ. Microbiol.* **7**, 1593–1601.
31. Riveros, R., Haun, M., Campos, V., and Duran, N. (1988). Bacterial chemistry IV: Complete characterization of violacein. *Arq. Biol. Technol.* **31**, 475–487.
32. August, P. R., Grossman, T. H., Minor, C., Draper, M. P., MacNeil, I. A., et al. (2000). Sequence analysis and functional characterization of the violacein biosynthetic pathway from *Chromobacterium violaceum*. *J. Mol. Microbiol. Biotechnol.* **2**, 513–519.
33. Sanchez, C., Brana, A. F., Mendez, C., and Salas, J. A. (2006). Reevaluation of the violacein biosynthetic pathway and its relationship to indolocarbazole biosynthesis. *Chem. Biochem.* **7**, 1231–1240.
34. Huang, S. Y. and Hadfield, M. G. (2003). Composition and density of bacterial biofilms determine larval settlement of the polychaete *Hydroides elegans*. *Mar. Ecol. Progr. Ser.* **260**, 161–172.
35. Halda-Alija, L. and Johnston, T. C. (1999). Diversity of culturable heterotrophic aerobic bacteria in pristine stream bed sediments. *Can. J. Microbiol.* **45**, 879–84.
36. Corpe, W. (1953). Variation in pigmentation and morphology of colonies of gelatinous strains of *Chromobacterium* species from soil. *J. Bacteriol.* **66**, 470–477.
37. Rusch, D. B., Halpern, A. L., Sutton, G., Heidelberg, K. B., Williamson, S., et al. (2007). The sorcerer II global ocean sampling expedition: Northwest Atlantic through Eastern Tropical Pacific. *PLoS Biol.* **5**, e77.
38. Seshadri, R., Kravitz, S. A., Smarr, L., Gilna, P., and Frazier, M. (2007). CAMERA: A community resource for metagenomics. *PLoS Biol.* **5**, e75.
39. McClean, K. H., Winson, M. K., Fish, L., Taylor, A., Chhabra, S. R., et al. (1997). Quorum sensing and *Chromobacterium violaceum*: Exploitation of violacein production and inhibition for the detection of N-acylhomoserine lactones. *Microbiology* **143**, 3703–3711.
40. Nakamura, Y., Sawada, T., Morita, Y., and Tamiya, E. (2002). Isolation of a psychrotrophic bacterium from the organic residue of a water tank keeping rainbow trout and antibacterial effect of violet pigment produced from the strain. *Biochem. Engineer.* **12**, 79–86.
41. Matz, C., Deines, P., Boenigk, J., Arndt, H., Eberl, L., et al. (2004). Impact of violacein-producing bacteria on survival and feeding of bacterivorous nanoflagellates. *Appl. Environ. Microbiol.* **70**, 1593–1599.
42. Bidle, K. D. and Falkowski, P. G. (2004). Cell death in planktonic, photosynthetic microorganisms. *Nat. Rev. Microbiol.* **2**, 643–655.
43. Ferreira, C. V., Bos, C. L., Versteeg, H. H., Justo, G. Z., Duran, N., et al. (2004). Molecular mechanism of violacein-mediated human leukemia cell death. *Blood* **104**, 1459–1464.
44. Kodach, L. L., Bos, C. L., Duran, N., Peppelenbosch, M. P., Ferreira, C. V., et al. (2006). Violacein synergistically increases 5-fluorouracil cytotoxicity, induces apoptosis and inhibits Akt-mediated signal transduction in human colorectal cancer cells. *Carcinogenesis* **27**, 508–516.
45. Väätänen, P. (1976). Microbiological studies on coastal waters of the Northern Baltic Sea, I. Distribution and abundance of bacteria and yeasts in the Tvarminne area. *Walter Andre Nottback Found Sci. Rep.* **1**, 1–58.
46. Leist, M. and Jaattela, M. (2001). Four deaths and a funeral: From caspases to alternative mechanisms. *Nat. Rev. Mol. Cell. Biol.* **2**, 589–598.

47. Egan, S., James, S., Holmström, C., and Kjelleberg, S. (2002). Correlation between pigmentation and antifouling compounds produced by *Pseudoalteromonas tunicata*. *Environ. Microbiol.* **4**, 433–442.

48. Markowitz, V. M., Korzeniewski, F., Palaniappan, K., Szeto, E., Werner, G., et al. (2006). The integrated microbial genomes (IMG) system. *Nucleic Acids Res.* **34**, D344–348.

49. Blosser, R. S. and Gray, K. M. (2000). Extraction of violacein from *Chromobacterium violaceum* provides a new quantitative bioassay for N-acyl homoserine lactone autoinducers. *J. Microbiol. Methods* **40**, 47–55.

50. Hild, E., Takayama, K., Olsson, R. M., and Kjelleberg, S. (2000). Evidence for a role of rpoE in stressed and unstressed cells of marine *Vibrio angustum* strain S14. *J. Bacteriol.* **182**, 6964–6974.

51. Gao, L. Y. and Abu Kwaik, Y. (2000). The mechanism of killing and exiting the protozoan host *Acanthamoeba polyphaga* by *Legionella pneumophila*. *Environ. Microbiol.* **2**, 79–90.

10

1. Nasci, R. S. and Miller, B. R. (1996). Culicine mosquitoes and the agents they transmit. In *The Biology of Disease Vectors*. B. J. Beaty and W. C. Marquardt (Eds.). University Press of Colorado, Niwot, pp. 85–97.

2. Anonymous (2008). California mosquito-borne virus surveillance and response plan. *California Dept. Public Health, Mosq. Vector Cont. Assoc. Calif., and University of California*.

3. Reisen, W. K. and Pfuntner, A. R. (1987). Effectiveness of five methods for urban sampling adult *Culex* mosquitoes in rural and urban habitats in San Bernadino County, California. *J. Am. Mosq. Contr. Assoc.* **3**, 601–606.

4. Allan, S. A. and Kline, D. (2004). Evaluation of various attributes of gravid female traps for collection of *Culex* in Florida. *J. Vector Ecol.* **29**, 285–294.

5. Hardy, J. L., Houk, E. J., Kramer, L. D., and Reeves, W. C. (1983). Intrinsic factors affecting vector competence of mosquitoes for arboviruses. *Ann. Rev. Entomol.* **28**, 229–262.

6. Ishida, Y., Cornel, A. J., and Leal, W. S. (2002). Identification and cloning of a female antenna-specific odorant-binding protein in the mosquito *Culex quinquefasciatus*. *J. Chem. Ecol.* **28**, 867–871.

7. Wojtasek, H. and Leal, W. S. (1999). Conformational change in the pheromone-binding protein from *Bombyx mori* induced by pH and by interaction with membranes. *J. Biol. Chem.* **274**, 30950–30956.

8. Wogulis, M., Morgan, T., Ishida, Y., Leal, W. S., and Wilson, D. K. (2006). The crystal structure of an odorant binding protein from *Anopheles gambiae*: Evidence for a common ligand release mechanism. *Biochem. Biophys. Res. Commun.* **339**, 157–164.

9. Biessmann, H., Walter, M. F., Dimitratos, S., and Woods, D. (2002). Isolation of cDNA clones encoding putative odourant binding proteins from the antennae of the malaria-transmitting mosquito, *Anopheles gambiae*. *Insect. Mol. Biol.* **11**, 123–132.

10. Ishida, Y., Cornel, A. J., and Leal, W. S. (2003). Odorant-binding protein from *Culex tarsalis*, the most competent vector of West Nile virus in California. *J. Asia-Pacific Entomol.* **6**, 45–48.

11. Ishida, Y., Chen, A. M., Tsuruda, J. M., Cornel, A. J., Debboun, M., et al. (2004). Intriguing olfactory proteins from the yellow fever mosquito, *Aedes aegypti*. *Naturwissenschaften* **91**, 426–431.

12. Zhou, J. -J, He, X. -L., Pickett, J. A., and Field, L. M. (2008). Identification of odorant-binding proteins of the yellow fever mosquito *Aedes aegypti*: genome annotation and comparative analyses. *Insect. Mol. Biol.* **17**, 147–163.

13. Clements, A. N. (1999). *The Biology of Mosquitoes: Sensory Reception and Behaviour*. CAB International, New York.

14. Hwang, Y. -H., Mulla, M. S., Chaney, J. D., Lin, G. -G., and Xu, H. -J. (1987). Attractancy and species specificity of 6-acetoxy-5-hexadecanolide, a mosquito oviposition attractant pheromone. *J. Chem. Ecol.* **13**, 245–252.

15. Laurence, B. R., Mori, K., Otsuka, T., Pickett, J. A., and Wadhams, L. J. (1985). Absolute configuration of mosquito oviposition attractant pheromone, 6-acetoxy-5-hexadecanolide. *J. Chem. Ecol.* **11**, 643–648.
16. Laurence, B. R. and Pickett, J. A. (1982). erythro-6-Acetoxy-5-hexadecanolide, the major component of a mosquito attractant pheromone. *J. Chem. Soc., Chem. Commun.* 59–60.
17. Du, Y. -J. and Millar, J. G. (1999). Electroantennogram and oviposition bioassay responses of *Culex quinquefasciatus* and *Culex tarsalis* (Diptera: Culicidae) to chemicals in odors from Bermuda grass infusions. *J. Med. Entomol.* **36**, 158–166.
18. Mboera, L. E. G., Takken, W., Mdira, K. Y., Chuwa, G. J., and Pickett, J. A. (2000). Oviposition and behavioral responses of *Culex quinquefasciatus* to skatole and synthetic oviposition pheromone in Tanzania. *J. Chem. Ecol.* **26**, 1193–1203.
19. Braks, M. A. H., Leal, W. S., and Cardé, R. T. (2007). Oviposition responses of gravid female *Culex quinquefasciatus* to egg rafts and low doses of oviposition pheromone under semifield conditions. *J. Chem. Ecol.* **33**, 567–578.
20. Dawson, G. W., Mudd, A., Pickett, J. A., Pile, M. M., and Wadhams, L. J. (1990). Convenient synthsesis of mosquito oviposition pheromone and a highly fluorinated analog retaining biological activity. *J. Chem. Ecol.* **16**, 1779–1789.
21. Olagbemiro, T. O., Birkett, M. A., Mordue, A. J., and Pickett, J. A. (1999). Production of (5R,6S)-6-acetoxy-5-hexadecanolide, the mosquito oviposition pheromone, from the seed oil of the summer cypress plant, *Kochia scoparia* (Chenopodiaceae). *J. Agric. Food Chem.* **47**, 3411–3415.
22. Olagbemiro, T. O., Birkett, M. A., Mordue Luntz, A. J., and Pickett, J. A. (2004). Laboratory and field responses of the mosquito, *Culex quinquefasciatus*, to plant-derived *Culex* spp. oviposition pheromone and the oviposition cue skatole. *J. Chem. Ecol.* **30**, 965–976.
23. Syed, Z. and Leal, W. S. (2007). Maxillary palps are broad spectrum odorant detectors in *Culex quinquefasciatus*. *Chem. Senses* **32**, 727–738.
24. Damberger, F., Nikonova, L., Horst, R., Peng, G., Leal, W. S., et al. (2000). NMR characterization of a pH-dependent equilibrium between two folded solution conformations of the pheromone-binding protein from *Bombyx mori*. *Protein Sci.* **9**, 1038–1041.
25. Leal, W. S., Chen, A. M., and Erickson, M. L. (2005). Selective and pH-dependent binding of a moth pheromone to a pheromone-binding protein. *J. Chem. Ecol.* **31**, 2493–2499.
26. Leal, W. S., Chen, A. M., Ishida, Y., Chiang, V. P., Erickson, M. L., et al. (2005). Kinetics and molecular properties of pheromone binding and release. *Proc. Natl. Acad. Sci. USA* **102**, 5386–5391.
27. Ban, L., Zhang, L., Yan, Y., and Pelosi, P. (2002). Binding properties of a locust's chemosensory protein. *Biochem. Biophys. Res. Commun.* **293**, 50–54.
28. Horst, R., Damberger, F., Luginbuhl, P., Guntert, P., Peng, G., et al. (2001). NMR structure reveals intramolecular regulation mechanism for pheromone binding and release. *Proc. Natl. Acad. Sci. USA* **98**, 14374–14379.
29. Lautenschlager, C., Leal, W. S., and Clardy, J. (2005). Coil-to-helix transition and ligand release of *Bombyx mori* pheromone-binding protein. *Biochem. Biophys. Res. Commun.* **335,** 1044–1050.
30. Sandler, B. H., Nikonova, L., Leal, W. S., and Clardy, J. (2000). Sexual attraction in the silkworm moth: Structure of the pheromone-binding-protein-bombykol complex. *Chem. Biol.* **7**, 143–151.
31. Xu, W. and Leal, W. S. (2008). Molecular switches for pheromone release from a moth pheromone-binding protein. *Biochem. Biophys. Res. Commun.* **372**, 559–564.
32. Wojtasek, H., Hansson, B. S., and Leal, W. S. (1998). Attracted or repelled? A matter of two neurons, one pheromone binding protein, and a chiral center. *Biochem. Biophys. Res. Commun.* **250**, 217–222.
33. Kline, D. L., Allan, S. A., Bernier, U. R., and Welch, C. H. (2007). Evaluation of the

enantiomers of 1-octen-3-ol and 1-octyn-3-ol as attractants for mosquitoes associated with a freshwater swamp in Florida, U.S.A. *Med. Vet. Entomol.* **21**, 323–331.

34. Christiansen, J. A., Smith, C., Madon, M. B., Albright, J., Hazeleur, W., et al. (2005). Use of gravid traps for collection of California West Nile Virus vectors. *Proc. Papers Calif. Mosq. Control Assoc. Ann. Conf.* 89–93.
35. Trexler, J. D., Apperson, C. S., Gemeno, C., Perich, M. J., Carlson, D., et al. (2003). Field and laboratory evaluations of potential oviposition attractants for *Aedes albopictus* (Diptera: Culicidae). *J. Am. Mosq. Control. Assoc.* **19**, 228–234.
36. Millar, J. G., Chaney, J. D., Beehler, J. W., and Mulla, M. S. (1994). Interaction of the *Culex quinquefasciatus* egg raft pheromone with a natural chemical associated with oviposition sites. *J. Am. Mosq. Contr. Assoc.* **10**, 374–379.
37. Leal, W. S. (2000). Duality monomer-dimer of the pheromone-binding protein from *Bombyx mori*. Biochem. *Biophys. Res. Commun.* **268**, 521–529.
38. Dobritsa, A. A., van der Goes van Naters, W., Warr, C. G., Steinbrecht, R. A., and Carlson, J. R. (2003). Integrating molecular and cellular basis of odor coding in the *Drosophila* antenna. *Neurons* **37**, 827–841.
39. Pitts, R. J., Fox, A. N., and Zwiebel, L. J. (2004). A highly conserved candidate chemoreceptor expressed in both olfactory and gustatory tissues in the malaria vector *Anopheles gambiae*. *Proc. Natl. Acad. Sci. USA* **101**, 5058–5063.
40. Kotsuki, H., Kadota, I., and Ochi, M. (1990). A novel carbon-carbon bond-forming reaction of triflates with copper(1)-catalyzed Grignard reagents. A new concise and enantiospecific synthesis of (+)-exo-brevicomin, (5R,6S)-(-)-6-acetoxy-5-hexadecanolidea, and L-factor. *J. Org. Chem.* **55**, 4417–4422.
41. Barbosa, R. M. R., Souto, A., Eiras, A. E., and Regis, L. (2007). Laboratory and field evaluation of an oviposition trap for *Culex quinquefasciatus* (Diptera: Culicidae). *Mem. Inst. Oswaldo. Cruz* **102**, 523–529.
42. McAbee, R. D., Green, E. N., Holeman, J., Christiansen, J., Frye, N., et al. (2008). Identification of *Culex pipiens* complex mosquitoes in a hybrid zone of West Nile Virus transmission in Fresno County, California. *Am. J. Trop. Med. Hyg.* **78**, 303–310.

11

1. Bock, K. W. and Kohle, C. (2006). Ah receptor: Dioxin-mediated toxic responses as hints to deregulated physiologic functions. *Biochem. Pharmacol.* **72**, 393–404.
2. Pelclova, D., Fenclova, Z., Preiss, J., Prochazka, B., Spacil, J., Dubska, Z., Okrouhlik, B., Lukas, E., and Urban, P. (2002). Lipid metabolism and neuropsychological follow-up study of workers exposed to 2,3,7,8-tetrachlordibenzo-p-dioxin. *Int. Arch. Occup. Environ. Health* **75**(Suppl.), S60–66.
3. Baccarelli, A., Mocarelli, P., Patterson, D. G. Jr., Bonzini, M., Pesatori, A. C., Caporaso, N., and Landi, M. T. (2002). Immunologic effects of dioxin: new results from Seveso and comparison with other studies. *Environ. Health Perspect.* **110**, 1169–1173.
4. Pesatori, A. C., Consonni, D., Bachetti, S., Zocchetti, C., Bonzini, M., Baccarelli, A., and Bertazzi, P. A. (2003). Short- and long-term morbidity and mortality in the population exposed to dioxin after the "Seveso accident". *Ind. Health* **41**, 127–138.
5. Birnbaum, L. S. (1995). Developmental effects of dioxins. *Environ. Health Perspect.* **103**(Suppl. 7), 89–94.
6. Peterson, R. E., Theobald, H. M., and Kimmel, G. L. (1993). Developmental and reproductive toxicity of dioxins and related compounds: Cross-species comparisons. *Crit. Rev. Toxicol.* **23**, 283–335.
7. Birnbaum, L. S. and Tuomisto, J. (2000). Non-carcinogenic effects of TCDD in animals. *Food Addit. Contam.* **17**, 275–288.
8. Pocar, P., Fischer, B., Klonisch, T., and Hombach-Klonisch, S. (2005). Molecular interactions of the aryl hydrocarbon receptor and its biological and toxicological relevance for reproduction. *Reproduction* **129**, 379–389.
9. Barker, D. J., Gluckman, P. D., and Robinson, J. S. (1995). Conference report: Fetal origins of adult disease—report of the First

International Study Group, Sydney, October 29–30, 1994. *Placenta* **16**, 317–320.

10. Wynn, M. and Wynn, A. (1988). Nutrition around conception and the prevention of low birthweight. *Nutr. Health* **6**, 37–52.
11. Kwong, W. Y., Wild, A. E., Roberts, P., Willis, A. C., and Fleming, T. P. (2000). Maternal undernutrition during the preimplantation period of rat development causes blastocyst abnormalities and programming of postnatal hypertension. *Development* **127**, 4195–4202.
12. Hunt, P. A., Koehler, K. E., Susiarjo, M., Hodges, C. A., Ilagan, A., Voigt, R. C., Thomas, S., Thomas, B. F., and Hassold, T. J. (2003). Bisphenol a exposure causes meiotic aneuploidy in the female mouse. *Curr, Biol.* **13**, 546–553.
13. Lo, C. W. and Gilula, N. B. (1979). Gap junctional communication in the post-implantation mouse embryo. *Cell* **18**, 411–422.
14. Reeve, W. J. (1981). The distribution of ingested horseradish peroxidase in the 16-cell mouse embryo. *J. Embryol. Exp. Morphol.* **66**, 191–207.
15. Johnson, M. H. and Maro, B. (1984). The distribution of cytoplasmic actin in mouse 8-cell blastomeres. *J. Embryol. Exp. Morphol.* **82**, 97–117.
16. Houliston, E., Pickering, S. J., and Maro, B. (1987). Redistribution of microtubules and pericentriolar material during the development of polarity in mouse blastomeres. *J. Cell Biol.* **104**, 1299–1308.
17. Ziomek, C. A. and Johnson, M. H. (1980). Cell surface interaction induces polarization of mouse 8-cell blastomeres at compaction. *Cell* **21**, 935–942.
18. Johnson, M. H. and McConnell, J. M. (2004). Lineage allocation and cell polarity during mouse embryogenesis. *Semin. Cell Dev. Biol.* **15**, 583–597.
19. Blankenship, A. L., Suffia, M. C., Matsumura, F., Walsh, K. J., and Wiley, L. M. (1993). 2,3,7,8-Tetrachlorodibenzo-p-dioxin (TCDD) accelerates differentiation of murine preimplantation embryos *in vitro*. *Reprod. Toxicol.* **7**, 255–261.
20. Tsutsumi, O., Uechi, H., Sone, H., Yonemoto, J., Takai, Y., Momoeda, M., Tohyama, C., Hashimoto, S., Morita, M., and Taketani, Y. (1998). Presence of dioxins in human follicular fluid: their possible stage-specific action on the development of preimplantation mouse embryos. *Biochem. Biophys. Res. Commun.* **250**, 498–501.
21. Shi, Z., Valdez, K. E., Ting, A. Y., Franczak, A., Gum, S. L., Petroff, B. K. (2007). Ovarian Endocrine Disruption Underlies Premature Reproductive Senescence Following Environmentally Relevant Chronic Exposure to the Aryl Hydrocarbon Receptor Agonist 2,3,7,8-Tetrachlorodibenzo-p-Dioxin. *Biol. Reprod.* **76**, 198–202.
22. Wu, Q., Ohsako, S., Baba, T., Miyamoto, K., and Tohyama, C. (2002). Effects of 2,3,7,8-tetrachlorodibenzo-p-dioxin (TCDD) on preimplantation mouse embryos. *Toxicology* **174**, 119–129.
23. Moran, F. M., Vande Voort, C. A., Overstreet, J. W., Lasley, B. L., and Conley, A. J. (2003). Molecular target of endocrine disruption in human luteinizing granulosa cells by 2,3,7,8-tetrachlorodibenzo-p-dioxin: Inhibition of estradiol secretion due to decreased 17alpha-hydroxylase/17,20-lyase cytochrome P450 expression. *Endocrinology* **144**, 467–473.
24. Pocar, P., Nestler, D., Risch, M., and Fischer, B. (2005). Apoptosis in bovine cumulus-oocyte complexes after exposure to polychlorinated biphenyl mixtures during *in vitro* maturation. *Reproduction* **130**, 857–868.
25. Li, B., Liu, H. Y., Dai, L. J., Lu, J. C., Yang, Z. M., and Huang, L. (2006). The early embryo loss caused by 2,3,7,8-tetrachlorodibenzo-p-dioxin may be related to the accumulation of this compound in the uterus. *Reprod. Toxicol.* **21**, 301–306.
26. Eichenlaub-Ritter, U., Winterscheidt, U., Vogt, E., Shen, Y., Tinneberg, H. R., and Sorensen, R. (2007). 2-methoxyestradiol induces spindle aberrations, chromosome congression failure, and nondisjunction in mouse oocytes. *Biol. Reprod.* **76**, 784–793.
27. Lattanzi, M. L, Santos, C. B., Mudry, M. D., and Baranao, J. L. (2003). Exposure of bovine oocytes to the endogenous metabolite 2-methoxyestradiol during *in vitro* maturation inhibits early embryonic development. *Biol. Reprod.* **69**, 1793–1800.

28. Brison, D. R. and Schultz, R. M. (1997). Apoptosis during mouse blastocyst formation: Evidence for a role for survival factors including transforming growth factor alpha. *Biol. Reprod.* **56**, 1088–1096.

29. Murray, A. (1994). Cell cycle checkpoints. *Curr. Opin. Cell Biol.* **6**, 872–876.

30. Artus, J., Babinet, C., and Cohen-Tannoudji, M. (2006). The cell cycle of early mammalian embryos: lessons from genetic mouse models. *Cell Cycle* **5**, 499–502.

31. Harrison, R. H., Kuo, H. C., Scriven, P. N., Handyside, A. H., and Ogilvie, C. M. (2000). Lack of cell cycle checkpoints in human cleavage stage embryos revealed by a clonal pattern of chromosomal mosaicism analysed by sequential multicolour FISH. *Zygote* **8**, 217–224.

32. Chatzimeletiou, K., Morrison, E. E., Prapas, N., Prapas, Y., and Handyside, A. H. (2005). Spindle abnormalities in normally developing and arrested human preimplantation embryos *in vitro* identified by confocal laser scanning microscopy. *Hum. Reprod.* **20**, 672–682.

33. Varmuza, S., Prideaux, V., Kothary, R., and Rossant, J. (1988). Polytene chromosomes in mouse trophoblast giant cells. *Development* **102**, 127–134.

34. Burke, B. and Stewart, C. L. (2006). The laminopathies: The functional architecture of the nucleus and its contribution to disease. *Annu. Rev. Genomics Hum. Genet.* **7**, 369–405.

35. Torres-Padilla, M. E., Parfitt, D. E., Kouzarides, T., and Zernicka-Goetz, M. (2007). Histone arginine methylation regulates pluripotency in the early mouse embryo. *Nature* **445**, 214–218.

36. Anway, M. D., Cupp, A. S., Uzumcu, M., and Skinner, M. K. (2005). Epigenetic transgenerational actions of endocrine disruptors and male fertility. *Science* **308**, 1466–1469.

37. Wu, Q., Ohsako, S., Ishimura, R., Suzuki, J. S., and Tohyama, C. (2004). Exposure of mouse preimplantation embryos to 2,3,7,8-tetrachlorodibenzo-p-dioxin (TCDD) alters the methylation status of imprinted genes H19 and Igf2. *Biol. Reprod.* **70**, 1790–1797.

38. Susiarjo, M., Hassold, T. J., Freeman, E., and Hunt, P. A. (2007) Bisphenol A exposure in utero disrupts early oogenesis in the mouse. *PLoS Genet.* 3(1), e5.

39. Zarrow, M., Yochim, J., and McCarthy, J. (1964). *Experimental endocrinology: A source book of basic techniques*. Academic Press, New York.

40. Ataniyazova, O. A., Baumann, R. A., Liem, A. K., Mukhopadhyay, U. A., Vogelaar, E. F., and Boersma, E. R. (2001). Levels of certain metals, organochlorine pesticides and dioxins in cord blood, maternal blood, human milk and some commonly used nutrients in the surroundings of the Aral Sea (Karakalpakstan, Republic of Uzbekistan). *Acta. Paediatr.* **90**, 801–808.

41. Hooper, K., Petreas, M. X., Chuvakova, T., Kazbekova, G., Druz, N., Seminova, G., Sharmanov, T., Hayward, D., She, J., Visita, P., Winkler, J., McKinney, M., Wade, T. J., Grassman, J., and Stephens, R. D. (1998). Analysis of breast milk to assess exposure to chlorinated contaminants in Kazakstan: High levels of 2,3,7, 8-tetrachlorodibenzo-p-dioxin (TCDD) in agricultural villages of southern Kazakstan. *Environ. Health Perspect.* **106**, 797–806.

42. Kociba, R. J., Keyes, D. G., Beyer, J. E., Carreon, R. M., Wade, C. E., Dittenber, D. A., Kalnins, R. P., Frauson, L. E., Park, C. N., Barnard, S. D., Hummel, R. A., and Humiston, C. G. (1978). Results of a two-year chronic toxicity and oncogenicity study of 2,3,7,8-tetrachlorodibenzo-p-dioxin in rats. *Toxicol. Appl. Pharmacol.* **46**, 279–303.

43. Combelles, C. M., Cekleniak, N. A., Racowsky, C., and Albertini, D. F. (2002). Assessment of nuclear and cytoplasmic maturation in in-vitro matured human oocytes. *Hum. Reprod.* **17**, 1006–1016.

12

1. Leslie, A. and Cuperus, G. W. (1993). *Successful Implementation of Integrated Pest Management for Agricultural Crops.* Lewis/ CRC Press, Florida.

2. Thrupp, L. A. (1996). *New Partnerships for Sustainable Agriculture*. World Resources Institute, Washington, DC.

3. Premazzi, G. and Ziglio, G. (1995). Regulations and Management. In *Pesticide Risk in Groundwater.* M. Vighi and E. Funari (Eds.). CRC Lewis Publishers, Boca Raton, Chapter 10, pp. 203–240.
4. London, L. (1992). An overview of agrochemical hazards in the South African farming sector. *S. Afr. Med. J.* **81**, 560–564.
5. London, L. (1995). An investigation into the neurological and neurobehavioral effects of long-term agrochemical exposure amongst deciduous fruit farm workers in the Western Cape, South Africa. Doctoral Thesis. *Cape Town, Department of Community Health, University of Cape Town.*
6. London, L. and Myers, J. E. (1995). General patterns of agrochemical usage in the Southern Region of South Africa. *SA J. Sci.* **91**, 509–514.
7. London, L. and Rother, A. (1998). People, pesticide and the environment: Who bears the brunt of backward policy in South Africa? In *Conference Proceedings: Environmental Justice and the Legal Process.* Environmental Law Unit, University of Cape Town, Cape Town, South Africa and Environmental Law Centre, Macquarie University, Sydney, Australia.
8. Maroni, M. and Fait, A. (1993). Health Effects in man from long-term exposure to pesticides. A review of the 1975–1991 literature. *Toxicology* **78**, 1–174.
9. Dalvie, M. A., White, N., Raine, R., Myers, J. E., London, L., Thompson, M., and Christiani, D. C. (1999). The long-term respiratory health effects of the herbicide, paraquat, among workers in the Western Cape. *Occup. Environ. Med.* **56**, 391–396.
10. Myers, J. E. (1990). Occupational health of farm workers. *S. Afr. Med. J.* **78**, 562–563.
11. London, L. and Myers, J. E. (1995). Critical issues in agrochemical safety in South Africa. *Am. J. Ind. Med.* **27**, 1–14.
12. Hassett, A. J., Viljoen, P. T., and Liebenberg, J. J. E. (1987). An assessment of chlorinated pesticides in the major surface water resources of the Orange Free State during the period September 1984 to September 1985. *Water SA* **13**(3), 133–136.
13. Schultz, R., Peall, S. K. C., Dabrowski, J. M., and Reinecke, A. J. (2001). Current-use insecticides, phosphates and suspended solids in the Lourens River, Western Cape, during the first rainfall event of the wet season. *Water SA* **27**(1), 65–70.
14. Greichus, Y. A., Greichus, A., Amman, B. D., Call, D. J., Hamman, D. C. D., and Pott, R. M. (1977). Insecticides polychlorinated biphenyl and metals in African ecosystems.1. Hartebeespoort Dam, Transvaal and Voëlvlei Dam, Cape Province, Republic of South Africa. *Arch. Environ. Contam. Toxicol.* **6**, 371–383.
15. Davies, H. (1997). An assessment of the suitability of a series of Western Cape Farm Dams as water bird habitats. *MSc (Conservation Biology) Thesis.* Zoology Department, University of Cape Town.
16. Grobler, D. F. (1994). A note on PCBs and chlorinated hydrocarbon pesticide residues in water, fish and sediment from the Olifants River, Eastern Transvaal, South Africa. *Water SA* **20**(3), 187–194.
17. Weaver, J. M. C. (1993). A preliminary survey of pesticide levels in groundwater from a selected area on intensive agriculture in the Western Cape. *Report to the Water Research Commission.* Pretoria, Division of Water Technology, CSIR.
18. London, L., Dalvie, M. A., Cairncross, E., and Solomon, A. (2001). The quality of surface and groundwater in the rural Western Cape with regard to pesticides. *WRC Report No: K5/795/00.* Pretoria, WRC 2001.
19. World Wild Life Fund (1997). *Known and Suspected Hormone Disruptors List.* World Wild Life, Canada, Toronto.
20. Schettler, T., Solomon, G., Burns, P., and Valenti, M. (1996). Generations at Risk. How Environmental Toxins may affect reproductive health in Massachusetts. *Greater Boston Physicians for Social Responsibility, Massachusetts Public Interest Research Group (MASSPIRG) Education Fund,* Cambridge.
21. EPA (1995). *Solid Phase Extraction Method 3535.* U.S. Environmental Protection Agency, Washington, DC.
22. EPA (1995). *Organochlorine Pesticides by Capillary Column Gas Chromatography Method 8081A.* U.S. Environmental Protection Agency, Washington, DC.

23. McGregor, F. (1999). The mobility of endosulfan and chlorpyrifos in the soil of the Hex River Valley. *Thesis Submitted in Partial Fulfilment of the Requirements for the Degree of Masters of Science in Environmental Geochemistry*. Department of Geological Sciences, University of Cape Town.
24. California Environmental Protection Agency (1997). Sampling for pesticide residues in California well water. 1996 update of the well inventory database. *California Environmental Protection Agency.* Department of pesticide regulation, California.
25. Council of the European Community (1980). Directive relating to the quality of water intended for human consumption (80/778/EEC). *EEC.*
26. EPA (1992). *Guidelines for Drinking Water Quality. 202, 260–7572.* U.S. Environmental Protection Agency, Washington, DC.
27. WHO (1993). *Guidelines for Drinking Water Quality. Recommendations.* WHO2 Edition, Geneva, p. 1.
28. DWAF (1996). *South African Water Quality Guidelines.* Pretoria, DWAF, pp. 1–7.
29. Dallas, H. F. and Day, J. A (1993). The effect of water quality variables on riverine ecosystems: A review. *Report Prepared for the Water Research Commission.* Rondebosch, Freshwater Research Unit, University of Cape Town.
30. International Programme on Chemical Safety (IPCS) (1993). Summary of Toxicological Evaluations performed by the joint WHO/FAO meeting on pesticide residues (JMPR). WHO, Geneva.
31. Soto, A. M., Chung, K. L., and Sonnenschein, C. (1994). The pesticides endosulfan, toxaphene, and dieldrin have estrogenic effects on human estrogen-sensitive cells. *Environ. Health Perspect.* **102**(4), 380–383.
32. Funari, E., Donati, L., Sandroni, D., and Vighi, M. (1995). Pesticide levels in groundwater: Value and limitations of monitoring. In *Pesticide Risk in Groundwater*. M. Vighi and E. Funari (Eds.). CRC Lewis Publishers, Boca Raton, Chapter 1, pp. 3–44.
33. Oskam, G., Van Genderen, J., Hopman, R., Noij, T. H. M., Noordsij, A., and Piuker, L. M. (1993). A general view of the problem, with special reference to the Dutch situation. *Water Supply* **11**, 1–17.
34. Espigares, M., Coca, C., Fernandez-Crehuet, M. O., Bueno, A., and Galvez, R. (1997). Pesticide concentrations in the waters from a section of the Guadal river basin, Spain. *Environ. Toxicol. Water Qual.* **12**, 249–256.
35. Domagalski, J. (1997). Results of a prototype surface water network design for pesticides developed for the San Joaquin River Basin, California. *J. Hydrology* **192**, 33–50.
36. Gustafson, D. I. (1989). Groundwater ubiquity score: A simple method for assessing pesticide leachability. *Environ. Toxi. Chem.* **8**, 339–357.
37. Jabber, A., Masud, S. Z., Parveen, Z., and Ali, M. (1993). Pesticide residues in cropland soils and shallow groundwater in Punjab Pakistan. *Bull. Environ. Contam. Toxicol.* **51**, 268–273.

13

1. Pratt, G. C., Palmer, K., Wu, C. Y., Oliaei, F., Hollerbach, C., and Fenske, M. J. (2000). An assessment of air toxics in Minnesota. *Environ. Health Perspect.* **108**, 815825.
2. Woodruff, T. J., Axelrad, D. A., Caldwell, J., Morello-Frosch, R., and Rosenbaum, A. (1998). Public health implications of 1990 air toxics concentrations across the United States. Environ. *Health Perspect.* 106, 245251.
3. Clean Air Act Amendments: Part A, Section 112. Public Law 1990, pp. 101549.
4. Leikauf, G. D. (2002). Hazardous Air Pollutants and Asthma. *Environ. Health Perspect.* **110**(S4), 505526.
5. Rumchev, K., Spickett, J., Bulsara, M., Phillips, M., and Stick, S. (2004). Association of domestic exposure to volatile organic compounds with asthma in young children. *Thorax* **59**, 746751.
6. International Agency for Research on Cancer (1982). IARC Monograph on the Evaluation of the Carcinogenic Risk of Chemicals: Some Industrial Chemicals and Dyestuffs, Volume 29, IARC, Lyon, France.
7. Lin, M., Chen, Y., Villeneuve, P. J., Burnett, R. T., Lemyre, L., Hertzman, C., Mcgrail, K. M., and Krewski, D. (2004). Gaseous

air pollutants and asthma hospitalization of children with low household income in Vancouver, British Columbia. *Am. J. Epidemiol.* **159**, 294303.

8. U.S. Department of Health and Human Services (2007). The Agency for Toxic Substances and Disease Registry (ATSDR). [http://www.atsdr.cdc.gov/toxprofiles/tp3.pdf] Toxicological Profile for Benzene.
9. Glass, D. C., Gray, C. N., Jolley, D. J., Gibbons, C., Sim, M. R., Fritschi, L., Adams, G. G., Bisby, J. A., and Manuell, R. (2003). Leukemia risk associated with low-level benzene exposure. *Epidemiology* **14**, 569577.
10. U.S. Department of Health and Human Services (2000). The Agency for Toxic Substances and Disease Registry (ATSDR). [http://www.atsdr.cdc.gov/toxprofiles/tp3.pdf] Toxicological Profile for Toluene.
11. Chang, S., Chen, C., Lien, C., and Sung, F. (2006). Hearing Loss in Workers Exposed to Toluene and Noise. *Environ, Health Perspect,* **114**, 12831286.
12. Gerin, M., Siemiatychi, J., Desy, M., and Krewski, D. (1998). Associations between several sites of cancer and occupational exposure to benzene, toluene, xylene, and styrene: Results of a case-control study in Montreal. *Am. J. Ind. Med.* **34**, 144156.
13. Antilla, A., Pukkala, E., Riala, R., Sallmén, M., and Hemminki, K. (1998). Cancer incidence among Finnish workers exposed to aromatic hydrocarbons. *Int. Arch. Occup. Environ. Health* **71**, 187193.
14. International Agency for Research on Cancer (1999). Monographs on the evaluation of carcinogenic risks to humans. Xylenes. In Part Three. Re-evaluation of some organic chemicals, hydrazine, and hydrogen peroxide. Volume 71. World Health Organization, Lyon, France, pp. 11891208.
15. Integrated Risk Information System (2001). Benzene, Toluene, Ethylbenzene, and Xylenes. Integrated Risk Information System, U.S. Environmental Protection Agency.
16. Vyskocil, A., Leroux, T., Truchon, G., Lemay, F., Gendron, M., Gagnon, F., El Majidi, N., and Viau, C. (2008). Ethyl benzene should be considered ototoxic at occupationally relevant exposure concentrations. Toxicol. Ind. **24**, 241246.
17. U.S. Environmental Protection Agency (1999). *Integrated Risk Information System (IRIS) on Xylenes.* National Center for Environmental Assessment, Office of Research and Development, Washington, DC.
18. U.S. Department of Health and Human Services (2007). The Agency for Toxic Substances and Disease Registry (ATSDR). [http://www.atsdr.cdc.gov/tfacts71.pdf]. Toxicological Profile for Xylene (Update) U.S. Department of Public Health and Human Services, Public Health; Service, Atlanta, GA.
19. Aguilera, I., Sunyer, J., Fernandez-Patier, R., Hoek, G., Aguirre-Alfaro, A., Meliefste, K., Bomboi-Mingarro, M. R., Nieuwenhuijsen, M. J., Herce-Garraleta, D., and Brunekreef, B. (2008). Estimation of outdoor NOx, NO_2, and BTEX exposure in a cohort of pregnant women using land use regression modelling. *Environ. Sci. Technol.* **42**, 815821.
20. Sexton, K., Adgate, J. L., Ramachandran, G., Pratt, G. C., Mongin, S. J., Stock, T. H., and Morandi, M. T. (2004). Comparison of personal, indoor, and outdoor exposures to hazardous air pollutants in three urban communities. *Environ. Sci. Technol.* **38**, 423430.
21. Adgate, J. L., Church, T. R., Ryan, A. D., Ramachandran, G., Fredrickson, A. L., Stock, T. H., Morandi, M. T., and Sexton, K. (2004). Outdoor, indoor, and personal exposure to VOCs in Children. *Environ. Health Perspect.* **112**, 13861392.
22. Lee, S. (1997). Comparison of indoor and outdoor air quality at two staff quarters in Hong Kong. *Environ. Int.* **23**(6), 791797.
23. Rava, M., Verlato, G., Bono, R., Ponzio, M., Sartori, S., Blengio, G., Kuenzli, N., Heinrich, J., Götschi, T., and de Marco, R. (2007). A predictive model for the home outdoor exposure to nitrogen dioxide. *Sci, Total Environ.* **384**, 163170.
24. Brunekreef, B. and Holgate, S. T. (2002). Air pollution and health. *Lancet* **360**, 12331242.
25. Pope, C. A. and Dockery, D. W. (2006). Health effects of fine particulate air pollution:

Lines that connect. *J, Air Waste Manag, Assoc.* **56**, 709742.

26. Hoek, G., Beelen, R., de Hooh, K., Vienneau, D., Gulliver, P. F., and Briggs, D. (2008). A review of land use regression models to assess spatial variation of outdoor air pollution. *Atmos. Environ.* **42**(33), 75617578.
27. Ryan, P. and LeMasters, G. (2007). A review of land-use regression models for characterising intraurban air pollution exposure. *Inhal. Toxicol.* **19**(Suppl. 1), 127133.
28. Briggs, D. (2005). The role of GIS: Coping with space (and time) in air pollution exposure assessment. *J. Toxicol. Environ. Health* A **68**, 12431261.
29. Dockery, D. W. and Stone, P. H. (2007).Cardiovascular risks from fine particulate air pollution. *N. Eng. J. Med.* **356**, 511513.
30. Burnett, R. T., Stieb, D., Brook, J. R., Cakmak, S., Dales, R., Raizenne, M., Vincent, R., and Dann, T. (204). Associations between short-term changes in nitrogen dioxide and mortality in Canadian cities. *Arch. Environ. Health* **59**, 228236.
31. Pope, C. A., Burnett, R. T., Thun, M. J., Calle, E. E., Krewski, D., Ito, K., and Thurston, G. D. (2002). Lung cancer, cardiopulmonary mortality, and long-term exposure to fine particulate air pollution. *J. Am. Med. Assoc.* **287**, 11321141.
32. Jerrett, M., Arain, A., Kanaroglou, P., Beckerman, B., Crouse, D., Gilbert, N. L., Brook, J. R., Finkelstein, N., and Finkelstein, M. M. (2007). Modelling the intra-urban variability of ambient traffic pollution in Toronto, Canada. *J. Toxicol. Environ. Health A* **70**, 200212.
33. Levy, J., Houseman, E. A., Ryan, L., Richardson, D., and Spengler, J. D. (2000). Particle concentration in urban microenvironments. *Environ. Health Persp.* **108**, 10511057.
34. Madsen, C., Carlsen, K. C., Hoek, G., Oftedal, B., Nafstad, P., Meliefste, K., Jacobsen, R., Nystad, W., Carlsen, K., and Brunekreef, B. (2007). Modelling the intraurban variability of outdoor traffic pollution in Oslo, Norway–A GA2 LEN project. *Atmos. Environ.* **41**, 75007511.
35. Jerrett, M., Arain, A., Kanaroglou, P., Beckerman, B., Potoglou, D., Sahsuvaroglu, T., Morrison, J., and Giovis, C. (2005). A review and evaluation of intra-urban air pollution exposure models. *J. Expo. Anal. Environ. Epidemiol.* **15**, 185204.
36. Briggs, D., Collins, S., Elliott, P., Kingham, S., Lebret, E., Pryl, K., van Reeuwijk, H., Smallbone, K., and Laan, A. (1997). Mapping urban air pollution using GIS: A regression-based approach. *Int. J. Geogr. Info. Syst.* **11**, 699718.
37. Ryan, P., LeMasters, G., Biswas, P., Levin, L., Hu, S., Lindsey, M., Bernstein, D., Lockey, J., Villareal, M., Hershey, G. K., and Grinshpun, S. A. (2006). A comparison of proximity and land use regression traffic exposure models and wheezing in infants. *Environ. Health Perspect.* **115**, 278284.
38. Fischer, P., Hoek, G., van Reeuwijk, H., Briggs, D. J., Lebret, E., van Wijnen, J. H., Kingham, S., and Elliott, P. (2000). Traffic-related differences in outdoor and indoor concentrations of particles and volatile organic compounds in Amsterdam. *Atmos. Environ.* **34**, 37133722.
39. Linaker, C., Chauhan, A., Inskip, H., Holgate, S., and Coggon, D. (2000). Personal exposures of children to nitrogen dioxide relative to concentrations in outdoor air. *J. Occup. Environ. Med.* **57**, 472476.
40. Fung, K., Luginaah, I., and Gorey, K. (2007). Impact of air pollution on hospital admissions in Southwestern Ontario, Canada: Generating hypotheses in sentinel high-exposure places. *Environ. Health* **6**, 18.
41. Brauer, M., Hoek, G., van Vliet, P., Meliefste, K., Fischer, P., Gehring, U., Heinrich, J., Cyrys, J., Bellander, T., Lewne, M., and Brunekreef, B. (2003). Estimating long-term average particulate air pollution concentrations: Application of traffic indicators and geographic information systems. *Epidemiology* **14**, 228239.
42. Statistics Canada (2006). [http://www12.statcan.ca/english/census06/data/profiles/community/Index.cfm?Lang=E] Community Profiles.
43. Gilbertson, M. and Brophy, J. (2001). Community health profile of Windsor, Ontario, Canada: Anatomy of a great lakes area of

concern. *Environ. Health Persp.* **109**(Suppl. 6), 827843.

44. Curren, K. C., Dann, T. F., and Wang, D. K. (2006). Ambient air 1,3-butadiene concentration in Canada (1995–2003): Seasonal, day of week variations, trends, and source influences. *Atmos. Environ.* **40**, 171181.
45. Health Canada (2000). *Health Data and Statistics Compilations for Great Lakes Areas of Concern.* Health Canada, Ottawa, Ontario.
46. Kanaroglou, P. S., Jerrett, M., Morrison, J., Beckerman, B., Arain, M. A., Gilbert, N. L., and Brook, J. (2005). Establishing an air pollution monitoring network for intra-urban population exposure assessment: A location-allocation approach. *Atmos. Environ.* **39**, 23992409.
47. Miller, L., Xu, X., and Luginaah, I. (2009). Spatial variability of VOC concentrations in Sarnia, Ontario, Canada. *J. Toxicol. Environ. Health A* **72**, 115.
48. SPSS 15.0 for Windows SPSS Inc. Headquarters, 233 S. Wacker Drive, 11th floor, Chicago, IL 60606 2007.
49. Hamilton, L. (1992). Regression with Graphics: A Second Course in Applied Statistics. Duxbury Press, Belmont, California.
50. Odland, J. (1998). Spatial Autocorrelation. Sage Publications, New Delhi, India.
51. Griffith, D. A. (1987). Spatial Autocorrelation: A Primer. Resource publications in geography. Association of American Geographers, Washington, DC.
52. Isaaks, E. and Srivastava, R. (1989). An Introduction to Applied Geostatistics. Oxford University Press, New York, NY.
53. Sahsuvaroglu, T., Arain, A., Kanaroglou, P. S., Finkelstein, N., Newbold, B., Jerrett, M., Beckerman, B., Brook, J. R., Finkelstein, M., and Gilbert, N. L. (2006). A land use regression model for predicting ambient concentrations of nitrogen dioxide in Hamilton, Ontario, Canada. *J. Air Waste Manag. Assoc.* **56**, 10591069.
54. Chow, G. (1960). Tests of equality between sets of coefficients in two linear regressions. *Econometrica* **28**, 591605.
55. Atari, D. O., Luginaah, I., Xu, X., and Fung, K. (2008). Spatial variability of ambient nitrogen dioxide and sulphur dioxide in Sarnia, "Chemical Valley", Ontario, Canada. *J. Toxicol. Environ. Health A* **71**, 110.
56. Gilbert, N. L., Goldberg, M. S., Beckerman, B., Brook, J. R., and Jerrett, M. (2005). Assessing Spatial Variability of Ambient Nitrogen Dioxide in Montreal, Canada, with a Land-Use Regression Model. *J. Air Waste Manag. Assoc.* **55**(8), 10591063.
57. Monod, A., Sive, B. C., Avino, P., Chen, T., Blake, D. B., Rowland, F. S. (2001). Monoaromatic compounds in ambient air of various cities: A focus on correlations between the xylenes and ethylbenzene. *Atmos. Environ.* **35**, 135149.
58. Environment Canada (2001). National Air Pollution Surveillance (NAPS) Network. [http://www.etc-cte.ec.gc.ca/NAPS/naps_summary_e.html] NAPS Network Summary.
59. Carr, D., von Ehrestein, O., Weiland, S., Wagner, C., Wellie, O., Nicolai, T., and von Mutius, E. (2002). Modelling annual benzene, toluene, NO_2, and soot concentrations on the basis of road traffic characteristics. *Environ. Res. Sect.* **90**, 111118.
60. Smith, L., Mukerjee, S., Gonzales, M., Stallings, C., Neas, L., and Norris, G. H. (2006). Use of GIS and ancillary variables to predict volatile organic compound and nitrogen dioxide levels at unmonitored locations. *Atmos. Environ.* **40**, 37733787.
61. Wheeler, A. J., Smith-Doiron, M., Xu, X., Gilbert, N. L., and Brook, J. R. (2008). Intra-urban variability of air pollution in Windsor, Ontario-measurement and modeling for human exposure assessment. *Environ. Res.* **106**, 716.
62. Beckerman, B., Jerrett, M., Brook, J. R., Verma, D. K., Arain, M. A., and Finkelstein, M. M. (2008). Correlation of nitrogen dioxide with other traffic pollutants near a major expressway. *Atmos. Environ.* **42**, 275290.
63. Ross, Z., English, P. B., Scalf, R., Gunier, R., Smorodinsky, S., Wall, S., and Jerrett, M. (March, 2006). Nitrogen dioxide prediction in Southern California using land use regression modeling: potential for environmental health analyses. *J. Expo. Sci. Environ. Epidemiol.* **16**(2), 106114.
64. Pankow, J. F., Luo, W., Bender, D. A., Isabelle, L. M., Hollingsworth, J. S., Chen, C.,

Asher, W. E., and Zogorski, J. S. (2003). Concentration and co-occurrence correlations of 88 volatile organic compounds (VOC) in the ambient air of 13 semi-rural to urban locations in the United States. *Atmos. Environ.* **37**, 50235046.

65. Lebret, E., Briggs, D., van Reeuwijk, H., Fischer, P., Smallbone, K., Harssema, H., Kriz, B., Gorynski, P., and Elliott, P. (2000). Small area variations in ambient NO_2 concentrations in four European areas. *Atmos. Environ.* **34**, 177185.

14

1. Hammer, U. T. (1978). The saline lakes of Saskatchewan, III. Chemical characterization. *Inter. Revue Der. Ges. Hydrobiol.* **63**, 311–3335.
2. Cowardin, L. M., Carter, V., Golet, F. C., and LaRoe, E. T. (1979). *Classification of Wetlands and Deepwater Habitats in the United States*. USFW Service.
3. LaBaugh, J. W. (1989). Chemical characteristics of water in northern prairie wetlands. In *Northern Prairie Wetlands*. A. V. Ames Valk (Ed.). Iowa University Press, pp. 57–90.
4. Hammer, U. T. and Haynes, R. C. (1978). The saline lakes of Saskatchewan. II. Locale, hydrology, and other physical aspects. *Inter.Revue Der Ges. Hydrobio.* **63**, 179–1203.
5. Last, W. and Ginn, F. (2005). Saline systems of the Great Plains of western Canada: An overview of the limnogeology and paleolimnology. *Saline Systems* **1**(1), 10.
6. Wollheim, W. M. and Lovvorn, J. R. (1995). Salinity effects on macroinvertebrate assemblages and waterbird food webs in shallow lakes of the Wyoming high-plains. *Hydrobiologia* **310**(3), 207–223.
7. Donald, D. B. and Syrgiannis, J. (1995). Occurrence of pesticides in prairie lakes is Saskatchewan in relation to drought and salinity. *J. Env. Quality* **24**(2), 266–270.
8. Hammer, U. T. (1978). The saline lakes of Saskatchewan. I Background and rationale for saline lakes research. *Inter. Revue Der Gesamten Hydrobiolgie* **63**, 173–1177.
9. Lieffers, V. J. and Shay, J. M. (1983). Ephemeral saline lakes in the Canadian prairies—Their classification and management for emergent macrophyte growth. *Hydrobiologia* **105**, 85–894.
10. Redberry Lake Biosphere Rerserve. Retrieved from [http://www.redberrylake.ca/].
11. Hammer, U. T. and Parker, J. (1984). Limnology of a perturbed highly saline Canadian lake. *Verhandlungen Internationale Vereinigung für theoretische und angewandte Limnologie* **102**, 31–342.
12. *Mineral Resource Map of Saskatchewan Regina* (2006). Saskatchewan Department of Industry and Resources.
13. Hammer, U. T., Shamess, J., and Haynes, R. C. (1989). The distribution and abundance of algae in saline lakes of Saskatchewan, Canada. *Hydrobiologia* **105**, 1–26.
14. Haynes, R. C. and Hammer, U. T. (1978). The saline lakes of Saskatchewan. IV Primary production of phytoplankton in selected saline ecosystems. *Inter. Revue. Der. Ges. Hydrobio.* **63**, 337–3351.
15. Bierhuizen, J. F. H. and Prepas, E. E. (1985). Relationships between nutrients, dominant ions, and phytoplankton standing crop in prairie saline lakes. *Canadian J. Fisher. Aquatic. Sci.* **42**, 1588–11594.
16. Evans, J. C. and Prepas, E. E. (1996). Potential effects of climate change on ion chemistry and phytoplankton communities in prairie saline lakes. *Limnol. Oceanogr.* **41**(5), 1063–1076.
17. Ferreyra, G. A., Demers, S., delGiorgio, P., and Chanut, J. P. (1997). Physiological responses of natural plankton communities to ultraviolet-B radiation in Redberry Lake (Saskatchewan, Canada). *Canadian J. Fisher. Aquatic. Sci.* **54**(3), 705–714.
18. Sorokin, D. Y. and Kuenen, J. G. (2005). *Chemolithotrophic haloalkaliphiles* from soda lakes. *FEMS Microbiol. Eco.* **52**(3), 287–295.
19. Hammer, U. T. (1986). Saline Lake Ecosystems of the World. In *Monographiae Biologicae*, Volume 59. H. J. Dumont and Dr. W. Dordrecht (Eds.). Junk Publishers.
20. Grasby, S. E. and Londry, K. L. (2007). Biogeochemistry of hypersaline springs supporting a mid-continent Marine Ecosystem:

An analogue for Martian springs? *Astrobiology* **7**(4), 662–683.

21. Sorensen, K. B., Canfield, D. E., Teske, A. P., and Oren, A. (2005). Community composition of a hypersaline endoevaporitic microbial mat. *Appl. Environ. Microbiol.* **71**(11), 7352–7365.
22. Sorokin, D. Y., Tourova, T. P., Lysenko, A. M., and Muyzer, G. (2006). Diversity of culturable halophilic sulfur-oxidizing bacteria in hypersaline habitats. *Microbiology* **152**(10), 3013–3023.
23. Hammer, U. T., Sheard, J. S., and Kranabetter, J. (1990). Distribution and abundance of littoral benthic fauna in Canadian prairie saline lakes. *Hydrobiologia* **197**, 173–192.
24. Oren, A. (1999). Microbiology and biogeochemistry of halophilic microorganisms—An overview. In *Microbiolgy and Biogeochemistry of Hypersaline Environments*. A. Oren (Ed.). New York, CRC.
25. Timms, B. V., Hammer, U. T., and Sheard, J. W. (1986). A study of benthic communities in some saline lakes in Saskatchewan and Alberta, Canada. *Inter. Revue. Der. Ges. Hydrobio*. **71**(6), 759–777.
26. Oren, A. (2002). Diversity of halophilic microorganisms: Environments, phylogeny, physiology, and applications. *J. Industrial Microbiol. Biotechnol.* **28**, 56–63.
27. Litchfield, C. D. and Gillevet, P. M. (2002). Microbial diversity and complexity in hypersaline environments: A preliminary assessment. *J. Ind. Microbiol. Biotechnol.* **28**(1), 48–55.
28. Wobeser, G. and Howard, J. (1987). Mortality of waterfowl on a hypersaline wetland as a result of salt encrustation. *J. Wildl. Dis.* **23**(1), 127–134.
29. Schmutz, J. K. (1999). *Community Conservation Plan for the Redberry Lake Important Bird Area*. Center for studies in agriculture, Law, and the Environment, Saskatoon.
30. Traylor, J. J., Alisauskas, R. T., and Kehoe, F. P. (2004). Nesting ecology of white—Winges scoters (*Melanitta fusca* deglandi) at Redberry Lake, Saskatchewan. In *Auk*, Volume 121. American Ornithologists Union, pp. 950–962.
31. Sachs, J. P., Pahnke, K., Smittenberg, R., and Zhang, Z. (2007). Paleoceanography, biological proxies; biomarkers. *Encyclo. Quarter. Sci.* **2**, 1627–1634.
32. Hammer, U. T. (1990). The effects of climate change on the salinity, water levels, and biota of Canadian prairie lakes. *Verhandlungen Internationale Vereinigung Für Theoretische Und Angewandte Limnologie* **24**, 321–326.
33. Last, W. M. (1993). Geolimnology of Freefight Lake: An unusual hypersaline lake in the northern Great Plains of western Canada. *Sedimentology* **40**, 431–4448.
34. Birks, S. J. and Remenda, V. H. (1999). Hydrogeological investigation of Chappice Lake, southeastern Alberta: Groundwater inputs to a saline basin. *J. Paleolimnol.* **21**, 235–2255.
35. Waiser, M. J. and Robarts, R. D. (1995). Microbial nutrient limitation in prairie saline lakes with high sulfate concentrations. *Limno. Oceano*. **40**(3), 566–574.
36. Conly, F. M. and Van der Kamp, G. (2001). Monitoring the hydrology of Canadian prairie wetlands to detect the effects of climate change and land use changes. *Environ. Monitor. Asses*. **67**(1–2), 195–215.
37. Covich, A. P., Fritz, S. C., Lamb, P. J., Marzolf, R. D., Matthews, W. J., Poiani, K. A., Prepas, E. E., Richman, M. B., and Winter, T. C. (1997). Potential effects of climate change on aquatic ecosystems of the Great Plains of North America. *Hydrol. Proc.* **11**(8), 993–1021.
38. Last, W. M. (1991). Sedimentology, geochemistry, and evolution of saline lakes of the Northern Great Plains. *Saskatoon, Post-conference Fieldtrip Guidebook, Sedimentary and Paleolimnological Records of Saline Lakes*.
39. Environment Canada Monthly Data Report. Retrieved from [http://www.climate.weatheroffice.ec.gc.ca/climatedata/monthlydata_e.html].
40. Mitsch, W. J. (1993). *Wetlands*. Van Nostrand Reinhold, New York.
41. Anati, D. A. (1999). The salinity of hypersaline brines: Concepts and misconceptions. *Inter. J. Salt Lake Res.* **8**(1), 55–70.
42. Development, O. R. (Ed.) (1983). *Methods of Analysis of Water and Wastes*. United

States Environmental Protection Agency, Washington, DC.

43. Richards, S. R., Rudd, J. W. M., and Kelley, C. A. (1994). Organic volatile sulfur in lakes ranging in sulfate and dissolved salt concentration over five orders of magnitude. *Limnol. Oceanogr.* **39**, 562–572.

15

1. Boschi-Pinto, C., Velebit, L., and Shibuya, K. (2008). Estimating child mortality due to diarrhea in developing countries. *Bull. WHO* **86**, 710–717.
2. Bhutta, Z. A., Black, R. E., Brown, K. H., Gardner, J. M., Gore, S., et al. (1999). Prevention of diarrhea and pneumonia by zinc supplementation in children in developing countries: Pooled analysis of randomized controlled trials. Zinc Investigators Collaborative Group. *J. Pediatr.* **135**, 689–697.
3. Aggarwal, R., Sentz, J., and Miller, M. A. (2007). Role of zinc administratioon in prevention of childhood diarrhea and respiratory illness: Ameta analysis. *Pediatr* **119**, 1120–1130.
4. Baqui, A. H., Black, R. E., El Arifeen, S., Yunus, M., Chakraborty, J., et al. (2002). Effect of zinc supplementation started during diarrhea on morbidity and mortality in Bangladeshi children: Community randomised trial. *BMJ* **325**, 1059.
5. Jones, G., Steketee, R. W., Black, R. E., and Bhutta, Z. A. (2003). Morris SSBellagio Child Survival Study Group. How many child deaths can we prevent this year? *Lancet* **362**, 65–71.
6. World Health Organization (2004). WHO/UNICEF Joint Statement. *Clinical management of acute diarrhea.* World Health Organization, Geneva.
7. Ellis, A. A., Winch, P. J., Daou, Z., Gilroy, K. E., and Swedberg, E. (2007). Home management of childhood diarrhea in southern Maliimplications for the introduction of zinc treatment. *Soc. Sci. Med.* **64**, 701–712.
8. Winch, P. J., Gilroy, K. E., and Fischer-Walker, C. L. (2008). The effect of HIV/AIDS and malaria on the context for introduction of zinc treatment and low-osmolarity ORS for childhood diarrhea. *J. Health Popul. Nutr.* **26**, 1–11.
9. Bennett, S., Woods, T., Liyanage, W. M., and Smith, D. L. (1991). A simplified general method for cluster-sample surveys of health in developing countries. *World Health Statist. Quart.* **44**, 98–106.
10. Khan, M. A., Larson, C. P., Faruque, A. S. G., Saha, U. R., Hoque, A. M., Alam, N. U., and Salam, M. A. (2007). Introduction of routine zinc therapy for children with diarrhea: Evaluation of safety. *J. Health Popul. Nutr.* **25**, 127–133.
11. Larson, C. P., Hoque, M., Larson, C.P., and Khan, A. M. (2005). Initiation of zinc treatment for acute childhood diarrhea and the risk for vomiting or regurgitation: Arandomized, double-blind, placebo-controlled trial. *J. Health Popul. Nutr.* **23**, 311–318.
12. Nasrin, D., Larson, C. P., Sultana, S., and Khan, T. U. (2005). Acceptability and adherence to zinc dispersible tablet treatment of acute childhood diarrhea. *J. Health Popul. Nutr.* **23**, 215–221.
13. Larson, C. P., Saha, U. R., Islam, R, and Roy, N, (2006), Childhood diarrhea management practices in Bangladesh: Private sector dominance and continued inequities in care. *Int. J. Epidemiol.* **35**, 1430–1439.
14. Gwatkin, D., Rustein, S., Johnson, K., Prande, R., and Wagstaff, A. (2000). *Socio-economic differences in health, nutrition and population in Bangladesh.* World Bank, Human Development Network, Washington, D.C. [2007], p. 139. Retrieved from [http://poverty.worldbank.org/library/view/4212/. Accessed 28/08/2009].
15. Filmer, D. and Pritchard, L. (1988). *Estimating wealth effects without expenditure data: An application to educational enrollments in states of India.* World Bank Policy Research Working Paper number 1994. Development Research Group (DECRG), World Bank, Washington (D.C.).
16. Quantitative Techniques for Health Equity Analysis—technical note number 7. The Concentration Index. Retrieved from [http://siteresources.worldbank.org/EXTEDSTATS/Resources/3232763-1171296378756/concentration.pdf Accessed 06/10/2009].

17. Rogers, E. (1962). *Diffusion of innovation.* The Free Press, New York.

18. Bandura, A. (1982). Self-efficacy mechanism in human agency. *Am. Psycho.* **37**, 122–147.

19. Rao, K. V., Mishra, V. K., and Retherford, R. D. (1998). Mass media can help improve treatment of childhood diarrhea. *Natl. Fam. Health Surv. Bull.* **11**, 1–4.

20. Greenhalgh, T., Robert, G., MacFarlane, F., Bate, P., and Kyriakidou, O. (2004). Diffusion of innovations in service organizations: Systematic review and recommendations. *Milbank Quart.* **82**, 581–629.

Index

F

U

V

W

Y

Z